Human Anatomy and Physiology Workbook

Human Anatomy and Physiology Workbook

Traditional and Innovative Exercises *to accompany* Tortora: Introduction to the Human Body

Second Edition

Kathryn E. Malone
Westchester Community College

Jane M. Schneider
Westchester Community College

HarperCollins*Publishers*

Sponsoring Editor: Bonnie Roesch
Project Management: Publishing Synthesis, Ltd.
Text Design: Publishing Synthesis, Ltd.
Cover Design: Heather A. Ziegler
Cover Illustration/Photo: David Stoecklein / The Stock Market
Production: Jimmy Spillane
Compositor: Publishing Synthesis, Ltd.
Printer and Binder: Malloy Lithographing
Cover Printer: Malloy Lithographing

HUMAN ANATOMY AND PHYSIOLOGY WORKBOOK: Traditional and Innovative Exercises

For information, address HarperCollins Publishers, Inc. 10 E. 53rd St., New York, NY 10022

ISBN: 0–06–046662–6
91 92 93 94 95 9 8 7 6 5 4 3 2 1

For our dads, with deep affection

Contents

Preface

This workbook has been formulated with you in mind. We know that Anatomy and Physiology is a fact-filled course, and that often the mere terminology can make you feel that you are filling a foreign language requirement. Thus, we have tried to mix the traditional with some nontraditional material so that your studying, comprehension, and reviewing might border on being enjoyable. It is our feeling that you might find "playing" with the facts of human anatomy and physiology an additional (hopefully pleasant!) way of tackling the concepts. We have included many types of review materials so that you may find and use those exercises that are most valuable for you.

The work has been organized in a pattern that corresponds to the typical systematic approach to human structure and function. The chapters parallel those in Tortora's *Introduction to the Human Body,* 2nd edition. This workbook can accompany any traditional Anatomy and Physiology course, however.

Each chapter features traditional question formats as well as various word games that incorporate factual information and theoretical constructs. The traditional material includes multiple choice, matching, completion, true/false questions, diagrams, and concept categories. Word games are in the form of addagrams, double crosses, word scrambles, crosswords and word cages.

We hope that this unique approach allows you to challenge yourself and further hone your knowledge and comprehension of the subject matter. Working at, and successfully completing, the activities provided should give you confidence in your grasp of Anatomy and Physiology.

We express our appreciation to those whose assistance has helped bring this workbook to fruition. Our thanks to our family and friends for their understanding, encouragement and good humor during the preparation of this work. We are grateful to the HarperCollins staff, particularly Bonnie Roesch, for their support and enthusiasm. We also thank Justin McShea for his efforts at making the writing of this workbook go smoothly.

Finally, we thank the students—past, present, and future—who serve as a lively inspiration in sustaining our efforts.

Kathryn Malone
Jane Schneider

Preface

Directions

Multiple Choice

Select the word or phrase that *best* completes each statement or question. Place the letter that precedes your choice on the line to the left of the statement.

True/False

Indicate whether the statements are true or false by placing a T or F at the start of the line to the left of the statement. If the statement is false, change the underlined word in the statement so that the statement reads correctly. Place the change after the F on the line to the left of the statement.

Completion

Complete each question or statement by filling in the *best* word or phrase.

Lost Sheep

Underline the word or phrase that does not "fit" in each of the grouped words or phrases.

Matching

Each of the terms/phrases in the left-hand column refers to an item in the right-hand column. Insert the letter from the appropriate terms in the right-hand column next to the corresponding numbered term in the left-hand column. Not all matching columns are even. In some cases a choice may be used more than once and in some there is more than one applicable answer.

Double Crosses

Complete each puzzle by using the clues provided and write in the correct word(s).

Word Cage

In the maze of letters find the words that refer to the clues stated below the cage. *Words can go in any direction: up, down, forward, backward, and diagonal.* Circle the word in the maze.

Sleuthing

In the cases given, you are the "medical" sleuth who tries to answer the questions. The cases deal with some aspect of the anatomy and/or physiology of the system under consideration.

Word Scrambles

In these puzzles, there is a list of jumbled words with corresponding clues. Once all the words have been unscrambled, the circled letters will spell out another scrambled term for which there is a total clue.

Addagrams

Each puzzle consists of a series of clues, the answers to which "add up" to solve the puzzle. The individual clues are followed by a series of numbers, which correspond to the letters in the total puzzle. Place the letters in the box that corresponds to the number. If a letter has a 0, that indicates it is not in the answer and should be ignored.

Crosswords

Complete these puzzles as you would a typical crossword puzzle. The difference here is that the majority of terms are related to the system you are studying.

Anatomic Artwork

For the diagrams, there are a series of tasks, which include labeling, shading, and drawing in structures related to the anatomy of the system.

Chapter 1

Organization of the Human Body

Introduction

The human body, like many successful companies, is well organized and relatively stable despite the changing environments to which it is exposed. There is, characteristically, a division of labor, central control and regulation, and the ability to adapt to all the variations that occur. This ability allows the human body to remain stable and healthy, a state called *homeostasis.*

After studying Chapter 1 in your text, you should be aware of these concepts, levels of organization, and descriptive terminology that applies to the study of the human body. Additionally, you should be able to answer the following questions and complete the varied activities.

Multiple Choice

___ 1. A midsagittal section will divide the body into which of the following?
  a. right and left halves
  b. dorsal and ventral halves
  c. anterior and posterior halves
  d. none of these

___ 2. Which term describes the location of the fingers in reference to the arm in the anatomic position?
  a. medial
  b. proximal
  c. distal
  d. lateral

___ 3. A section through the body which separates it into anterior and posterior portions is called which of the following?
  a. frontal
  b. sagittal
  c. transverse
  d. ventral

___ 4. The cavity that contains the brain only is called which of the following?
  a. ventral
  b. cranial
  c. dorsal
  d. thoracic

___ 5. Which of the following pairs is correctly matched?
a. anterior–dorsal
b. cranial–posterior
c. medial–side
d. none of these is correctly matched

___ 6. Loss of the spinal nerve supply to the brachial region might affect your ability to receive sensory information from and send motor information to the:
a. leg
b. chest
c. hand
d. lumbar area

___ 7. The processes that permit the body to maintain a dynamic equilibrium between the external and internal environments are collectively called:
a. life processes
b. homeostatic mechanisms
c. metabolism
d. physiologic gradients

___ 8. Most of the body's control mechanisms are regulated by which of the following?
a. positive exchanges
b. negative feedback
c. hormones
d. none of these

___ 9. Homeostatic mechanisms:
a. are self-regulating
b. have negative feedback control
c. regulate ongoing activities
d. all of these

___ 10. In the anatomic position, the thumb would be:
a. dorsal
b. lateral
c. proximal
d. sagittal

___ 11. An organism is in a homeostatic state when the internal environment has optimal:
a. levels of chemicals
b. temperature
c. pressure outside and inside the cells
d. all of these

___ 12. Fluid within cells is called:
a. interstitial
b. intercellular
c. intracellular
d. ECF

___ 13. Which of the following regions is included, at least in part, in the left upper quadrant (LUQ) of the abdominopelvic cavity?
a. left hypochondriac
b. umbilical
c. epigastric
d. all of these

___ 14. Which of the following organs would *not* be found in the ventral cavity?
- a. lungs
- b. intestine
- c. brain
- d. ovaries

___ 15. If the kidneys were cut longitudinally into anterior and posterior portions, it would be a ________ section.
- a. sagittal
- b. midsagittal
- c. transverse
- d. coronal

True/False

__________ 1. Standing erect, upper limbs at the sides and palms *inward,* describes the body in the anatomic position.

__________ 2. In humans, the terms inferior and *caudal* could be used to describe the same structure.

__________ 3. The phalanges are *proximal* to the elbow.

__________ 4. The *ventral* body cavity is subdivided into the cranial cavity and vertebral canal.

__________ 5. The collective name for the organs in the *ventral* body cavity is the viscera.

__________ 6. The RLQ would include part of the *hypogastric* and umbilical region.

__________ 7. The fluid filling the microscopic spaces between cells is referred to as *interstitial* fluid.

__________ 8. Extracellular fluid is found in two major areas: as plasma and *intercellular* fluid.

__________ 9. In a negative feedback system, the outcome is in the *same* direction as the initial condition.

__________ 10. The body's *homeostatic* mechanisms are controlled by the nervous and endocrine systems.

__________ 11. The characteristics of conductivity and *excitability* are life processes expressed by all living organisms.

__________ 12. The term *medial* means toward the midline of the body or structure.

Completion

1. The field of ________________ is concerned with the functions of the body and its parts.
2 The lowest level of organization within the body is the ________________ level.
3. ________________ are structures that have a definite shape and function and are composed of two or more tissues.
4. The total of all the chemical reactions that occur in the body is called ________________.
5. The brain is ________________ to the heart, in humans.
6. If a structure is away from the body surface, it would be described as ________________.
7. The term ________________ could be substituted for transverse section.
8. In humans, the dorsal body cavity is located toward the ________________ of the body.

9. There are ________________ regions and ________________ quadrants in the abdominopelvic cavity.
10. Homeostasis in all organisms would be altered by any stimulus that creates an imbalance in the ________________ environment.
11. In homeostatic control systems, another term that can be used to describe the input is ________________.
12. In a positive feedback system, the input is ________________ by the output.

Lost Sheep

1. dorsal body cavity, spinal cord, viscera, posterior
2. gallbladder, right hypochondriac, RUQ, appendix
3. mediastinum, heart, esophagus, lungs
4. ECF, plasma, interstitial fluid, ICF
5. negative feedback, blood pressure regulation, output intensifies input, blood glucose regulation
6. frontal, sagittal, longitudinal, transverse
7. thoracic cavity, pleural cavities, ventral cavity, vertebral cavity
8. nervous system, regulation, endocrine system, lymphatic system
9. cranial, superior, medial, cephalic
10. thoracic, cranial, abdominopelvic, brachial

Matching

Set 1

___ 1. contains all structures of the thoracic cavity except lungs
___ 2. has the cranial cavity and vertebral canal
___ 3. is divided into quadrants
___ 4. a vertical plane through a structure creating left and right halves
___ 5. frontal plane
___ 6. at right angle to a midsagittal section

a. dorsal cavity
b. ventral cavity
c. abdominopelvic cavity
d. thoracic cavity
e. mediastinum
f. diaphragm
g. coronal and regions
h. sagittal
i. parasagittal
j. midsagittal

Set 2

___ 1. toward the surface
___ 2. between two structures
___ 3. farther from the point of attachment
___ 4. toward the back
___ 5. toward the head
___ 6. away from the midline

a. ventral
b. dorsal
c. cranial
d. medial
e. proximal
f. distal
g. intermediate
h. lateral
i. sagittal
j. superficial

Double Crosses

1. *Across:*
   coronal

   Down:
   cephalic

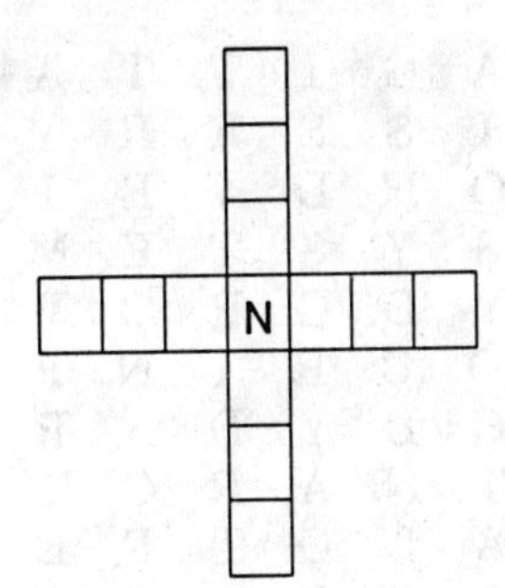

2. *Across:*
   a> toward the front
   <b toward the side

   Down:
   near to

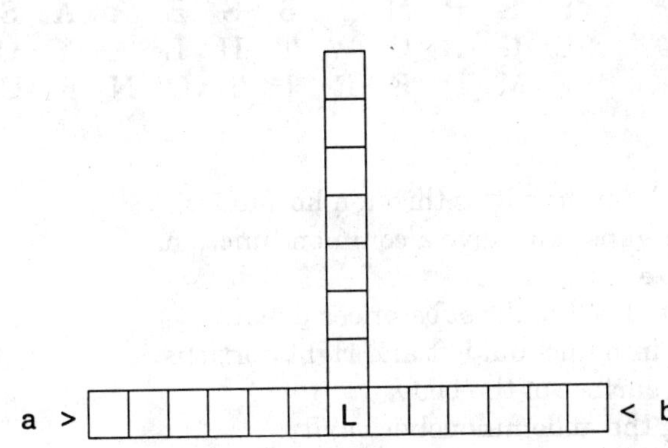

3. *Across:*
   area between lungs

   DOWN:
   steady state

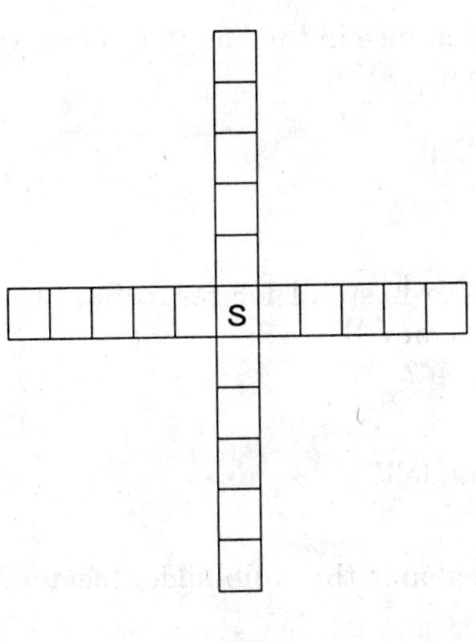

4. *Across:*
   has ability to shorten

   DOWN:
   central pubic region

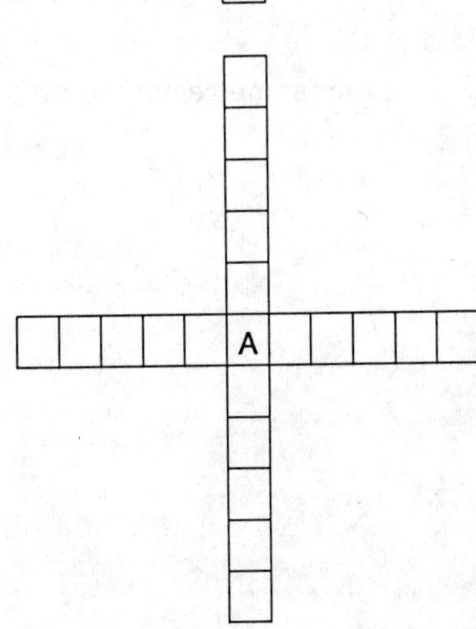

Word Cage

```
P A R A S A G I T T A L
C M M D R I S S E R T S
F I S F F O H D Y E I Y
L N T I G S Y S T E M R
A P R Y L V D G R O V E
R U E N D O C R I N E I
B T S P O C B Y D A R A
E I N P U T N A R C T D
T D O R S A L O T F E K
R E R N F S S E R A S L
E G R O W T H L A Y C Z
V M P E R I T O N E U M
```

1. System that regulates activities through hormones.
2. Several related organs that have a common function.
3. An increase in size.
4. Metabolic process that breaks substances down.
5. Divides the body into unequal left and right portions.
6. Toward the back surface of the body.
7. Membrane lining the abdominopelvic cavity.
8. Canal containing spinal cord.
9. Any stimulus that creates an imbalance in the body's internal environment.
10. The stimulus in a feedback system.

Sleuthing

1. Mrs. Jackson was admitted to the hospital for tests. She was complaining of abdominal pain. Ultimately, her gallbladder was removed.
   a. What side of the body was "sore"?

   b. In what quadrant was the pain felt?

   c. In what specific abdominal region is the gallbladder located?

   d. If she had appendicitis, would the exact same region be involved? Why?

Crossword

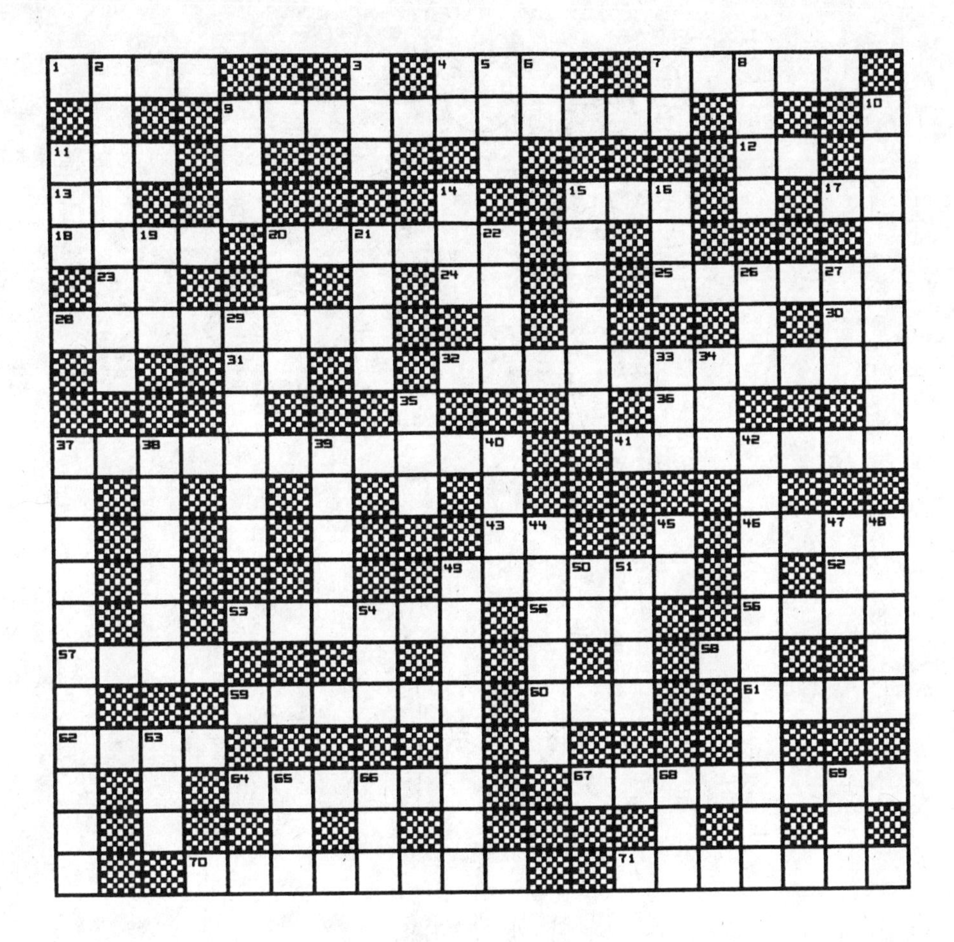

Across

1 Top of pyramid
4 Man's name in Scotland
7 Superior
9 Dynamic equilibrium
11 Animal of hunt
12 Type of radiation (abbr.)
13 Animal of burden
15 Green vegetable
17 Branch of medicine
18 Ferris wheel, for example
20 Toward the back
23 "Heart attack" (abbr.)
24 City of Angels
25 Toward the midline
28 Longitudinal section that creates left and right portions
30 Southern state (abbr.)
31 Helium (abbr.)
32 Abdominal region
36 Short form of mother
37 Toward the surface
41 Away from the midline
43 Neomorphic (abbr.)
46 Poetic preposition
49 Homeostatic state
52 Goes with either
53 Response in a feedback system
55 Direction when traveling from Chicago to Jacksonville
56 Found in thoracic cavity
57 Ingredient for punch
58 Part of the formula for the radius of a circle
59 Organization level composed of organs
60 High-energy molecule
61 What sun's rays do to the skin
62 Basic unit of living matter
64 Opposite of cranial
67 Level of organization between atoms and cells
70 Output in a feedback system
71 Secreted by endocrine glands

Down

2 Near to
3 Soldier
4 I am, you are, it ______
5 Produced in mitochondria
6 Sodium (abbr.)
7 Like
8 Egg
9 Found in pelvic region
10 Central abdominal region
11 Preposition
14 Police Athletic League (abbr.)
15 Time slot
16 Goal
19 Done with a shovel
20 Care for
21 Goes with coffee in the a.m.
22 Woman
26 Spot
27 French friend
29 Chest cavity
33 Association of MDs
34 One of the seven days (abbr.)
35 Secretion of sebaceous gland
37 Not deep
38 Awards
39 Stimulus
40 Fuzz on clothing
42 What is achieved between the internal and external cellular environments
44 Opposite lateral
45 Possessive pronoun
47 2000 pounds
48 Composed of two or more tissues
49 Will elicit a response
50 Good ____ gold
51 Not toward the surface
54 Fruit's stone
63 Lower extremity
65 Autonomic Nervous System (abbr.)
66 God (Latin)
68 Personality component
69 Girl's name

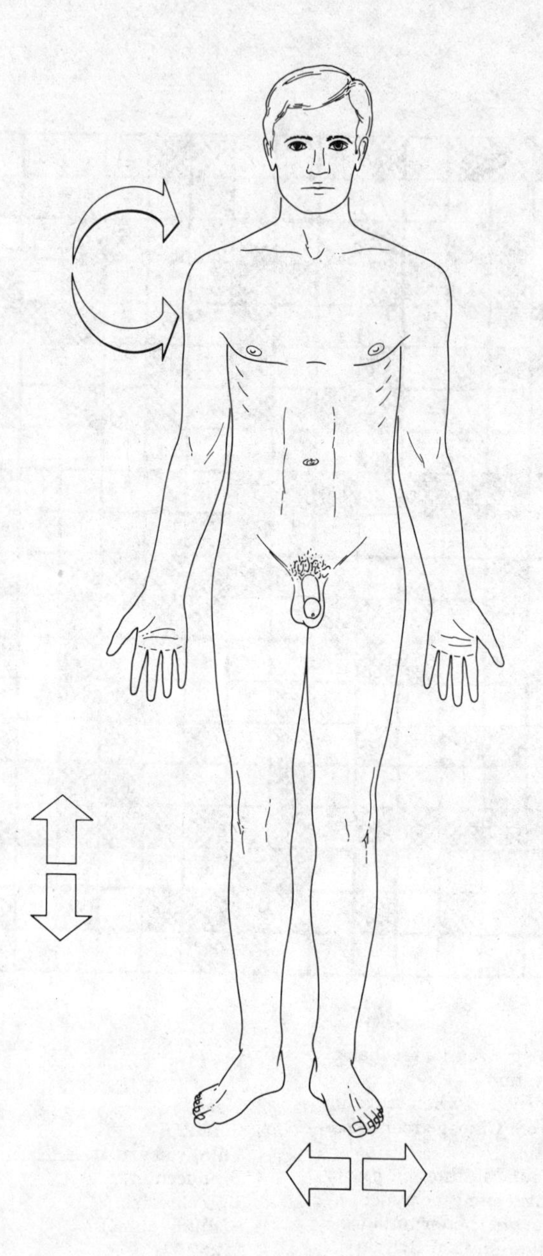

Anatomic Artwork

1. Label the specific abdominal region where the appendix is located.
2. Label the following anatomic regions:
   a. brachial
   b. cranial
   c. thoracic
3. Draw a line through the thigh which would make a transverse section through that structure.
4. Shade in and label the arrows that indicate each of the following directions:
   a. superior
   b. lateral
   c. dorsal
5. Use the appropriate term to describe the relationship of the fingers to the shoulder and label the fingers accordingly.

Chapter 2

Introductory Chemistry

Introduction

Whether examining wood, or seaweed, or a piece of meat, we are going to come across the same elemental makeup, for all living materials have a commonality: their chemistry. Certain elements appear repeatedly in living systems, and the most abundant are oxygen, carbon, hydrogen, and nitrogen. In fact, these four elements comprise about 96 percent by weight of all living organisms. If calcium and phosphorus are added to the list, the total is 99 percent. The remainder consists of trace elements, for example, iron, copper, sodium, magnesium, sulfur, chloride, and potassium. It is interesting to note that the fluids of the human body contain many of the same elements as seawater, although the concentration may be different.

While much oxygen exists chemically uncombined in the body as O_2 gas, it also enters into chemical union with other elements. Water (H_2O) and glucose ($C_6H_{12}O_6$) are such examples. In fact, in living organisms, many of the elements are not found free, but in chemical combination. In such a union, elements either share their innate charges (forming *covalent* bonds) or completely donate or receive these charges (forming *ionic* bonds).

In combination, the elements are united into *compounds*. Generally, those compounds that do not contain carbon together with hydrogen are classed as *inorganic*. The abundant biological inorganic compound is water. Compounds that contain carbon together with hydrogen are classified as *organic*. The most abundant biological organic compounds are proteins, lipids, carbohydrates, nucleic acid, and ATP (adenosine triphosphate).

After you have studied the textbook regarding Introductory Chemistry, you should be able to answer the following questions and carry out the activities of this chapter.

Multiple Choice

___ 1. Bases contained in RNA include:
- a. alanine and thymine
- b. adenine and thymine
- c. uracil and guanine
- d. alanine and uracil

___ 2. Which is characteristic of proteins?
- a. large size
- b. peptide bonds
- c. amino acid subunits
- d. all of these

___ 3. Which of these is related to the structure or function of proteins?
  a. enzyme
  b. antibodies
  c. muscle contraction
  d. all of these

___ 4. Mildly alkaline biologic solutions like pancreatic juice would have a pH of about:
  a. 5.5–6.0
  b. 7.5
  c. 12.5–13.0
  d. 14

___ 5. Which of the following is a covalent bond?
  a. C—O
  b. Ca—Cl
  c. Na—F
  d. all of these

___ 6. A molecule that serves in an energy storage function in living organisms is:
  a. DNA
  b. ATP
  c. RNA
  d. NaCl

___ 7. Which is not soluble in water?
  a. sodium chloride
  b. oxygen
  c. steroids
  d. glucose

___ 8. The four most abundant elements in living organisms are:
  a. C,H,O,P
  b. C,H,S,P
  c. S,N,O,P
  d. C,H,O,N

___ 9. Glycogen:
  a. consists of monosaccharide subunits
  b. forms by a synthesis reaction
  c. is a polysaccharide
  d. all of these

___ 10. Which is true concerning the properties of water?
  a. It requires little heat to raise its temperature
  b. It is a neutral compound regarding pH
  c. It can break down into hydrogen and hydroxide ions
  d. all of these

___ 11. Which compound releases hydrogen ions in water?
  a. HCl
  b. NaCl
  c. H_2O
  d. NaOH

___ 12. Which species would make a water solution basic?
a. HCl
b. NaOH
c. NaCl
d. KF

___ 13. Which ion would make the pH rise if its concentration fell?
a. OH^-
b. Cl^-
c. H^+
d. none of these

___ 14. Which compound has an equal number of hydrogen and hydroxide ions?
a. H_2O
b. $NaHCO_3$
c. HCl
d. NaOH

___ 15. Which compound is a salt?
a. HCl
b. NaOH
c. H_2O
d. NaCl

___ 16. When a terminal phosphate group is removed from an ATP molecule, what molecule is formed?
a. AMP
b. ADP
c. phosphate
d. none of these

___ 17. The sequence of bases on a DNA molecule directs the production of:
a. inorganic molecules
b. proteins
c. salts
d. all of these

___ 18. Which is incorrectly paired?
a. enzyme–protein
b. cholesterol–lipid
c. glucose–amino acid
d. all are correct

___ 19. A saturated fat contains ________ between its carbon atoms.
a. no double bonds
b. one double bond
c. multiple double bonds
d. none of these

___ 20. The bonds formed between amino acids are known as:
a. amino bonds
b. lipid bonds
c. peptide bonds
d. none of these

___ 21. Which of these products is formed by a synthesis reaction?
a. proteins
b. fats
c. polysaccharides
d. all of these

___ 22. Amino acids: proteins as glucose: ________ .
a. lipids
b. phospholipids
c. glycogen
d. fructose

___ 23. Water is considered:
a. a major constituent of lubricants
b. organic
c. a good lipid solvent
d. all of these

___ 24. The sugar unit in RNA is:
a. glucose
b. deoxyribose
c. ribose
d. none of these

True/False

__________ 1. Ribosomal RNA is the *only* nucleic acid in humans which is single stranded.
__________ 2. When four amino acids join together by a synthesis reaction, a *polypeptide* results.
__________ 3. A pH of 2 indicates a strongly *acidic* solution.
__________ 4. Water is a neutral solution with a pH of about *14*.
__________ 5. An acid in solution releases *H^+ ion*.
__________ 6. One of the properties of water is the *low* amount of heat needed to raise its temperature.
__________ 7. All proteins contain *C,H,O, and N*.
__________ 8. All lipids contain *C,H,O, and P*.
__________ 9. Most lipids are *fairly* soluble in water.
__________ 10. An *electron* is a positively charged particle found in the nucleus of an atom.
__________ 11. The "letters" in the protein molecule are *amino acids*.
__________ 12. The bonds between carbon and hydrogen in a carbohydrate molecule are *covalent* bonds.
__________ 13. When NaCl dissolves in water, the *electrolytes* Na^+ and Cl^- are formed.
__________ 14. The most abundant *organic* molecule in living organisms is water.
__________ 15. Fats, a class of lipid, consist of fatty acids linked to a *glycerol* molecule.
__________ 16. *Nucleic acids* consist of repeating units called amino acids.
__________ 17. In the human being, the hereditary molecules, called genes, are made up of *nucleic acids*.
__________ 18. Fe^{2+}, Zn^{2+}, and Mg^{2+} are *cations* (positively charged ions) which are present in trace amounts in living organisms.
__________ 19. The bond in NaCl is considered a *covalent* bond.

_______ 20. The breakdown of methane (CH_4) into carbon and hydrogen is considered a *catabolic* reaction.

_______ 21. The atomic *weight* of an element consists of the sum of its protons and neutrons.

_______ 22. If sugar is placed in a cup of tea, the tea water would be considered the *solvent*.

_______ 23. Chlorine with 17 electrons would be an electron *acceptor* in a chemical reaction.

_______ 24. Both proteins and polysaccharides are formed from subunits by *catabolic* reactions.

_______ 25. An ion with 17 protons in its nucleus and 18 electrons surrounding its nucleus would carry a *positive* charge of 1.

Completion

1. When electrons are lost or gained in the formation of a chemical bond, the bond is referred to as _______.
2. Water is an excellent medium for chemical reactions because it is _______.
3. All organic compounds by definition contain _______.
4. The four most common elements in living organisms are _______, _______, _______, and _______.
5. _______ and _______ are among the trace elements of living matter.
6. Proteins are composed of subunits called _______.
7. When elements share their electrons, the resulting bond is called _______.
8. A catabolic reaction is a _______ type of chemical reaction.
9. Solutions that are acidic have an excess of _______ ions.
10. Systems that help prevent changes in acid-base balance are called _______.
11. The base found only in RNA and not in DNA is _______.
12. A solution with a pH of 8.5 would be considered _______.
13. A high-energy organic molecule that is abundant in living material is _______.
14. The sugar molecule found in RNA is _______.
15. A common component of the backbone of both DNA and RNA is _______.
16. The basic subunit of a polysaccharide is a _______. An example of such a molecule is _______.
17. Genetic material is made up of molecules of _______ in the human.
18. A positively charged ion is known as a _______, whereas a negatively charged ion is known as a(n) _______.
19. Fats are also known by the chemical name _______, which describes the molecules comprising it.
20. A fatty acid that binds all the hydrogen that it can, and has no double bonds, is considered _______.
21. _______ is the term for the polysaccharide that is the storage form of carbohydrate in animals.
22. The sugar ribose in the RNA molecule is replaced by _______ in the DNA molecule.
23. Organic molecules of the human body which generally do not dissolve in water but do dissolve in solvents such as alcohol and ether are the _______.

24. Gastric juices are strongly acidic, having a pH of about ________________.
25. Proteins and polysaccharide polymers are formed by a process known as ________________.

Lost Sheep

1. ionic, salt, inorganic, carbon
2. amino acid, polypeptide, steroid, protein
3. OH^-, low pH, hydrochloric, H^+
4. steroid, water, cholesterol, triglyceride
5. sodium, iron, chlorine, calcium
6. synthesis, anabolism, formation, catabolism
7. sodium chloride, calcium phosphate, methane, ferrous sulfate
8. chloride ion, electron gain, anion, positive charge
9. fructose, amino acid, glycogen, glucose
10. guanine, phosphate, adenine, cytosine
11. electron, proton, neutron, photon
12. organic, carbon, hydrogen, phosphate
13. acidic, pH = 13, basic, alkaline
14. iron, hydrogen, copper, magnesium
15. electron sharing, covalent, ions, oxygens in O_2
16. ATP, O_2, high energy, great caloric release
17. DNA, phosphate, ribose, adenine
18. single stranded, uracil, thymine, ribose
19. carbon, compound, glucose, CCl_4
20. ribose, alanine, starch, glucose

Matching

Set 1

___ 1. steroids
___ 2. high-energy molecule
___ 3. enzymes
___ 4. glycogen
___ 5. contains cytosine

a. protein
b. carbohydrate
c. lipid
d. nucleic acid
e. ATP

Set 2

___ 1. protein
___ 2. NaCl
___ 3. carbohydrate
___ 4. lipid
___ 5. HCl

a. inorganic
b. organic

Set 3

___ 1. fat
___ 2. starch
___ 3. protein
___ 4. steroid
___ 5. triglyceride

a. peptide bond
b. polysaccharide
c. lipid

Double Crosses

1. *Across*:
   typical inorganic bond

   Down:
   DNA–RNA base

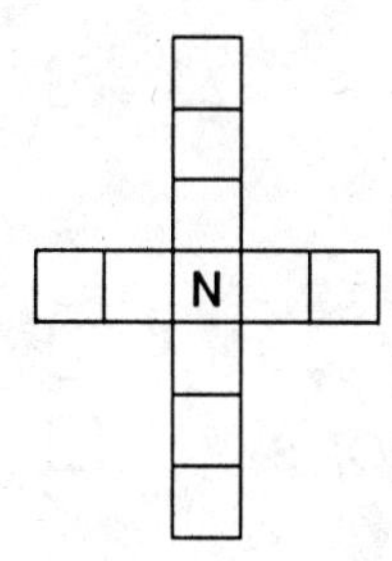

2. *Across:*
   in the presence of oxygen

   Down:
   a. pH=1
   b. what a salt does in water
   c. organic element

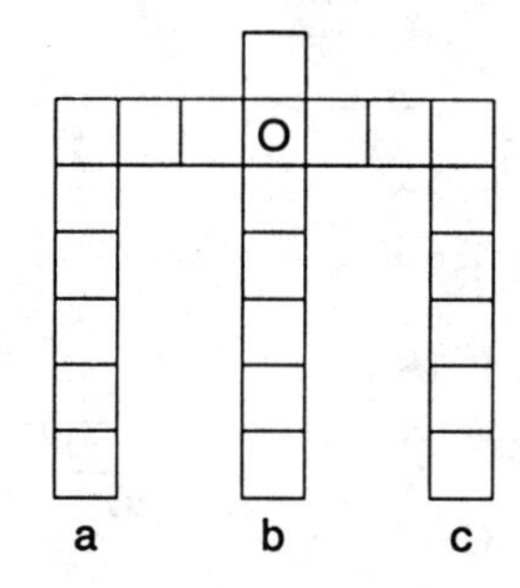

3. a. element typical of amino acid
   b. element carried on hemoglobin
   c. element of water
   d. element of organic compounds

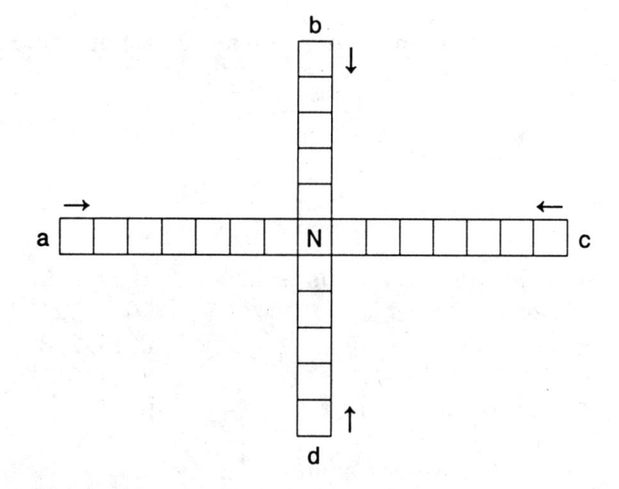

Sleuthing

1. Mary came in to the hospital with the pH of her arterial blood measuring 7.3.
   a. Was her blood more acidic or more basic than the normal range?

   b. What ion would her kidneys be trying to excrete to bring her body fluid back to the normal range?

Word Scrambles

1. a. glycogen subnits — LOSUGEC — □ _ □ _ _ □ _
   b. class of DNA bases — SURPINE — _ _ □ _ _ _ _
   c. good inorganic solvent — TWERA — _ □ _ _ _
   d. absent in RNA — HITYNME — □ _ _ _ _ _ □
   e. element making up 1 percent of body weight — HPSOPHROUS — □ □ □ □ □ □ _ _ _ _
   f. in the nucleus — AND — _ _ □

Total: genetic backbone

□□□□□–□□□□□□□□□

2. a. Na Na Na — ODUMSIS — _ □ _ □ _ _ _
   b. carbohydrate polymer — GONGLECY — □ _ _ _ _ _ _ _
   c. element characteristic of proteins — TRONGINE — _ _ _ □ _ _ _ □
   d. organic element — ABCORN — □ □ _ _ _ _

Total: type of molecule characteristic of living matter

□□□□□□□

Addagrams

1. a. units released in water when a salt dissociates — 12, 4, 3, 5
   b. solution with a pH of 1 — 10, 7, 12, 13
   c. inert chemical gas — 3, 14, 2, 3
   d. what not to do before A&P exam — 8, 11, 6, 1
   e. element bonded to oxygen in water — 9

Total: glucose, for example

1	2	3	4	5	6	7	8	9	10	11	12	13	14

2. a. smallest unit of matter — 2, 3, 6, 10
   b. alkaline — 5, 4, 9, 8, 1
   c. inorganic crystal — 9, 2, 7, 3

Total: decomposition reaction

1	2	3	4	5	6	7	8	9	10

Crossword

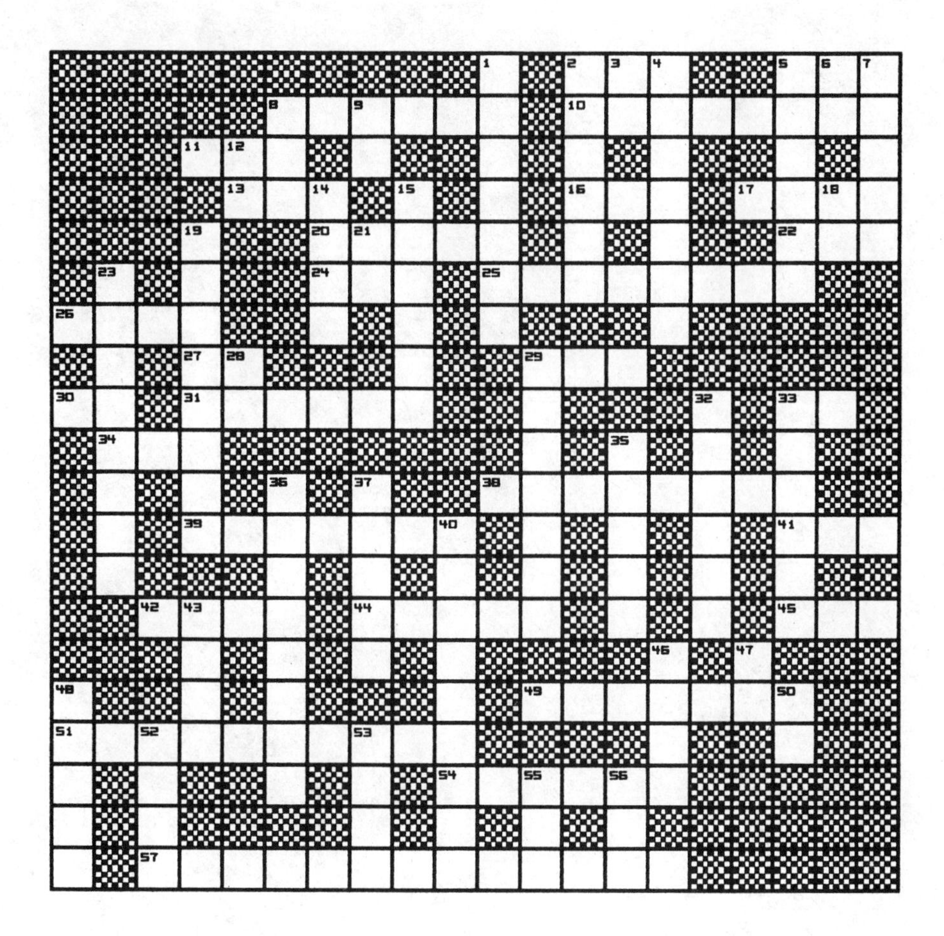

Across

2 Be a victor
5 Pounds per square inch (abbr.)
8 Abbreviation for elements
10 Negatively charged particle
11 South American capital
13 Genetic material
16 Obtain
17 Heavenly light
20 Element in body which is present in tiny amounts
22 Canadian capital (abbr.)
24 Over (poetic)
25 Element always found in protein
26 Chemical union
27 Egyptian sun god
29 Molecule that is high in energy
30 Symbol for iron
31 Element always present in an aerobic process
33 Symbol for chlorine
34 ________ in a wheel
38 A bond where electrons are shared
39 Contains protons and neutrons
41 Charged-up sodium
42 Material of nucleus in cell
44 Type of bond between sodium and chlorine
45 Prefix meaning "new"
49 Type of molecule that contains carbon and hydrogen
51 Repeating unit of RNA
54 Substance that dissolves
57 Breakdown

Down

1 A substance that cannot be broken down
2 Proton number and neutron number together equal the atomic ________
3 Abe's state (abbr.)
4 Neutral particle in an atom
5 Positively charged particle of nucleus of atom
6 ________ long!
7 Unreactive element
8 Gets my genes
9 Abbreviation for small metric weight unit
12 Freud's basic drives
14 Smallest unit of matter
15 Common to carbohydrates, proteins, fatty acids, and nucleic acids
18 Article before a vowel
19 Twice as much as the other in water
21 Regarding
23 A chemical reaction between atom and atom produces a ________
28 Chopper
29 In the presence of oxygen
32 Released when ATP splits
33 Positive ion
35 Abundant solvent in body
36 Chlorine is an electron ________
37 Configuration of DNA
40 Anabolic reaction
43 So bad
46 Alkaline
47 A sugarlike sucrose is a ________ saccaride
48 Negative ion
50 Symbol for calcium
52 Temperature that slows down diffusion rate
53 International House of Pancakes (abbr.)
55 He turned on the light
56 Also

Chapter 3

Cells

Introduction

Trillions of cells arranged into specialized tissues and organs compose the human body. The *cell* is the smallest whole unit, which, in itself, is a complex organization of *organelles* and chemicals adapted for specialized functions. The colloidal suspension of cytoplasm within the cell is kept separate from the external environment of body fluids by the *plasma membrane*. This same type of membrane forms the boundary of many organelles within the cell.

The nucleus acts as the conductor in the symphony of each cell's unique music. The ultimate sound is directed by the genetic material, or DNA in the nucleus, as it coordinates protein synthesis and the integrated activity of all members of the orchestra of the cell.

During the lifetime of most cells of the body, they undergo several cell divisions, producing identical copies of the original cell. Division of the nuclear material is called *mitosis*. It is accompanied by division of the cytoplasm, or *cytokinesis*.

After reading and working with the material in the textbook, you should be able to answer the following questions and do the activities of this chapter.

Multiple Choice

___ 1. Duplication of genetic material occurs during:
   a. interphase
   b. prophase
   c. metaphase
   d. telophase

___ 2. Spindle fibers emanate from which of the following?
   a. centrosome
   b. centrioles
   c. nucleolus
   d. chromatin

___ 3. Which of the following statements is/are incorrect?
   a. Messenger, transfer, and ribosomal RNA play a role in protein synthesis
   b. If the base sequence of DNA is AGGTCA, the messenger RNA template will be TCCAGT
   c. There is a specific type of t-RNA for each amino acid
   d. All statements are correct

___ 4. Aerobic cellular respiration occurs in which organelle?
a. Golgi complex
b. lysosomes
c. mitochondrion
d. nucleus

___ 5. Which of the following statements is/are correct?
a. The rate of diffusion can be altered by changing the temperature
b. The greater the concentration gradient, the faster the rate of diffusion
c. Molecular weight of a substance will affect the rate of diffusion
d. All statements are correct

___ 6. Facilitated diffusion occurs when:
a. a concentration gradient exists
b. the membrane is permeable to the diffusing substance
c. carrier molecules are present
d. all of these

___ 7. Which of the following factors will help determine the permeability of a membrane to a given substance?
a. pore size
b. electrical charge
c. lipid solubility
d. all of these

___ 8. If cells are placed in a hypotonic solution containing a solute to which the membrane is impermeable, what will happen?
a. The cells will swell and ultimately burst
b. The cells will shrink and ultimately dehydrate
c. The cells will shrink at first but ultimately will reach equilibrium with the surrounding solution
d. The cells will show no appreciable change due to adaptation and diffusion of solute and solvent

___ 9. A physiologic process involving expenditure of energy to move substances against the concentration gradient is called:
a. osmosis
b. active transport
c. diffusion
d. facilitated diffusion

___ 10. Which of the statements below is true concerning the most recent theoretical model of cell membrane structure?
a. The "bulk" of the membrane consists of a bilayer of lipid molecules with proteins scattered in an irregular pattern at various points on and through the bilayer
b. Nucleic acid molecules associated with protein and lipid layers form an integral part of the membrane structure
c. The predominant molecules in most membranes are phosphorylated proteins and cholesterol
d. All statements are true

___ 11. Which of the pairs below is correctly matched?
a. lysosome–respiratory enzymes
b. centrosome–centrioles
c. nucleolus–DNA
d. sugar molecule–thymine

___ 12. Experimental manipulation to increase cellular energy production might include implanting additional:
a. lysosome
b. mitochondria
c. Golgi bodies
d. ribosomes

___ 13. The organelle thought to be involved in packaging secretory materials such as glycoproteins is the:
a. mitochondrion
b. lysosome
c. Golgi complex
d. centrosome

___ 14. The series of membrane-bound channels between the nuclear and cell membranes, serving as a cellular transport system, is the:
a. lysosome
b. Golgi complex
c. endoplasmic reticulum
d. centrosome

___ 15. The nuclear membrane begins to disintegrate during:
a. prophase
b. interphase
c. metaphase
d. anaphase

___ 16. The division of the nucleus in the somatic cells of the body is:
a. meiosis
b. mitosis
c. cytokinesis
d. none of these

___ 17. The excess tissue that grows as a result of uncontrollable cell duplication is called a:
a. tumor
b. neoplasm
c. growth
d. all of these

___ 18. A malignant tumor
a. metastasizes
b. is invasive
c. secretes autocrine motility factor
d. all of these

___ 19. When nonmalignant cells divide, the growth stops because of:
a. angiogenesis
b. contact inhibition
c. metastasis
d. none of these

___ 20. Which of these is a possible cause for a cell to lose control of its normal growth limits?
a. viruses
b. oncogene activation
c. carcinogens
d. all of these

___ 21. Which of these cancer terms is/are properly matched?
a. sarcoma—connective tissue cancer
b. myeloma—cancer involving cells of the marrow
c. osteogenic sarcoma—involving bone tissue
d. all of these are correctly matched

___ 22. Removal of tissue for microscopic examination is called a:
a. necrosis
b. biopsy
c. endoscopy
d. none of these

___ 23. A solution that has a greater concentration than body cells and fluids is considered:
a. isotonic
b. hypotonic
c. hypertonic
d. none of these

___ 24. Distilled water would be considered ________ to body cells.
a. isotonic
b. hypotonic
c. hypertonic
d. none of these

___ 25. Necrotic cells are considered:
a. dead
b. spreading
c. malignant
d. tumors

True/False

__________ 1. Ribosomes are composed primarily of protein and *DNA*.

__________ 2. Fluid and dissolved particles are engulfed by *pinocytosis*.

__________ 3. If a membrane is impermeable to salt, then water will tend to *leave* the cell by osmosis if the salt concentration outside the cell is greater than inside the cell.

__________ 4. Mitosis and cytokinesis produce daughter cells that contain *half* the identical DNA information as the parent.

__________ 5. Hydrostatic pressure is necessary in order to move molecules across a membrane by *filtration*.

__________ 6. A cell membrane is composed chiefly of lipid and *protein*.

__________ 7. The *mitochondria* are largely responsible for packaging cell products to be secreted.

__________ 8. ATP is produced in the *lysosome*.

__________ 9. Proper sequencing of amino acids during protein synthesis is the function of *ribosomal RNA*.

__________ 10. *Aerobic* respiration occurs in the mitochondrion.

__________ 11. During *metaphase* the centromeres split and sister chromatids move toward opposite poles of the cell.

__________ 12. The *centrosome* contains digestive enzymes.

_______ 13. One major difference between DNA and RNA is that the latter is a single-stranded molecule containing uracil instead of *thymine.*

_______ 14. *Translation* is the process by which DNA produces RNA.

_______ 15. *Meiosis* is the term for somatic cells producing two daughter cells.

_______ 16. The term used to describe RNA producing protein is *transcription.*

_______ 17. The phase of the cell life cycle when the genetic material is condensed and coiled up is *interphase.*

_______ 18. A *peripheral* protein picks up a molecule like glucose to help carry it across the membrane.

_______ 19. Extracellular fluid is considered the body's *external* environment.

_______ 20. Plasma is considered an *extracellular* fluid.

_______ 21. *Peripheral* proteins penetrate through the membrane.

_______ 22. Lipid-soluble substances pass through *channels* in the cell membrane.

_______ 23. Substances are extruded from the cell by the process of *exocytosis.*

_______ 24. When a red blood cell is placed in an *isotonic* solution, it crenates.

_______ 25. When the contents of a lysosome are released, *digestion* of engulfed materials, for example, by the enzymes it contains, occurs.

Completion

1. The process by which the genetic material in the nucleus copies itself into the molecule RNA is called _______________.
2. The process by which RNA carries out protein synthesis in the cell is called _______________.
3. The molecule that possesses an anticodon to identify which amino acid to carry to the ribosome is called _______________.
4. Duplication of genetic material takes place during which phase of mitosis? _______________.
5. Reduction of the genetic material, which occurs during gamete formation, is called _______________.
6. During cell division, the mitotic spindle forms from the organelle known as the _______________.
7. A normal cell of the body, *not* a reproductive cell, is known as a _______________ cell.
8. The process of cytoplasmic division that accompanies mitosis is known as _______________.
9. An anticodon consists of a series of _______________ bases.
10. In the sequence of transcription DNA produces _______________, which then directs protein synthesis.
11. The cell organelle at which the different amino acids hook into a protein is the _______________.
12. The network of membranes within the cell which is continuous with both the nuclear membrane and the plasma membrane is called the _______________.
13. Lipids are synthesized on _______________ while proteins are synthesized on the _______________.
14. Microfilaments and microtubules within the cell are part of the structure of the _______________.

15. Slender microtubules, made of the protein tubulin, form slender extensions of the cell called ________________ and ________________.
16. The organelle of the cell where energy is packaged into molecules called ATP is known as the ________________.
17. The lysosome contains ________________, which are involved in the breakdown of worn-out cell parts.
18. The ________________ consists of a series of flattened sacs in the cell involved in sorting and delivering proteins.
19. Which organelle is generally the largest structure in the cell and controls cellular activities? ________________.
20. The structure that consists primarily of phospholipids and proteins and separates the internal contents from the external environment is the ________________.
21. Movement of substances across a selectively permeable membrane, with the aid of gravitational or hydrostatic pressure, is called ________________.
22. What structures "stud" the nuclear membrane and the endoplasmic reticulum and make it "rough" looking? ________________.
23. Enzymes capable of digesting bacteria brought into the cell by phagocytosis are contained in the ________________ of the cell.
24. Diffusion and osmosis are considered ________________ (active, passive) processes.
25. When substances are brought into the cell against their concentration gradient, an expenditure of energy from the breakdown of the molecule ________________ must occur.

Lost Sheep

1. anaphase, telophase, metaphase, interphase
2. active transport, diffusion, filtration, osmosis
3. ribosome, synthesis, codon, lysis
4. mitochondrion, digestion, ATP, cristae
5. lysosome, enzymes, lysis, centrosome
6. hypertonic, 10 percent sodium chloride, hemolysis, crenation
7. crenate, shrivel, dehydrate, swell
8. hemolysis, hypotonic, highly salty, distilled water
9. diffusion, against gradient, passive, oxygen entering blood
10. microfilaments, micrometer, microtubules, cytoskeleton
11. flagella, melanin, lipids, glycogen
12. phospholipid, cytoplasm, peripheral protein, integral protein
13. interstitial, extracellular, plasma, intracellular
14. passive, pinocytosis, phagocytosis, active transport
15. lipid, spindle, flagella, microfilament
16. mitosis, cytokinesis, meiosis, digestion
17. chromatin, chromosomes, chromomere, chromatids
18. lysis, equatorial plate, metaphase, chromosomal microtubule attachment
19. nucleus, cytokinesis, cleavage furrow, cytoplasm
20. metastasis, benign, autocrine motility factor, no contact inhibition

Matching

Set 1

___	1. site of protein synthesis	a. centrosome
___	2. site of carrier molecules	b. Golgi complex
___	3. hereditary information	c. ribosome
___	4. location of cristae	d. mitochondria
___	5. contains hydrolytic (digestive) enzymes	e. cell membrane
___	6. contains centrioles	f. chromatin
___	7. has flattened membranous sacs	g. lysosome
___	8. is selectively permeable	
___	9. made of r-RNA	
___	10. producer of ATP	

Set 2

___	1. process that requires carrier system	a. osmosis
___	2. process that requires energy	b. diffusion
___	3. process where substances move against gradient	c. active transport
___	4. movement of salt down concentration gradient	d. filtration
___	5. responsible for exchange of O_2 and CO_2 between lungs and blood	
___	6. movement of solvent from higher concentration to lower solvent concentration	
___	7. requires hydrostatic pressure	

Set 3

___	1. chromatids lined up at center of plate	a. interphase
___	2. production of new DNA	b. metaphase
___	3. centromeres duplicate	c. anaphase
___	4. centrioles separate	d. prophase
___	5. nuclear membrane disappears	e. telophase
___	6. two identical sets of chromosomes have reached opposite poles	
___	7. process of movement of chromatids to opposite poles	
___	8. new nuclear membrane begins to form	

Double Crosses

1. *Across*:
movement of solvent from higher concentration to lower concentration

Down:
somatic nuclear division

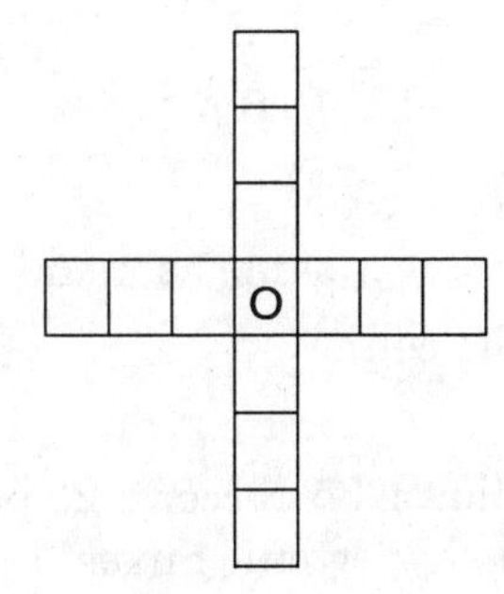

2. *Across:*
hair-like cell processes

Down:
flattened sacs that communicate with ER

3. *Across:*
discriminatory membrane permeability

Down:
membranous transport system within cell

4. *Across:*
diffuse nuclear DNA

Down:
a sister attached by centromere

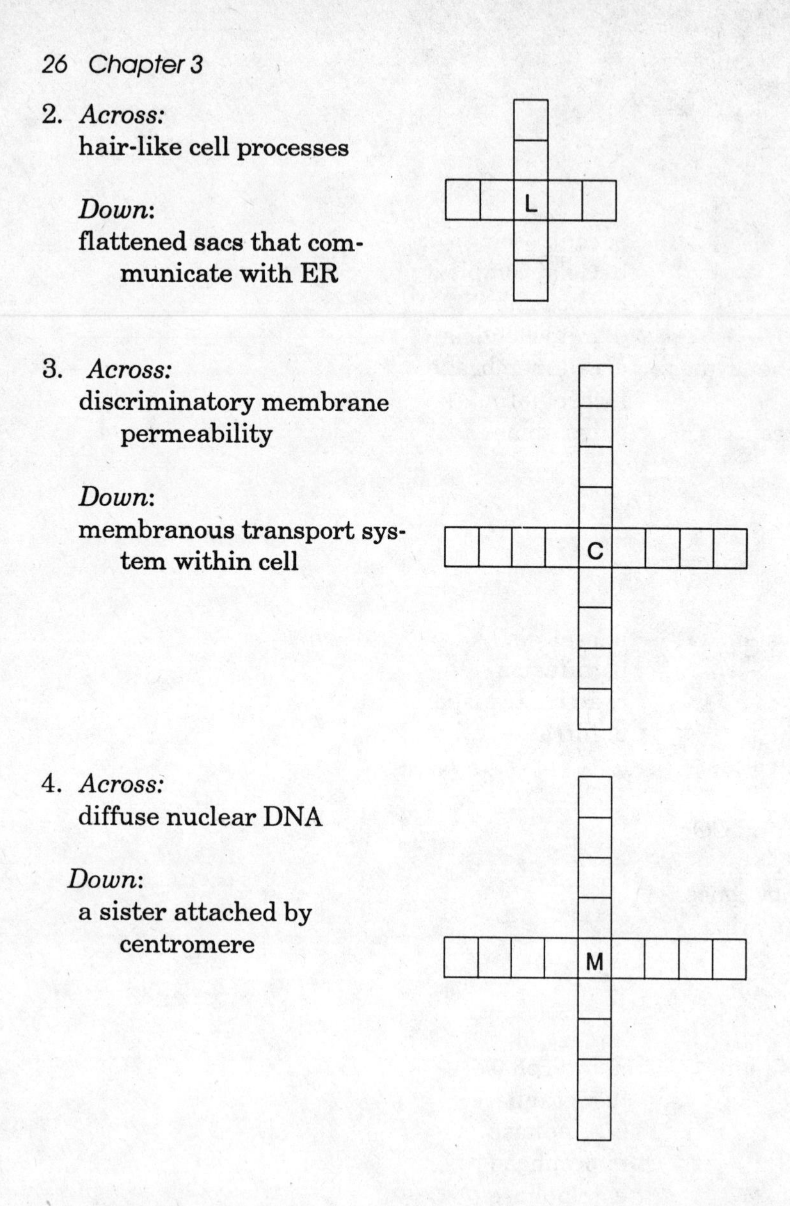

Sleuthing

1. Listed below are the outstanding characteristics of three unknown materials: A, B, and C. Using the information presented and your knowledge regarding the factors that affect diffusion, answer the questions below.

	A	B	C
diameter of molecule	8.6 Å	10.8 Å	3. 86 Å
lipid solubility	–	++	–
ionic charge	no charge	same as membrane	opposite to membrane

a. Which would diffuse most readily into the cell?

b. Three solutions made with B were as follows: 35 percent, 20 percent, and 10 percent. The cell was placed into solutions of each of these hypertonic mixes. In which of these three solutions would the cell change size most readily? How?

c. Which molecule would pass most readily through the pores or channels of the membrane of the cell? Through the lipid bilayer? How, in each case?

d. Which molecule would probably have to be transported through the cell membrane by a carrier molecule?

Word Scrambles

1. a. adjective meaning "body" — T O M A S I C — ☐ ☐ ☐ _ _ _ _
   b. bursting apart of red blood cell — S H I M O L E S Y — _ ☐ _ ☐ ☐ ☐ ☐ _ _

Total: busy after phagocytosis

☐☐☐☐☐☐☐☐

2. a. endoplasmic — T R I C E M U L U — ☐ _ ☐ ☐ ☐ _ _ _ _
   b. type of RNA — F R E N S T A R — _ ☐ ☐ ☐ ☐ _ _ _
   c. charged element — O N I — ☐ ☐ ☐
   d. made in cristae — P A T — _ ☐ ☐

Total: double stranded makes single stranded

☐☐☐☐☐☐☐☐☐☐☐☐☐

Addagrams

1. a. cytoplasmic organelle with digestive enzymes — 6, 2, 8, 4, 8, 4, 12, 19
   b. sum of chemical processes that cells carry out — 9, 13, 3, 7, 15, 4, 6, 10, 8, 14
   c. active process of cell taking in liquid material — 5, 10, 18, 4, 11, 2, 3, 4, 8, 10, 8
   d. mitochondrial infoldings — 1, 16, 10, 8, 3, 17, 19

Total: regulates movement of substances in and out of cell

☐☐☐☐☐☐☐☐☐☐☐ ☐☐☐☐☐☐☐☐

1 2 3 4 5 6 7 8 9 10 11 12 13 14 15 16 17 18 19

2. a. molecules made in mitochondrion — 3, 8, 1, 12
   b. solution less concentrated is ——tonic — 2, 7, 1, 5
   c. in the vapor state — 4, 3, 10
   d. what one chromatid of a pair is to the other — 10, 11, 12
   e. prefix meaning "cell" — 6, 7, 8, 9

Total: process of engulfment

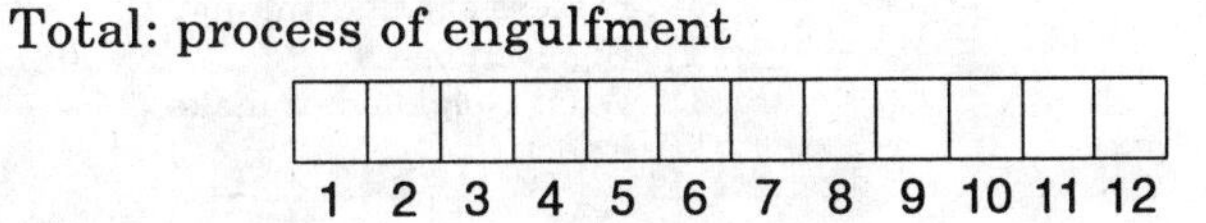

Crossword

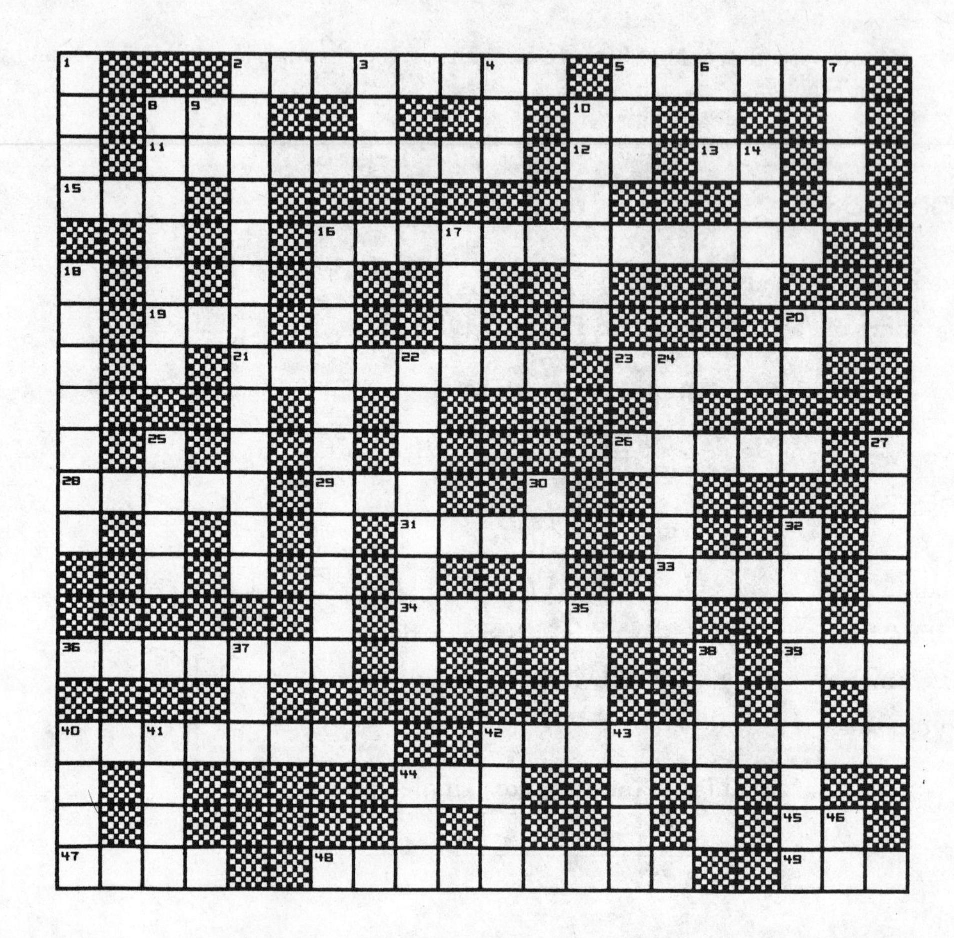

Across

2 Semipermeable cover
5 Packaged in mitochondrion
8 Cell-mediated Immunity (abbr.)
10 Abbreviation for energy
11 Endoplasmic ________
12 Compass direction
13 Fourth note on scale
15 Driving While Intoxicated (abbr.)
16 Process of engulfing
19 Much ________ about nothing
20 Small point
21 Excessive growth
23 ________ acid of protein
26 The bilayer of the cell membrane is made of this
28 ________ complex where carbohydrates are packaged
29 How many centrioles there are in a centrosome
31 Remain
33 What you scratch
34 Nuclear division
36 Organelle in charge
39 "Hi" to Jose
40 Type of protein that is part of the cell membrane
42 Part of cell cycle when new DNA is made
44 Pull
45 Not out
47 Woodwind
48 From which the spindle fibers emanate
49 Negative

Down

1 Activity at the library
2 Organelle that packages energy into ATP molecules
3 Unit used to measure energy
4 "Name" to Pierre
5 Ending on an organic molecule meaning double bond
6 Tiny creature
7 What DNA looks like before it is condensed
8 Inner membrane of mitochondrion
9 Myself
10 Chemical that speeds up biological reactions
14 Gives off hydrogen ions in solution
16 Process of cell drinking
17 Big even
18 What Golgi bodies make
20 Enact
22 Organelle involved in digestion of engulfed material
24 Process of reduction division of nucleus in a sex cell
25 Applaud
27 Stage of mitosis
30 Prefix meaning "cell"
32 Uncoiled genetic material
35 Little play
37 Unit of energy
38 ________-tonic solution that makes cells shrivel
40 Phagocytosis involves taking a substance ________ the cell
41 Dorothy's dog
42 Russian fellow
43 Bad
44 What ultraviolet light can do for the skin
46 Yes and ________

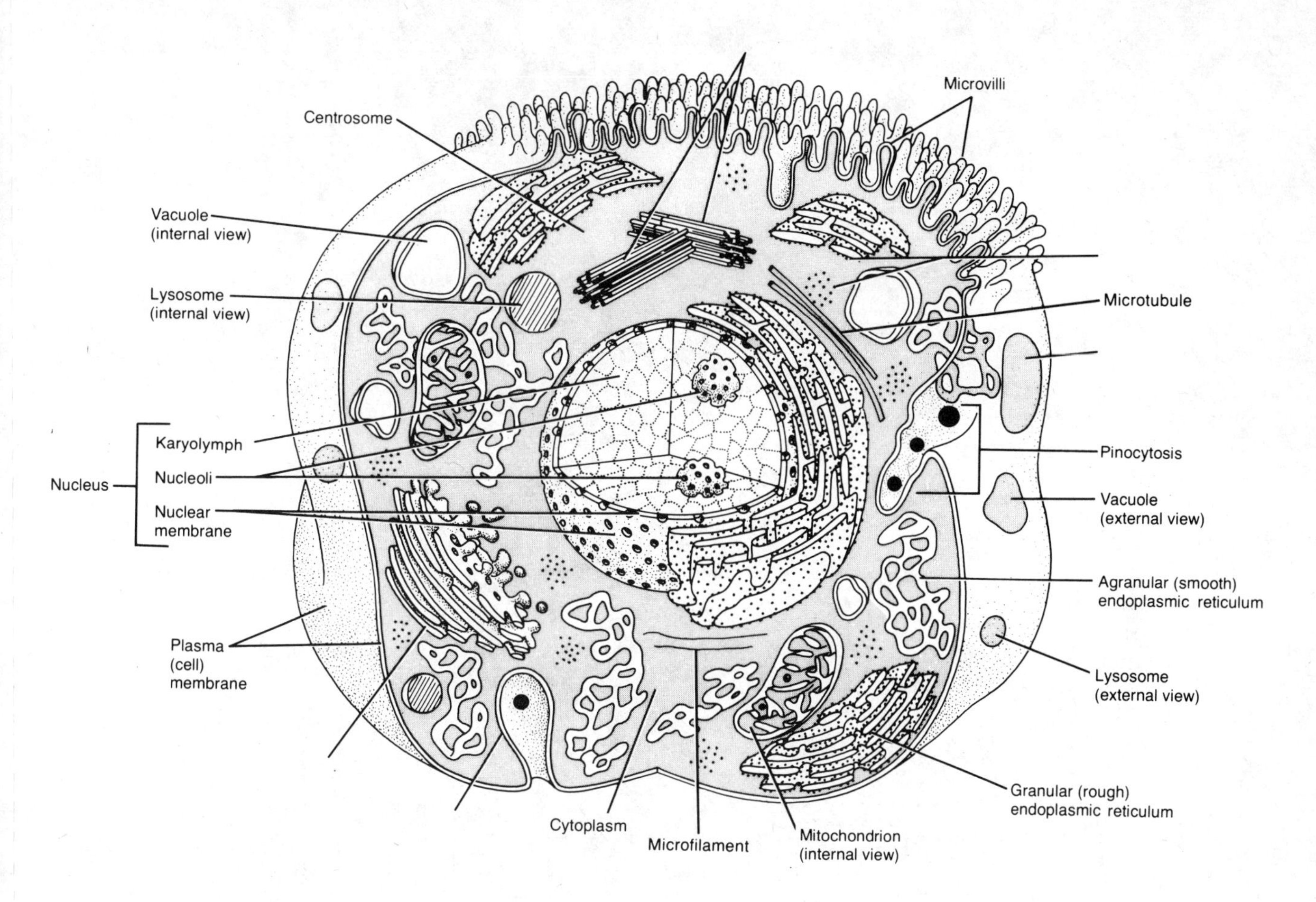

Anatomic Artwork

1. Label the organelle where protein is made.
2. Label the vesicle where fluid is being taken into the cell.
3. Label the perpendicular organelles that give rise to the mitotic spindles.
4. Label the organelle where energy is packaged.
5. Label the organelle where carbohydrates are combined with proteins for secretion.

Chapter 4

Tissues

Introduction

Trillions of cells arranged into specialized tissues and organs comprise the human body. During embryological development, cells that at first are all alike become differentiated, guided by nuclear instructions and environmental factors.

The first level of organization in the hierarchy of living things is the *cellular* level, which was covered in Chapter 3. Cells of a similar structure with the same function become organized into the next level, which is that of a *tissue*. Tissues, in turn, are organized into *organs*, which form the *systems*, and all of the systems combined make up the total *organism*.

The human body has a myriad of tissues, all of which have their own special adaptations and functions. After careful study, these varied tissues can be classified into one of four major tissue types: *epithelium, connective, muscle,* and *nervous* tissue. In the text, the specific characteristics, functions, and locations of these tissues were outlined. You should therefore be familiar with this material and should be able to answer the following questions and complete the varied activities.

Multiple Choice

___ 1. Basement membranes are characteristically found in which of the following tissues?
- a. hyaline cartilage
- b. cardiac muscle
- c. pseudostratified columnar
- d. neuroglia

___ 2. Filtration, secretion, and absorption are specialized functions of which of the following tissues?
- a. dense connective tissue
- b. epithelial tissue
- c. skeletal and cardiac muscle
- d. none of these

___ 3. Which of the following connective tissue cells is involved in the production of antibodies?
- a. macrophages
- b. fibroblasts
- c. mast cells
- d. plasma cells

___ 4. Which of the following fiber types predominates in the cartilage of the external ear and epiglottis?
a. collagenous
b. elastic
c. reticular
d. keratin

___ 5. Which of the pairs below is/are correctly matched?
a. dense connective tissue–muscle sheaths
b. simple columnar–lining of intestine
c. macrophages–loose connective tissue
d. all pairs are correctly matched

___ 6. Several varieties of fibers, loosely interwoven with fibroblasts, mast cells, and macrophages within a gel-like matrix, is a description of which of the following tissues?
a. reticular tissue
b. hyaline cartilage
c. dense connective tissue
d. loose connective tissue

___ 7. Which of the following cells would be found in epithelial tissue?
a. adipocytes
b. goblet cells
c. mast cells
d. none of these

___ 8. Which of the following is specialized for contractility?
a. nervous tissue
b. dense connective tissue
c. muscle tissue
d. none of these

___ 9. Which of the following tissues is avascular?
a. cardiac muscle
b. stratified squamous
c. compact bone
d. none of these

___ 10. A major constituent of the matrix of loose connective tissue is:
a. fibrocytes
b. hyaluronic acid
c. perichondrium
d. goblet cells

___ 11. Which of the following constitutes an epithelial membrane?
a. epithelial tissue and underlying connective tissue
b. epithelium and its nerve supply
c. epithelium and its direct blood supply
d. none of these

___ 12. Which of the following membranes does not contain epithelial tissue?
a. mucous
b. serous
c. cutaneous
d. synovial

___ 13. Which of the following types of connective tissue has no blood vessels or nerves within it?
a. bone
b. areolar
c. cartilage
d. stratified squamous

___ 14. Which of the following is not a connective tissue fiber?
a. elastic
b. reticular
c. collagenous
d. keratin

___ 15. Which of the following is not associated with the osteon?
a. lamellae
b. canaliculi
c. chondrocyte
d. Haversian canal

___ 16. Which of the following is/are properly matched?
a. stratified squamous–basement membrane
b. pseudostratified columnar–upper respiratory tract
c. goblet cell–mucus
d. All are correctly matched

___ 17. Which of the following is/are properly matched?
a. areolar–loose connective tissue
b. elastic cartilage–articulating surfaces
c. canaliculi–fibrocartilage
d. reticular fibers–dense connective tissue

___ 18. Which of the following statements is/are true?
a. Sweat glands are examples of endocrine glands
b. Exocrine secretions may vary from very fluid, such as tears, to more viscous, such as mucus
c. Both a and b are true
d. Neither a nor b is true

___ 19. Which of the following statements is false?
a. Skeletal muscle is voluntary and multinucleate
b. Smooth muscle cells are separated by intercalated discs
c. Involuntary smooth muscle is found in the digestive viscera
d. Cardiac muscle is striated and involuntary

___ 20. Which set is properly matched?
a. neuron–axon
b. neuroglia–dendrite
c. mesenchymal cell–epithelium
d. serous membrane–mouth

___ 21. Which of the following is/are examples of extracellular materials?
a. mucus
b. plasma
c. interstitial fluid
d. all of these

___ 22. The term "matrix" refers to:
a. intercalated discs
b. ground substance
c. protein fibers
d. none of these

___ 23. Which of the following tissues has lost the capacity for mitosis?
a. epithelium and connective tissue
b. epithelium and muscle
c. muscle and nerve
d. connective tissue and muscle

___ 24. The rate at which an injured tissue heals can be affected by:
a. the tissue's regenerative capacity
b. the blood supply to the tissue
c. both a and b
d. neither a nor b

___ 25. Which of the following is/are correctly paired?
a. hyaluronic acid–viscous fluid that lubricates joints
b. chondroitin sulfate–jelly substance found in cartilage
c. mucous connective tissue–Wharton's jelly
d. all are correctly paired

True/False

__________ 1. Tendons and ligaments are composed primarily of *loose* connective tissue.

__________ 2. Internally, *cartilage* characteristically lacks a direct blood and nerve supply.

__________ 3. The most regenerative of the primary tissues is *epithelium.*

__________ 4. Epithelial tissue covers all body surfaces and is *avascular.*

__________ 5. Neuroglia are supportive and protective cells found within *nerve* tissue.

__________ 6. The jellylike material found within cartilage is called *chondroitin sulfate.*

__________ 7. The most common cell found in connective tissue proper is the *plasma cell.*

__________ 8. Transitional, microvilli, and goblet are terms that apply to *connective* tissue categories.

__________ 9. A characteristic of *epithelial* tissue is conductivity.

__________ 10. *Nerve* tissue is specialized for support.

__________ 11. The major antibody-producing cells are called *macrophages.*

__________ 12. The tiny spaces where osteocytes are located are the *canaliculi.*

__________ 13. *Exocrine* glands are ductless glands.

__________ 14. Serous membranes line *closed* body cavities.

__________ 15. The *intercellular material* in connective tissue is largely responsible for the qualities of the tissue.

__________ 16. Axons carry the nerve impulse *away* from the cell body.

__________ 17. *Cardiac* muscle is striated and voluntary.

__________ 18. Reticular fibers are short, branching fibers made of *collagen* with a glycoprotein coating.

__________ 19. Tears, secretions of glands, and mucus are examples of *extracellular* fluids.

__________ 20. Mesenchymal cells are found in *epithelial* tissues.

__________ 21. The body's coverings and linings are composed primarily of *connective* tissue.

__________ 22. Filtration, absorption, and secretion are the outstanding functions of *epithelial* tissue.

__________ 23. Adipose tissue is a *loose* connective tissue that has been filled with adipocytes.

__________ 24. The osteon is the structural unit of *compact* bone.

__________ 25. *Cutaneous* membranes typically lack epithelial tissue.

Completion

1. The cementing layer that attaches epithelial tissue to adjacent connective tissue is called __________.
2. __________ cells are specialized mucus secretors found in columnar epithelium.
3. When epithelial tissues have several layers of cells, the tissue is called __________.
4. The specialized epithelium that lines the urinary bladder and changes shape to accommodate varying urine volumes is called __________.
5. Simple __________ epithelium is found where diffusion and filtration are carried out.
6. Stratified __________ epithelium protects and secretes and is found in the male urethra.
7. Two of the least commonly found epithelial tissues are stratified __________ and __________.
8. The major protein fiber in hyaline cartilage is __________.
9. The major differences among the categories of connective tissue are the nature of the __________ which determines the tissue's qualities.
10. __________ are the cells in cartilage and __________ are the cells in osseous tissue.
11. The __________ is another name for the ground substance in which connective tissue cells are embedded.
12. __________ fibers consist of the protein collagen with an outer coating of glycoprotein.
13. Compact bone is comprised of osseous tissue arranged in microscopic units called __________.
14. The outer covering on cartilage is called the __________.
15. __________ tissue is the most abundant tissue in the body.
16. All connective tissues are formed from __________.
17. Muscle tissue that is under our conscious control is referred to as __________ muscle.
18. __________ is characterized by intercalated discs, which aid in transmitting impulses within the heart.
19. Highly branched processes that conduct nerve impulses toward the cell body are called __________.
20. __________ are supportive and protective, nonconducting cells in nerve tissue.
21. The conducting cells of nerve tissue are called __________.
22. Glands that secrete their products directly into the bloodstream are called __________ glands.
23. A __________ is a mass of epithelial cells adapted for secretion.

24. Membranes that line body cavities with an opening to the exterior are called ________________ membranes.
25. ________________ glands have ducts and secrete their products to the surface through these ducts.

Lost Sheep

1. squamous, areolar, cuboidal, columnar
2. lacunae, adipose, perichondrium, hyaline
3. collagen, cilia, microvilli, goblet
4. transitional, mast, plasma, macrophage
5. collagen production, secretion, diffusion, absorption
6. areolar, dense connective tissue, fibroblasts, tendon
7. epidermis, bone, blood, cartilage
8. lamellae, Haversian canal, compact, hyaline
9. striated, tapered ends, smooth, involuntary
10. cell body, neuroglia, dendrite, axon
11. matrix, chondrocytes, hyaline, Haversian canal
12. stroma, reticular, elastic, liver
13. mesenchyme, cuboidal, transitional, pseudostratified
14. hyaluronic acid, synovial membrane, elastic cartilage, joints
15. osteocytes, anticoagulant, connective tissue proper, mast cell

Matching

Set 1

___	1. flat, irregular cells	a. pseudostratified
___	2. one layer of cells that appears as many	b. mucous membrane
___	3. cell that secretes mucus	c. squamous
___	4. membrane that lines digestive, respiratory, reproductive tracts	d. goblet
		e. transitional
___	5. "anchors" epithelium	f. exocrine
___	6. lines urinary bladder	g. endocrine
___	7. membrane lining closed cavity	h. simple
___	8. ductless gland	i. basement membrane
___	9. single layer of cells	j. serous membrane
___	10. salivary and mammary glands	

Set 2

___	1. support and protection for neurons	a. axon
___	2. impulse toward cell body	b. neuroglia
___	3. voluntary, striated	c. smooth muscle
___	4. involuntary	d. dendrite
___	5. multinucleate	e. skeletal muscle

Set 3

___	1. phagocytosis	a. chondrocyte
___	2. produces anticoagulant	b. erythrocyte
___	3. cell found in hyaline cartilage	c. reticular tissue
___	4. common to bone and cartilage matrix	d. osteocyte
___	5. organ stroma	e. fibroblast
___	6. primitive cell that produces fibers	f. macrophage
___	7. osteonic cell	g. lacunae
___	8. blood	h. mast cell
___	9. type of dense collagenous tissue	i. tendon
___	10. produces antibodies	j. plasma cell

Double Crosses

1. *Across:*
protects neurons

Down:
stores fat

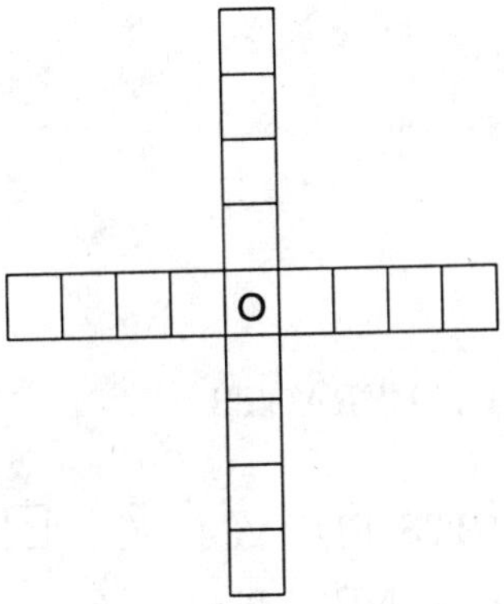

2. ACROSS:
delicate, branching fiber

DOWN:
primitive connective tissue
cell

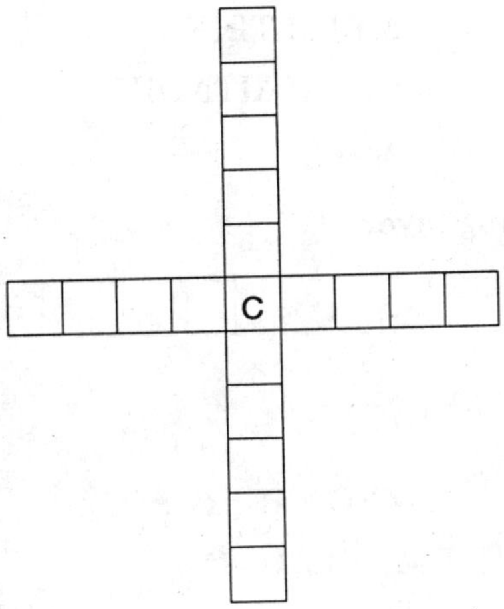

3. *Across:*
   a sulfate in cartilage

   Down:
   cell in lacunae

R

Word Scrambles

1. a. matrix producer of connective tissue — BLIBFRAOTS — _ _ ☐ _ _ ☐ _ _ _ _
   b. several layers — RITSIFDEAT — ☐ _ ☐ _ ☐ _ _ _ ☐ _
   c. cell type lining respiratory tract — CMANROLU — _ _ _ _ ☐ ☐ _ _
   d. histiocyte — PROAMCGEHA — ☐ ☐ _ _ _ _ _ ☐ _ ☐
   e. primary tissue type — MEIULTPIEH — ☐ _ _ _ _ ☐ _ _ _ ☐
   f. epithelium typical of urinary bladder — LATRSAITNOIN — _ _ _ ☐ _ _ _ _ _ _ _ _

   Total: anchoring layer

2. a. gland with duct — XECONEIR — _ _ ☐ ☐ _ _ _ ☐
   b. type of epithelial cell — BDILAUOC — _ _ ☐ _ _ _ _ ☐
   c. secretes into bloodstream: endocrine ________ — DNLAG — ☐ ☐ _ _ _
   d. group of similar cells — SITSEU — ☐ _ _ _ _ ☐
   e. loose connective tissue — REAROAL — _ _ _ _ ☐ _ _

   Total: mucus secretor

Addagrams

1. a. product of most common connective tissue cell — 7, 8, 0, 9, 3
   b. type of cartilage — 9, 13, 17, 1, 2, 6, 11
   c. band of dense connective tissue — 5, 9, 16, 10, 12, 16
   d. goblet's product — 15, 14, 11, 14, 1
   e. nervous tissue cell — 16, 9, 14, 18, 12, 16
   f. vascular — 4

Total: larger ducts of glands lined by this not too common tissue

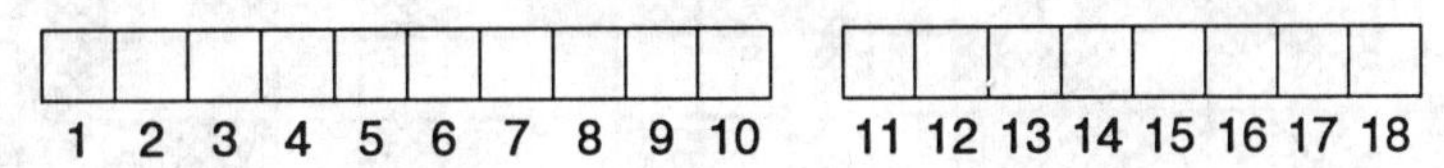

2. a. muscle type that shows branching and fusing — 16, 9, 5, 13, 14, 7, 6
   b. skeletal muscle characteristic resulting from arrangement of myofilaments — 15, 10, 5, 1, 7, 3, 4, 12
   c. fiber of cartilage in epiglottis — 11, 8, 9, 15, 10, 1,2

Total: unique to cardiac muscle

1 2 3 4 5 6 7 8 9 10 11 12   13 14 15 16

Crossword

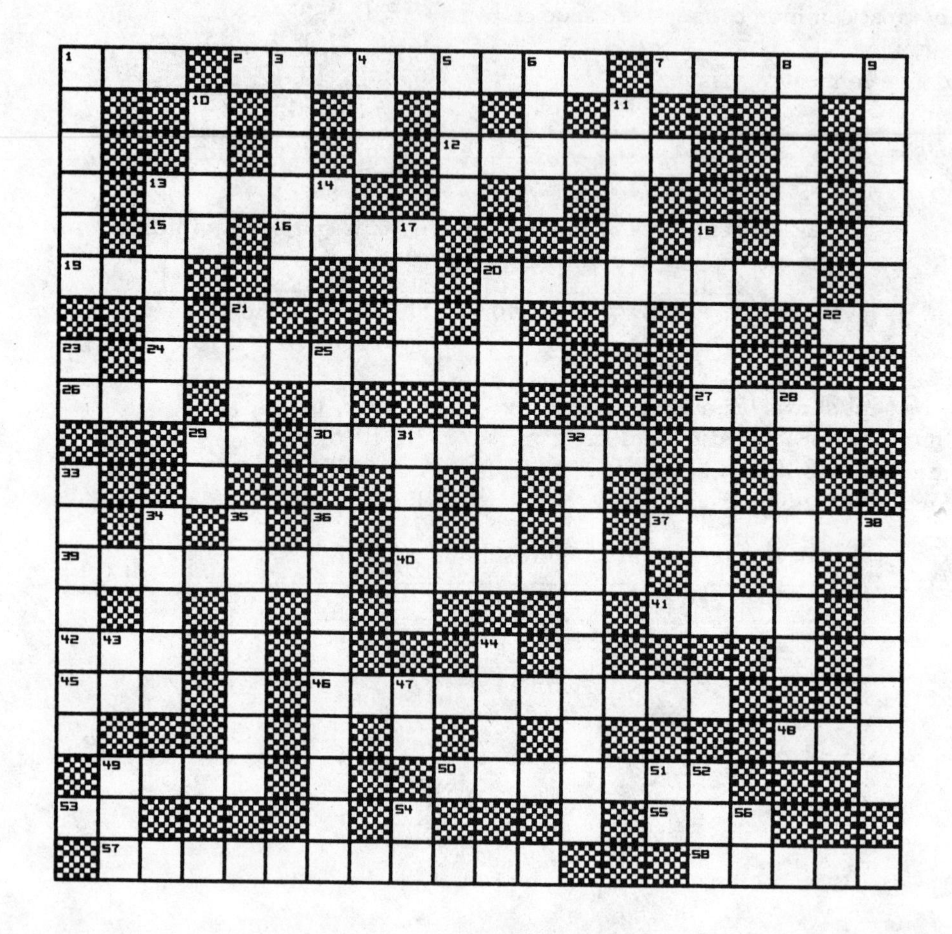

Across

1 Cup
2 Bone cell
7 Cell that produces antibodies
12 Intercellular ground substance
13 Epithelial secretor
15 Grand ________ Opri
16 Vice
19 Decline, fade
20 A function of glandular epithelium
22 Operating Room (abbr.)
24 Transitional is this type of tissue
26 Wager
27 One of five senses
29 Classified unit
30 Epithelial cells that line small intestine
37 Only one layer of cells
39 Cell that stores fat
40 Before long
41 Smell
42 Roman number three
45 Profits after expenses
46 Tissue specialized to transport and bind
48 Southern military college (abbr.)
49 Connective tissue cell
50 Loose connective tissue
53 Calcium (abbr.)
55 Not bright
57 Cartilage cell
58 Elastic tissue

Down

1 Specialized to contract
3 Cancellous
4 Consume food
5 Past tense of come
6 French head
8 Carbohydrates
9 Connective tissue with many cells and fibers
10 Grinds grain
11 Collagenous and elastic ________
13 Secretes mucus
14 Physician (abbr.)
17 Tablet
18 Has multiple layers
20 Broad, flat irregular cell
21 Function of connective tissue
23 Nota bene (abbr.)
25 Ad ________ committee
28 A function of bone
29 Amino acid (abbr.)
31 Areolar
32 A function of epithelium
33 Articular cartilage
34 Boundary
35 Bone with Haversian Systems
36 Connective tissue forming framework of many organs
38 Fibers that can recoil
43 That is (abbr.)
44 Proximal
47 Negative reply
49 Big ________ hamburger
51 Advertisement (abbr.)
52 To eliminate
54 Care unit where surgical patients receive special care
56 ________, myself and I

Chapter 5

The Integumentary System

Introduction

The skin is a tough protective covering over the surface of the body which weighs between 6 and 10 pounds. It functions to help regulate heat in the body, is a route for water loss, is involved in the production of vitamin D, and contains sensory cells and cells of the immune system. It is composed of an upper layer of keratinized stratified epithelium (the *epidermis*) and a lower layer of dense connective tissue (the *dermis*). The two layers of skin rest on a layer of loose connective tissue known as subcutaneous tissue.

After studying Chapter 5 in the textbook, you should be able to answer the following questions and complete the various activities.

Multiple Choice

___ 1. The arrector pili muscle is under ________ control.
   a. conscious
   b. voluntary
   c. autonomic
   d. none of these

___ 2. The stratum corneum consists of:
   a. cells producing melanin
   b. cells undergoing mitosis
   c. dead cells
   d. cells containing adipose

___ 3. Which two layers are involved in the joining of epidermis and dermis?
   a. germinativum and reticular
   b. basale and reticular
   c. germinativum and papillary
   d. basale and subcutaneous

___ 4. Which is/are properly matched?
   a. sudoriferous–sweat
   b. sebaceous–keratin
   c. adipocyte–triglyceride (fat)
   d. all are properly matched

___ 5. A third degree burn is not considered painful because:
  a. it does not reach down into the dermis
  b. it does not damage the nerve endings
  c. it destroys nerve endings in the dermis
  d. it usually involved second degree burns also, which are not painful

___ 6. Vitamin D:
  a. is produced in the deep layers of the epidermis
  b. is synthesized upon exposure to ultraviolet light
  c. needs to be further activated in the liver and kidney
  d. all of these

___ 7. Cells begin to die in the ________ layer of the epidermis.
  a. lucidum
  b. granulosum
  c. germinativum
  d. corneum

___ 8. Which of the following is involved in heat regulation?
  a. thermoreceptors
  b. perspiration
  c. varying blood flow to skin
  d. all of these

___ 9. A decubitus ulcer is also known as:
  a. a bedsore
  b. acne
  c. a blackhead
  d. all of these

___ 10. Acne is an inflammation of the
  a. oil glands
  b. sweat glands
  c. epithelium
  d. none of these

True/False

__________ 1. In Caucasians, melanin is produced in the stratum *lucidum* of the epidermis.

__________ 2. The difference in darkness of the skin is related to how much *keratin* is produced.

__________ 3. New cells are constantly being produced in the stratum *lucidum* and being pushed up to the stratum corneum in the epidermis.

__________ 4. The *subcutaneous* layer is a third layer of the skin.

__________ 5. The vitamin D produced in the skin is the *active* form of the vitamin.

__________ 6. The arrector pili muscles are under *involuntary* control in the body.

__________ 7. A strong sunburn represents a *third* degree burn.

__________ 8. Psoriasis is characterized by a *low* mitotic rate.

__________ 9. The stratum *lucidum* is found primarily in the soles of the feet and the palms of the hands.

__________ 10. The *color* of the skin is affected by the amount of melanin and carotene, as well as the amount of blood supply to the surface of the skin.

Completion

1. Dark and light skin differ in the amount of ________________ that is produced in the epidermis.
2. The second major layer of the skin is called ________________ .
3. Skin rests on a layer of ________________ tissue, which is a loose connective tissue.
4. The contours of the dermal papillae are projected onto the surface of the skin in a set of ridges on the tips of the digits known as ________________ .
5. The layer of flat dead cells and keratin fibers at the topmost layer of the epidermis is known as ________________ .
6. The most lethal form of skin cancer is called ________________ .
7. The excretory ducts of the sudoriferous glands of the skin release ________________ .
8. Sensitivity to heat, cold, and pain occurs as a result of sense receptors located for the most part in the ________________ of the skin.
9. An invagination of the ________________ layer of the epidermis produces the generating layer of the hair follicle.
10. The epidermis is "attached" to the ________________ layer of the dermis.
11. The layer of the epidermis which is constantly undergoing mitotic division is called the ________________ .
12. The muscle that attaches the shaft of the hair follicle to the surface of the skin is called the ________________ .

Lost Sheep

1. keratin, papillae, Pacinian corpuscles, reticular layer
2. skin of axillae, skin of forehead, skin of nipples of breast, skin of scrotum
3. arrector pili muscles, sebaceous glands, eccrine glands, root and follicle
4. keratin, melanin, carotene, subcutaneous blood supply
5. nail body, artery, lunula, cuticle
6. exocrine, sweat, sudoriferous glands, arrector pili
7. follicle, root, corpuscle, shaft
8. reticular, basale, germinativum, corneum
9. thermoreceptors, hair follicle, adipose cells, loose connective tissue
10. sweat pore, elastin, collagen, sebaceous gland

Matching

Set 1

___ 1. found only in thick skin
___ 2. capable of constant mitosis
___ 3. cells have spinelike projections
___ 4. first layer in epidermis to begin producing keratin
___ 5. flat, dead cells

a. stratum corneum
b. stratum lucidum
c. stratum granulosum
d. stratum spinosum
e. stratum basale

Set 2

___ 1. tan pigment
___ 2. protects against disease
___ 3. vitamin produced upon exposure to ultraviolet light
___ 4. fibers of dermis
___ 5. waterproof protein

a. melanin
b. vitamin D
c. . keratin
d. elastin and collagen
e. immune cells
f. vitamin C

Sleuthing

The Waldorf family was caught in a fire but escaped. Unfortunately, the father suffered second degree burns on his chest, abdomen, and the whole of one of his arms. He also suffered third degree burns on one of his lower extremities. Emergency treatment included IV fluids.

a. What percentage of the father's body was covered with third degree burns?

b. He experiences a good deal of pain in the area of his chest and abdomen, but little pain in the leg. Why?

c. Why was an IV infusion begun?

d. What complications might be expected from damage to large areas of skin?

Word Scrambles

1. a. type of gland — CERIENOX — ☐ _ ☐ ☐ ☐ _ ☐ _
   b. secretion of goblet cell — SUCUM — ☐ ☐ _ _ _

Total: uppermost layer of the epidermis

☐☐☐☐☐☐☐

2. a. elements found in very small quantities in the body — CRATE — ☐ ☐ _ ☐ ☐
   b. stratum — ERYLA — ☐ ☐ _ _ ☐
   c. layer under dermis — USCUBTOSUANE — _ _ _ _ ☐ _ _ _ _ _ _ _
   d. waxy, waterproofing protein of skin — EAKTRIN — _ _ _ _ _ ☐ _

Total: second layer of the dermis

☐☐☐☐☐☐☐☐☐

Addagrams

1. a. portion of hair below surface of skin — 2, 7, 7, 6
   b. nipplelike projection at junction of epidermis and dermis — 9, 1, 9, 10, 11, 11, 1
   c. water at 31 · F — 12, 5, 4
   d. animal's posterior — 3, 4, 1, 8

   Total: muscle causing "goose bumps"

1	2	3	4	5	6	7	8		9	10	11	12

2. a. what melanin does to skin — 6, 7, 8, 12
   b. encrustation on surface of healing wound — 1, 4, 7, 3
   c. what "AAA" on DNA would code for on RNA — 2, 5, 11
   d. inert gas — 8, 9, 10, 8

   Total: layer on which dermis rests

1	2	3	4	5	6	7	8	9	10	11	12

Crossword

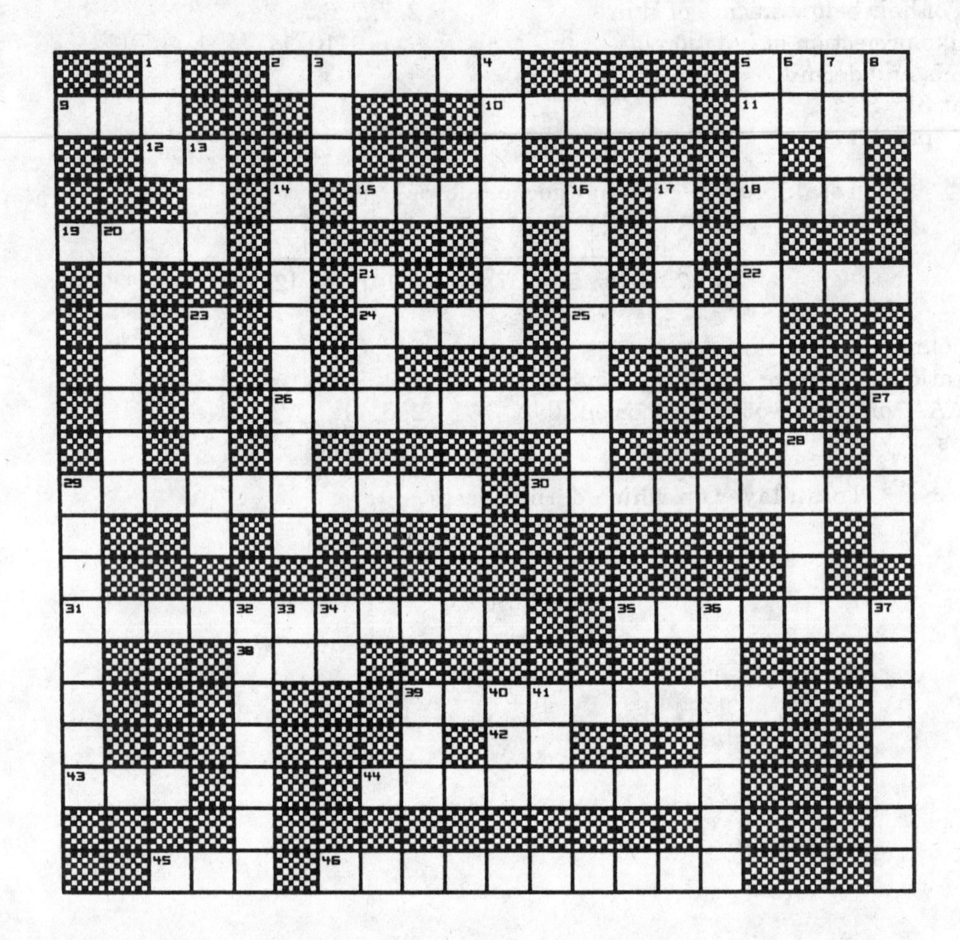

Across

2 Connective tissue layer of skin
5 Opening of sweat gland on surface
9 Cell-mediated immunity (abbr.)
10 What degree is a burn that destroys all of dermis?
11 Part of flower
12 One of four primary tissues (abbr.)
15 Lowest layer of epidermis
18 Color of a first degree burn
19 Integument
22 Inflammation and infection of oil glands
24 Thing
25 Basic functioning unit of life
26 Upper layer of skin
29 Skin
30 Lower layer of dermis
31 ________ reticulum
35 Layer of epidermis which is missing in thin skin
38 Decay
39 Layer of dermis which is connected to epidermis
42 Alternating current (abbr.)
43 Secretion of sebaceous gland
44 Nerve ending in dermis which is wrapped in connective tissue
45 Charged element
46 Layer of epidermis which starts producing keratin

Down

1 Nervous twitch
3 Extrasensory perception (abbr.)
4 Layer
5 Abnormality of skin characterized by high mitotic rate and flaking of skin
6 Occupational Therapist (abbr.)
7 Tear apart
8 Electron microscope (abbr.)
13 What melanocytes make for your skin
14 Oil secreting
16 Gland such as sudoriferous
17 Skin derivative
20 Tough protein of upper layer of skin
21 Arrector ________ muscle
23 What kind of endings are present in dermis?
27 Heat damage
28 A secretor
29 Characterized by rash of the skin
32 What kind of chemical is keratin?
33 ________ and behold!
34 Preposition
36 Topmost stratum of skin
37 Pigment produced by epidermis
39 Best player
40 Type of smear done by the gynecologist
41 Intensive Care Unit (abbr.)

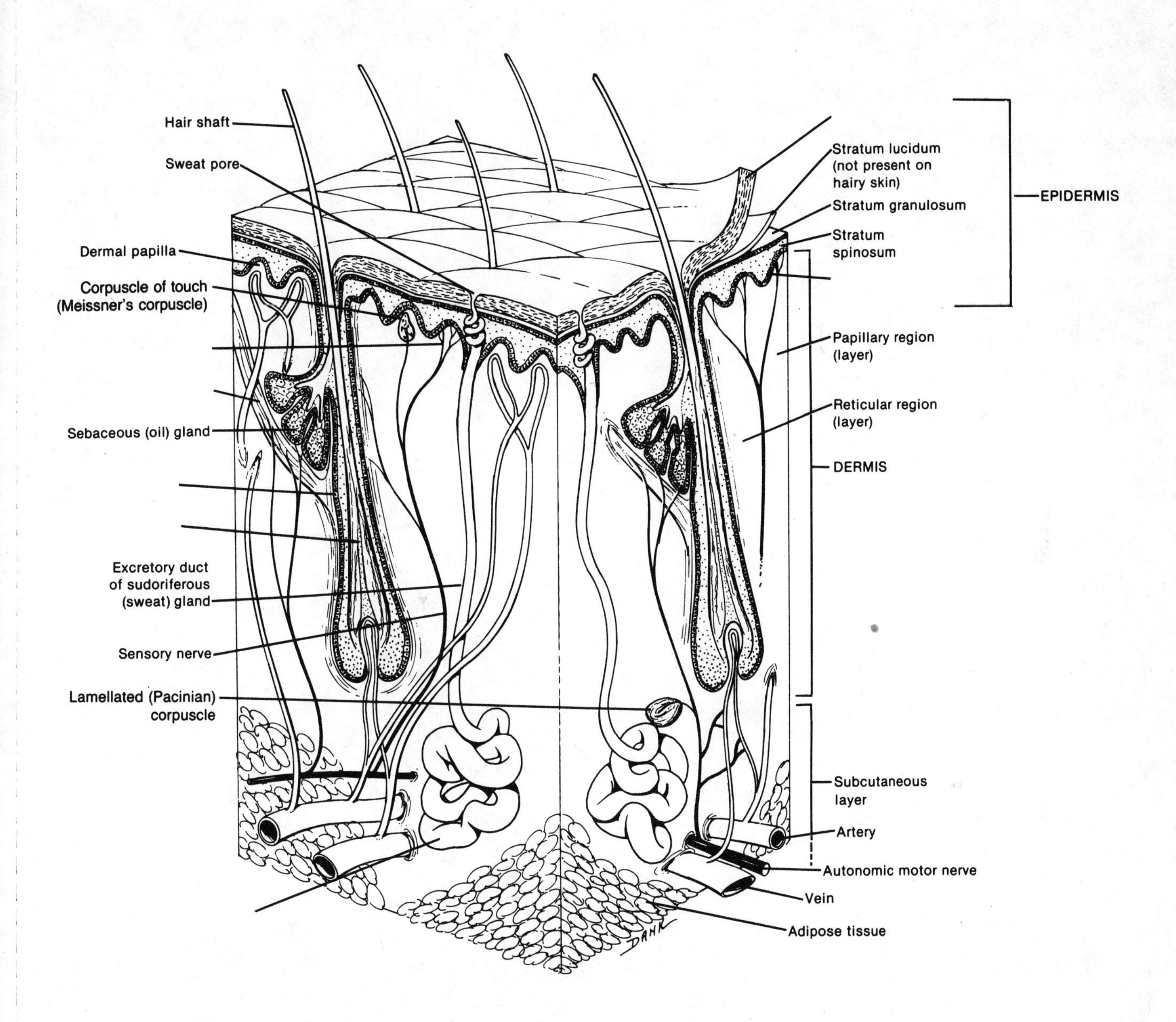

Anatomic Artwork

1. Label the layer of the epidermis that binds to the papillary layer of the dermis.
2. Label the sudoriferous gland.
3. Label the muscle that makes the shaft of the hair stand up vertically.
4. Label the layer of the epithelium which is constantly mitotic.
5. Label the stratum of epithelium which consists of dead cells that are completely filled with keratin.

Chapter 6

The Skeletal System

Introduction

The skeletal system is involved in the *support, protection,* and *movement* of the body. Calcium and phosphate salts make the framework of bones firm and strong for these functions. As a result of this makeup, the skeletal system plays an important role in the regulation and *storage of calcium* in the body.

Calcium ions are necessary for the formation of blood clots, the contraction of muscle, and the stabilization of the neuron cell membranes. Most of the body's calcium is stored in bone and the exchange between skeletal system and body fluids is mediated through two hormones: *calcitonin* and *parathyroid hormone* (PTH). This control mechanism works in conjunction with the gastrointestinal absorption of calcium and kidney regulation to achieve homeostasis.

A final function attributed to the skeletal system is *hematopoiesis,* since the *red marrow* spaces of bone constitute the prime areas of blood cell production. The rate of this function is greatest in the newborn and declines with age as the amount of red marrow decreases and is gradually replaced by yellow marrow.

After completing the reading in your text, you should be able to answer the following questions and complete the varied activities, which deal with the skeletal system, its histology, functions, and organization.

Multiple Choice

___ 1. The primary ossification center of long bone is located in which of the following structures?
   a. epiphysis
   b. epiphyseal plate
   c. diaphysis
   d. endosteum

___ 2. Abnormal bone development can result from vitamin D deficiency since this vitamin is required for the:
   a. differentiation of osteoblasts and osteoclasts
   b. proper absorption of calcium by the intestinal tract
   c. stimulation of cellular activity in the epiphyseal line
   d. maintenance of the matrix

___ 3. Which of the following functions can be attributed to the skeletal system?
   a. support
   b. mineral storage
   c. hematopoiesis
   d. all of these

___ 4. Which of the following is the primary *organic* constituent of bone tissue?
a. calcium phosphate
b. calcium carbonate
c. collagen
d. keratin

___ 5. Cells involved in bone resorption are:
a. osteoclasts
b. osteoblasts
c. fibroblasts
d. chondrocytes

___ 6. The hormone that would be released in the body to correct for low blood calcium levels is:
a. estrogen
b. calcitonin
c. parathyroid hormone
d. vitamin C

___ 7. The portion of bone laid down first in bone formation is the:
a. matrix
b. calcium salts
c. trabeculae
d. marrow

___ 8. Which of the following cell types has the ability to undergo mitosis?
a. osteoblast
b. osteoprogenitor
c. osteocyte
d. osteoclast

___ 9. Which of the following statements is/are true?
a. Compact bone is made of Haversian systems with few spaces
b. Trabeculae of bone are found in cancellous bone tissue
c. The process by which bone hardens is calcification
d. All statements are true

___ 10. Which of the following is/are correctly matched?
a. compact bone–osteon
b. spongy bone–trabeculae
c. cancellous bone–red marrow
d. all pairs are correctly matched

___ 11. A "soft spot" in an infant's skull is called:
a. suture
b. meatus
c. fossa
d. fontanel

___ 12. In the cranial cavity, the site of attachment for the meninges (the covering on the brain) is the:
a. sphenoid bone
b. frontal sinus
c. crista galli of ethmoid
d. mastoid process of temporal bone

___ 13. The union of the parietal and occipital bones forms which of the following sutures?
a. coronal
b. sagittal
c. lambdoidal
d. squamosal

___ 14. Which of the bones listed below does *not* contain a paranasal sinus?
a. sphenoid
b. frontal
c. ethmoid
d. temporal

___ 15. A cleft palate results from "failure to fuse" of which of the following bones?
a. maxillae
b. mandible
c. vomer
d. conchae

___ 16. Excessive lumbar curvature of the vertebral column is:
a. scoliosis
b. lordosis
c. kyphosis
d. "humpback"

___ 17. Which of the bones listed does *not* have a styloid process?
a. temporal
b. ethmoid
c. radius
d. ulna

___ 18. Which of the following pairs is/are mismatched?
a. sternum–xiphoid process
b. radius–olecranon
c. scapula–glenoid cavity
d. humerus–capitulum

___ 19. The point of articulation for the femur to the pelvic bone is the:
a. lateral condyle
b. acetabulum
c. trochanteric ridge
d. linea aspera

___ 20. Rickets is a condition resulting from deficiency of:
a. vitamin C
b. vitamins A and B
c. Fe^{2+} and K^{+}
d. vitamin D

___ 21. Flat bones are:
a. cube shaped and made primarily of spongy bone
b. found in the ankle
c. thin with a "core" of spongy bone covered with parallel plates of compact bone
d. none of these

___ 22. Which of the following bones is *not* part of the axial skeleton?
a. sternum
b. vomer
c. humerus
d. cervical vertebrae

___ 23. Which of the following pairs is/are mismatched?
a. sella turcica–sphenoid bone
b. pectoral girdle–clavicle
c. fibula–medial malleolus
d. manubrium–sternum

___ 24. Which of the following statements concerning the pelvis is *incorrect?*
a. The greater pelvis is deeper in the male than in the female
b. The pubic arch in females is less than a 90° angle
c. The male pelvis outlet is heart shaped
d. The iliac crest is less curved in females than in males

___ 25. Inflammation or infection in bone tissue is called:
a. osteomalacia
b. osteomyelitis
c. osteoporosis
d. none of these

True/False

___________ 1. All bone tissue consists basically of the same components, however, the *organization* of these components may vary.

___________ 2. Bone is an active tissue that is continuously being broken down and reconstructed *throughout an individual's life.*

___________ 3. Bones that develop from connective tissue membranes undergo *endochondral* ossification.

___________ 4. Calcitonin stimulates the *release* of calcium by bone.

___________ 5. *Red* marrow is adipose type tissue.

___________ 6. Hematopoiesis occurs in *all* bone marrow.

___________ 7. Short bones *are* miniature versions of long bones with a central marrow cavity.

___________ 8. PTH increases *osteoclastic* activity.

___________ 9. Absorption of calcium by the digestive tract is promoted by *vitamin C.*

___________ 10. In response to *mechanical stress,* bone produces minute currents of electricity, which can stimulate growth.

___________ 11. *Sex hormones* aid osteoblastic activity and promote the degeneration of chondrocytes in the epiphyseal plate.

___________ 12. Vertebrae are examples of *short* bones.

___________ 13. The *periosteum* has an outer fibrous and inner osteogenic layer.

___________ 14. All bones contain *both* compact and cancellous bone tissue.

___________ 15. The irregular latticework of spongy bone is made of thin bone plates called *canaliculi.*

___________ 16. *All bone* tissue when first formed is of the spongy type.

___________ 17. The lining of the *medullary cavity* has several cell types and is called the endosteum.

__________ 18. The *epiphyseal plate* is between the diaphysis and epiphysis and allows for elongation of the diaphysis into adulthood.

__________ 19. The *axial* skeleton consists of the skull, hyoid, vertebral column, and rib cage.

__________ 20. The *pectoral* girdle is formed by the hip and sacrum.

__________ 21 An opening in bone which serves as a passage for blood vessels and nerves is called a *fossa*.

__________ 22. Rounded processes on bones that usually articulate with other bones are called *condyles*.

__________ 23. A marked curvature of the vertebral column to the left or right is known as *kyphosis*.

__________ 24. The dens is part of the *scapula*.

__________ 25. The number of tarsals is the *same* as the number of carpals.

Completion

1. Blood cell formation, known as ______________, occurs in all bone marrow spaces of the newborn.
2. Parietal bones form by means of ______________ ossification.
3. Yellow marrow is primarily composed of ______________ cells.
4. A person with ______________ blood calcium levels may suffer from inability to clot blood.
5. Materials diffuse through compact bone tissue through channels called ______________.
6. Calcitonin ______________ (increases/decreases) blood calcium levels and parathyroid hormone has the ______________ effect on blood calcium levels.
7. The ______________ surrounds bone and is the source of its blood supply.
8. The production of a tiny electric current in bone as a result of mechanical stress is known as the ______________ effect.
9. The growth in length of a bone occurs at the ______________.
10. The daily cellular activities of bone tissue are maintained by cells called ______________ .
11. ______________ marrow is found in the medullary cavity of adult bones.
12. Demineralization of bone in adults due to vitamin D deficiency is known as ______________.
13. Failure of vertebral laminae to unite at the midline results in ______________.
14. In osseous tissue, the ratio of mineral salts to collagen fibers is ______________ : ______________.
15. Replacement of cartilage by bone is known as ______________ ossification.
16. During the repair of a fracture, the hematoma is ultimately replaced by spongy bone tissue referred to as a ______________.
17. The ______________ skeleton consists of those bones that form the support framework of the arms and legs.
18. The major inorganic constituent of bone is comprised of ______________.
19. The maintenance of calcium levels in the body is important since this ion is needed for physiological processes such as ______________ .
20. Resorption of bone is carried out by cells called ______________.
21. In endochondral bone formation, the first "skeletal" model is made of ______________ cartilage.

22. ________________ are fibrous membrane areas in the skull where ossification has not yet taken place.
23. Long bones form by means of ________________ ossification.
24. The ulna is a bone in the ________________ division of the skeleton.
25. The ________________ bone does not articulate directly with any other bone.

Lost Sheep

1. collagen, matrix, organic, calcium salts
2. leads to high blood calcium, increased parathyroid secretion, PTH, calcitonin
3. estrogen, growth hormone, parathyroid hormone, collagen
4. adipocyte, red marrow, hematopoiesis, blood cell
5. subluxation, greenstick, open/compound, pathologic
6. increased calcium absorption by GI tract, increased retention of calcium by kidney, increased osteoclastic activity, increased estrogen production
7. periosteum, diaphysis, articular cartilage, medullary cavity
8. trabeculae, cancellous, lattice, compact
9. canaliculi, Volkmann's canals, medullary cavity, Haversian canals
10. vitamin D, vitamin A, vitamin C, vitamin B
11. hematoma, Colle's, callus, osteomalacia
12. thoracic region, kyphosis, lateral curvature, hunchback
13. cartilage, fibrous tissue, epiphyseal plate, growth in length
14. sternum, rib, scapula, femur
15. radius, ulna, tibia, sternum
16. parietal, zygomatic, occipital, frontal
17. trapezoid, capitate, phalange, pisiform
18. sacrum, ilium, pubis, ischium
19. lambdoidal, squamosal, mandibular, coronal
20. tibia, fibula, radius, femur

Matching

Set 1

___	1 most primitive cell in the inner layer of periosteum	a.	osteoclast
		b.	Paget's disease
___	2. cell that maintains bone	c.	osteoblast
___	3. cell responsible for bone resorption	d.	lordosis
___	4. infection of bone caused by microorganisms	e.	rickets
		f.	osteocyte
___	5. irregular thickening and softening of bone	g.	Pott's disease
___	6. exaggerated lumbar curvature of vertebral canal	h.	osteomyelitis
		i.	kyphosis
___	7. condition resulting from childhood vitamin deficiency	j.	osteoprogenitor

Set 2

___	1. trochlea	a.	scapula
___	2. coronoid process	b.	pelvis
___	3. acromion	c.	tarsal
___	4. trapezium	d.	sphenoid
___	5. obturator foramen	e.	ethmoid
___	6. linea aspera	f.	tibia
___	7. cuneiform	g.	occipital
___	8. mandibular fossa	h.	temporal
___	9. sella turcica	i.	humerus
___	10. foramen magnum	j.	femur
		k.	carpal
		l.	ulna

Set 3

___	1. reduced levels of this hormone may promote osteoporosis	a.	vitamin A
___	2. stimulates osteoclastic activity	b.	parathyroid hormone
___	3. vitamin responsible for themaintenance of the bone matrix	c.	estrogen
___	4. hormone that promotes degeneration of chondrocytes at epiphyseal plate	d.	vitamin C
___	5. abnormal amounts of this hormone can cause giantism/dwarfism	e.	growth hormone
___	6. hormone that raises blood calcium level	f.	calcitonin
___	7. increases kidney retention of calcium		
___	8. vitamin associated with bone cell differentiation		

Double Crosses

1. *Across:*
   CPR tip

   Down:
   sit on it

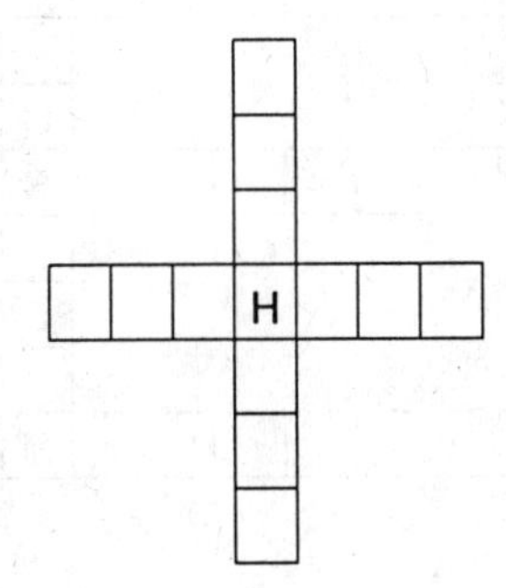

2. *Across*:
   bad breaks

   Down:
   knob for articulation between radius and humerus

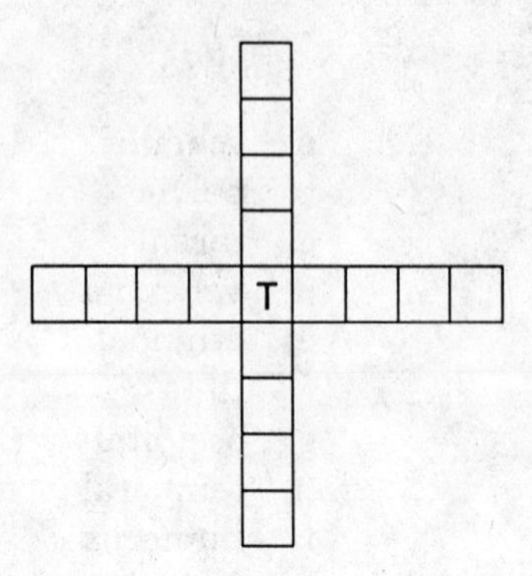

3. *Across*:
   blood cell formation

   Down:
   a. "funny bone"
   b. upper jaw
   c. bone projection
   d. pelvis bone
   e. a function of bone
   f. part of pectoral girdle

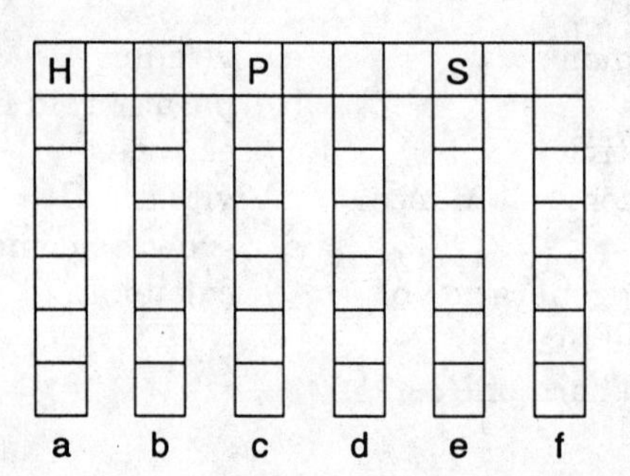

4. *Across*:
   a. unique to femur
   b. spongy
   c. sternum's "handle"
   d. digits
   e. maintainers
   f. growth plate

   Down:
   covering

Sleuthing

1. Mrs. Jones, a 56-year-old mother of six, was hospitalized for a hip fracture. Her husband had noticed that recently her posture had deteriorated. He informed the doctor of this and also of the fact that his wife had complained of acute, severe back pains during a recent camping trip. The doctor ordered a series of x-rays, which confirmed his suspicions.
   a. What disorder did the doctor suspect and why?

   b. What evidence would the x-ray show to confirm the diagnosis?

   c. What treatment would be appropriate to arrest the progress of the disease and reduce the risk of further fractures?

Word Scrambles

1. a. bone at base of skull — CIOCPTALI — ☐ _ _ _ _ _ _ _ _
   b. bat-shaped bone — HPONIEDS — ☐ _ _ _ _ _ _ _
   c. 26 in number — BAEVTRERE — _ ☐ _ _ _ _ _ _ _
   d. "chewing" bone — AMEDBNLI — ☐ ☐ _ _ ☐ _ ☐ _
   e. method by which tibia grows — ONDEDNOHCRLA — _ _ _ _ _ _ _ _ _ _ ☐ _
   f. bone formation — SICOSFIAIOTN — ☐ _ _ _ _ _ _ _ ☐ _ _ _
   g. essential bone element — ALICCMU — _ ☐ _ ☐ _ _ _

Total: softening of bone

☐☐☐☐☐☐☐☐☐☐☐☐

2. a. movable skull bone — EMLABNDI — _ _ _ ☐ ☐ _ _ ☐
   b. chest area — LPAERCTO — ☐ _ ☐ _ _ _ _ ☐
   c. fingers and toes — SEPAHANGL — _ _ _ _ _ _ ☐ ☐ _
   d. common to wrist and ankle — AICVLARNU — _ _ ☐ ☐ _ _ ☐ _ ☐

Total: "connected to the thigh bone"

☐☐☐☐☐☐ ☐☐☐☐☐☐

Addagrams

1. a. narrow ridge-like projection on bone — 1, 9, 15, 3, 10
   b. first vertebrae — 13, 4, 6, 8, 3
   c. "bone" prefix — 2, 3, 3, 15, 2
   d. cock's comb of ethmoid — 7, 9, 11, 3, 10, 8, 14, 5, 12, 12, 11

Total: some ribs have it, others don't

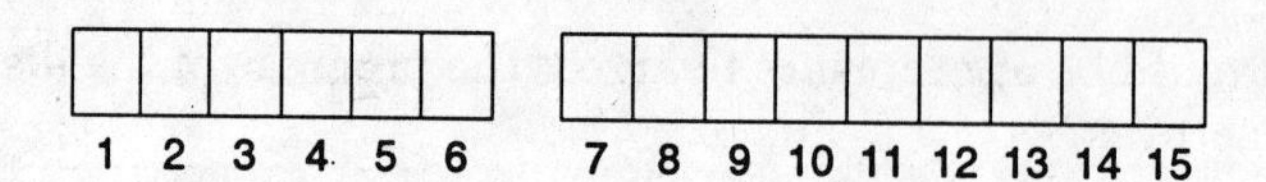

1 2 3 4 5 6 7 8 9 10 11 12 13 14 15

2. a. fossa on humerus — 13, 6, 2, 7, 14, 5, 9, 13, 9
   b. hip crest bone — 11, 6, 4, 8, 1
   c. hole in bone — 12, 13, 14, 5, 1, 10, 9
   d. trochanter found here — 12, 2, 15, 8, 14
   e. notch on ulna for "mate's" head — 14, 5, 3, 4, 5, 6

Total: one of the wedge-shaped tarsals

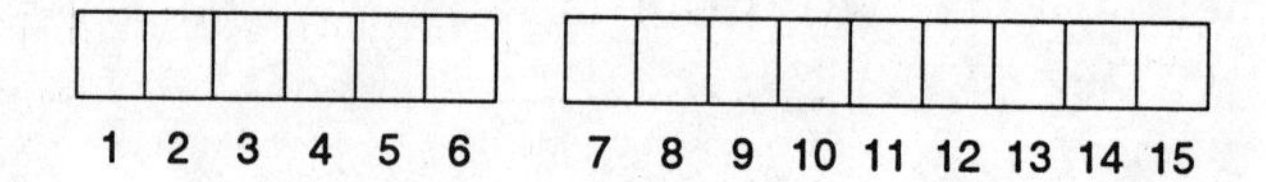

1 2 3 4 5 6 7 8 9 10 11 12 13 14 15

Crossword

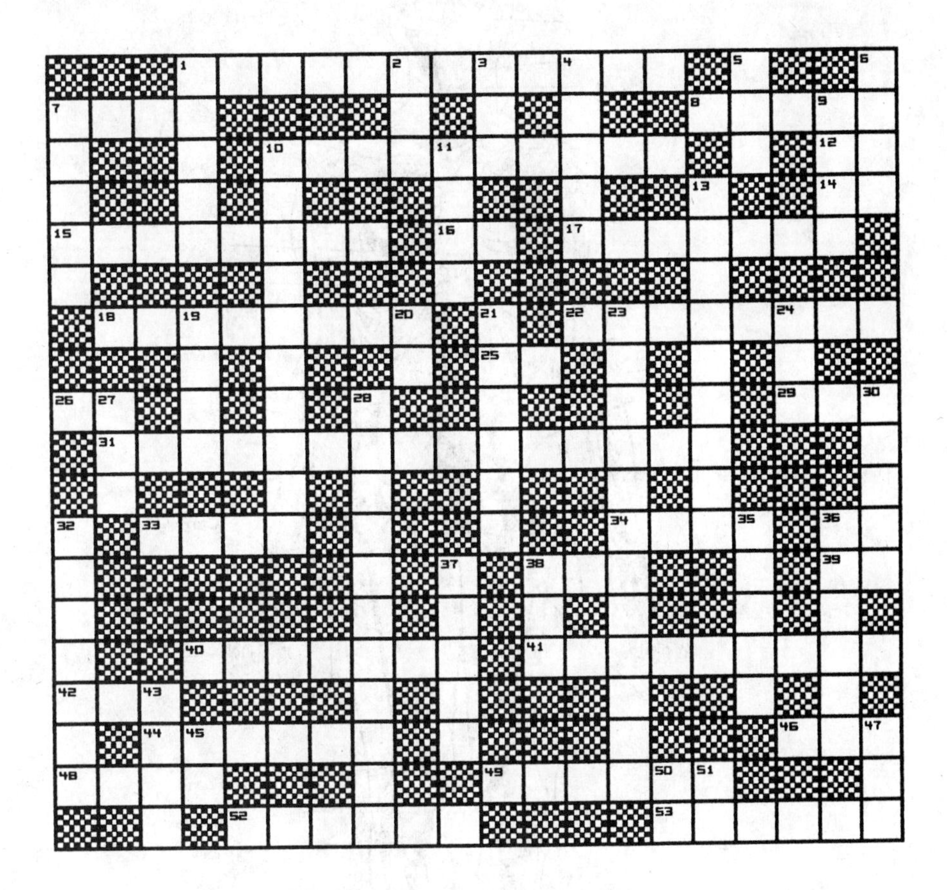

Across

1 Division of skeleton that has femur and humerus
7 Material for fabric
8 Linea aspera is here
10 Responsible for bone resorption
12 Musical note
14 The Dodgers and Giants are in this league (abbr.)
15 Comprised of 206 bones
16 Hydroxyl Ion (abbr.)
17 Home of the glenoid fossa
18 Causes closure of the epiphysis
22 Bone where inner ear is found
25 Sigh
26 Growth hormone (abbr.)
29 What a baby often wears
31 Flat bones form this way
33 Where phalanges and metatarsals
34 Firefighter's apparatus
36 Possessive pronoun
38 Time past
39 Preposition
40 Stored in bone
41 Cavity in long bones
42 Blend ingredients
44 Ulna's "mate"
46 Used for chopping
48 Not messy
49 Marrow of adult diaphysis
52 Navicular or cuboid
53 Do over

Down

1 Skull, rib cage and vertebrae form this skeleton
2 Female deer
3 Three hundred to a roman
4 "Springs" over
5 Marrow in cancellous bone
6 Refers to the mouth
7 Shallow depression in bone
9 Has a styloid process
10 Bone-forming cell
11 Aroma
13 Made of spongy bone
19 Slant
20 Symbol for ammonia
21 Has a hemotopoietic function
23 Long bones form this way
24 Associated with costal cartilage
27 Acetabulum is found here
28 Blood cell formation
30 Units of the skeletal system
32 Opening for a blood vessel or nerve
35 Not late
36 Intercellular material
37 Forms back of nasal septum
38 Goal
43 Used to "see" bones
45 Preposition
47 Emergency medical technician (abbr.)
50 Either ________
51 A group of us

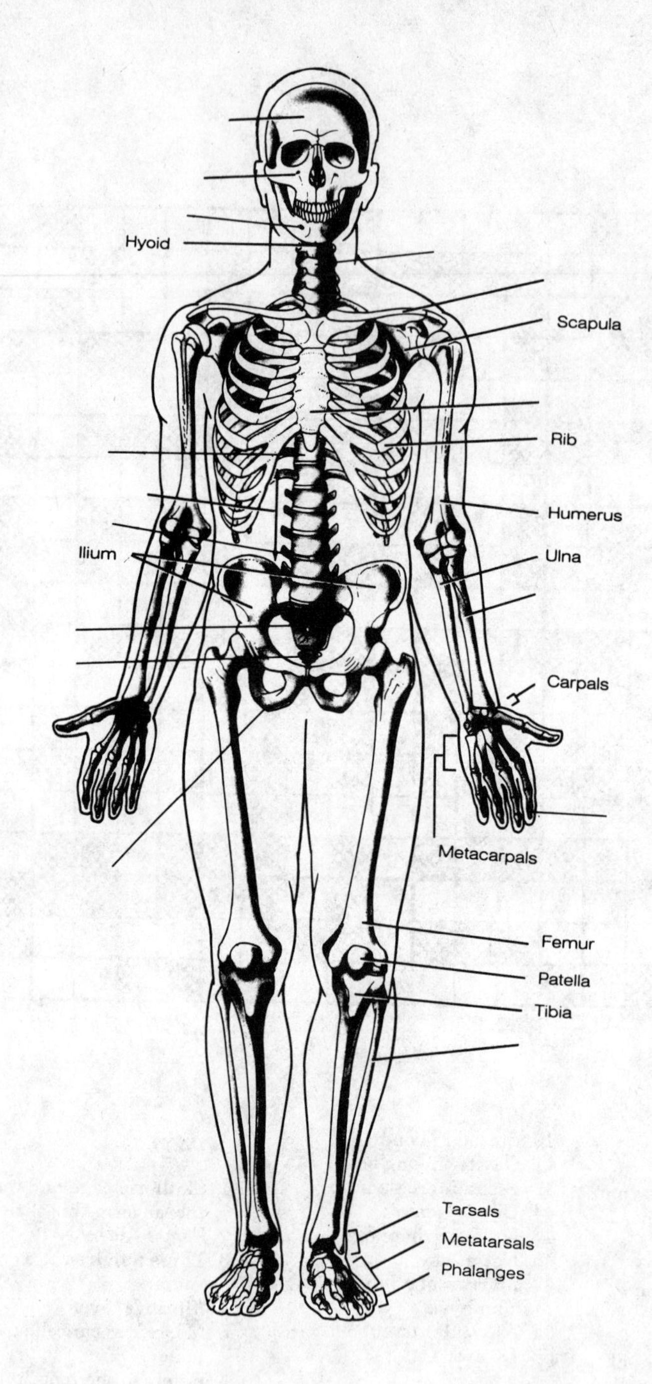

Anatomic Artwork

Figure 1

1. Label the bone lateral to the ulna.
2. Label the bone lateral to the tibia.
3. Label the bone pair that sits on either side of a necktie knot.
4. Label the five sections of the vertebral column.
5. Label the bones that grasp the pencil as you write.
6. Label the area of the hip bone on which you sit.
7. Label the bones (that are visible) that have a paranasal sinus.
8. Label the bone that is compressed during CPR.
9. Label the movable bone of the skull.
10. Label the largest sesamoid bone.

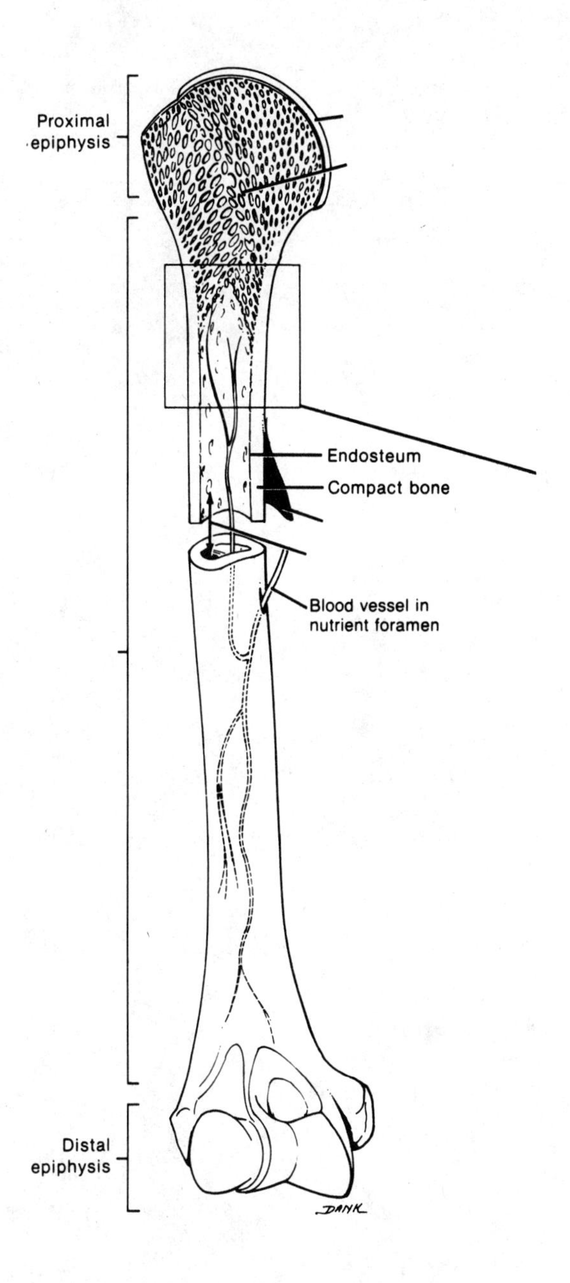

Figure 2

1. Shade in the area where there is articular cartilage.
2. Label the area where there is a predominance of red marrow and indicate the type of bone tissue.
3. Identify the covering on the bone and label it.
4. Identify and label the central elongated portion of the bone.
5. Identify and label the central cavity.
6. Label the area where the primary and secondary ossification centers meet.

Chapter 7

Articulations

Introduction

In the chapter on articulations, you studied the relationships of the bone to one another in terms of how they are held together. You saw that joints may be classified structurally or on the amount of motion possible at the joint. You saw that the structure of joints affects the degree of mobility and that in general there are three kinds of joints: *synarthritic, amphiarthritic,* and *diarthritic.* You also examined the various kinds of diarthritic or *synovial* joints as well as some of the diseases that affect these articulations.

After completing the reading in your text, you should be able to answer the following questions concerning the structure and functioning of joints. In addition, you should be able to complete the varied activities.

Multiple Choice

___ 1. Which of the following is/are true regarding synovial fluid?
  a. It is secreted by the synovial membrane
  b. It functions to lubricate and nourish articular cartilage
  c. It has the consistency of egg white
  d. All statements are true

___ 2. Joints in which articulating bone surfaces are separated by discs of fibrocartilage are known as:
  a. diarthroses
  b. symphyses
  c. pivots
  d. sutures

___ 3. Swollen, painful, and inflamed joints are characteristic of:
  a. luxation
  b. osteomalacia
  c. arthritis
  d. rickets

___ 4. Joints that are slightly movable are best referred to as:
  a. synovial
  b. amphiarthritic
  c. gliding
  d. diarthritic

___ 5. Which of the following statements is/are true?
a. Joint structure determines its mobility
b. Joint flexibility may be affected by hormones
c. Generally, greater joint mobility is achieved when there is a joint capsule
d. All of these statements are true.

___ 6. Which of the following is/are kinds of synarthroses?
a. sutures
b. gomphosis
c. synchondroses
d. all of these

___ 7. Synovial joints are characterized by:
a. articular cartilage
b. fibrous capsule
c. joint cavity
d. all of these

___ 8. Movement of the arm away from the midline of the body is:
a. adduction
b. rotation
c. abduction
d. circumduction

___ 9. Which type of joint permits movement in three planes?
a. ball and socket
b. gliding
c. pivot
d. symphysis

___ 10. Which of the following diseases is caused by a bacterium?
a. rheumatoid arthritis
b. Lyme disease
c. subluxation
d. gout

___ 11. Structurally, joints can be classified as:
a. amphiarthroses
b. fibrous
c. diarthroses
d. all of these

___ 12. The special motion that moves the sole of the foot outward is called:
a. pronation
b. dorsiflexion
c. eversion
d. depression

True/False

__________ 1. Sutures are *fibrous* synarthritic joints.

__________ 2. *Fibrocartilagenous discs* are present in synchondroses.

__________ 3. Synovial fluid lubricates *amphiarthritic* joints.

__________ 4. A *functional* classification of joints includes fibrous, cartilagenous and synovial varieties.

_______ 5. Movement of a bone around its axis is called *rotation.*

_______ 6. Moving the head from side to side as in saying "no" is an example of *circumduction.*

_______ 7. The range of movement is *greatest* in a ball and socket joint

_______ 8. Dorsiflexion and inversion are special movements of the *hand.*

_______ 9. There are several forms of *arthritis* which have different causes.

_______ 10. The term *pannus* is associated with rheumatoid arthritis.

Completion

1. Synchondrosis and symphysis are examples of structural joints called _______________.
2. A _______________ joint has a fluid-filled cavity and fibrous capsule.
3. A joint in which a cone-shaped peg fits into a socket is called a _______________.
4. A difference between an amphiarthrosis and synarthrosis is the degree of _______________.
5. _______________ are pads of fibrocartilage found in some synovial joints.
6. The primary movement at a pivot joint is _______________.
7. Small fluid-filled sacs that are located between the skin and bone and cushion the movement between them are called _______________.
8. The synovial joint at the elbow is an example of a _______________ joint.
9. Another term for joint is _______________.
10. Diarthroses are _______________ movable joints.
11. A partial dislocation is known as a _______________.
12. Increasing the angle between articulating surfaces of bone is called _______________.

Lost Sheep

1. pannus, bursitis, rheumatoid, arthritis
2. dislocation, luxation, displacement, Lyme
3. DNA, uric acid, gout, subluxation
4. inversion, plantar flexion, supination, eversion
5. saddle, hinge, ellipsoidal, symphysis
6. symphysis, synchondrosis, fibrocartilage, amphiarthrosis
7. synarthroses, amphiarthroses, gomphoses, suture
8. synovial, joint capsule, fibrocartilage, membrane
9. immovable, suture, symphysis, synarthritic
10. rotation, amphiarthrosis, pivot, radius-ulna

Matching

___ 1. gliding movement possible
___ 2. suture
___ 3. syndesmosis
___ 4. gomphosis
___ 5. ellipsoidal
___ 6. extension movement possible

a. fibrous joint
b. cartilagenous joint
c. synovial joint

___ 7. synchondrosis
___ 8. menisci
___ 9. symphysis
___ 10. protraction movement possible

Double Crosses

1. *Across:*
pubic
Down:
freely movable

H

2. Across:
luxation
Down:
movement toward the midline

C

3. *Across:*
oval-shaped condyle into elliptical cavity
Down:
amphiarthrosis with dense fibrous connective tissue

S

4. *Across:*
   sole of foot inward
   Down:
   inflamed joints

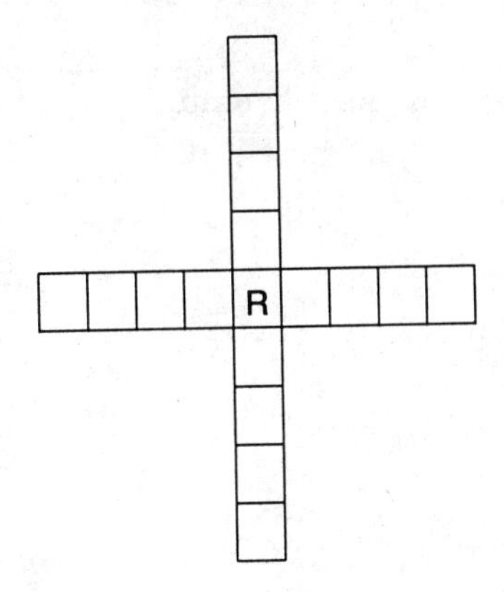

5. *Across:*
   special forward movement of mandible
   Down:
   type of joint lacking a cavity

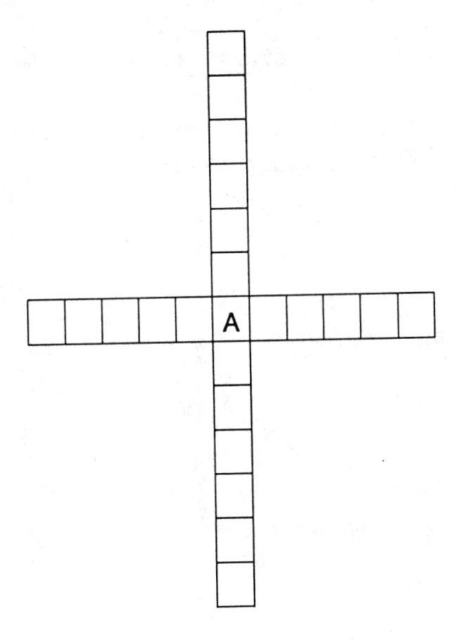

Word Cage

```
S C A R T I L A G E S K T
Y E N I L A Y H A I U H R
N R U O A J M J S L B G C
O T F L E X I O N O L O E
V B Q U T E R I K N U U M
I X B D I H M N A E X T Y
A O E P T J A T T B A L L
L B U R S A E D I E T I H
M A A W O N S P R A I N A
A I M X S U T U R E O L R
D R S I S O M S E D N Y S
S Y N C H O N D R O S I S
```

1. A movable joint.
2. A partial dislocation.
3. Reduction of the angle between two bones.
4. Saclike, fluid-filled structures.
5. Membrane that secretes fluid into joint cavity.
6. Common/lay term for articulation.
7. Predominant joint in skull.

7. Predominant joint in skull.
8. Connective tissue in symphysis and synchondritic joints.
9. Epiphyseal plate is this type of joint.
10. Distal articulation of tibia and fibula.
11. Fits into socket.
12. Disease transmitted by tick.
13. Caused by excessive uric acid production.
14. Forcible twisting of a joint results in this.
15. Articular cartilage in synovial joint.

Sleuthing

1. Katie has osteoarthritis in her hip. Osteoarthritis is considered to be a degenerative condition while rheumatoid arthritis is an autoimmune disease.
   a. Compare osteoarthritis and rheumatoid arthritis as to: usual age of onset, joint(s) involved, joint swelling, deformity, and disease progression.

   b. Why would physical therapy be an important part of treatment for both forms of arthritis?

   c. Are there other forms of arthritis? If so, what are they?

Word Scrambles

1. a. proximal ends of radius and ulna joint type — VOTIP — ☐ _ _ ☐ ☐
   b. inflamed joints — RIRTTIHAS — ☐ _ _ ☐ ☐ _ _ _ ☐
   c. fibrocartilagenous joint — SPSYHYMSI — _ _ ☐ _ ☐ _ _ ☐ _
   d. painful supporting structures — HUAISTMERM — ☐ _ ☐ _ _ ☐ _ _ ☐ _

Total: pubis and tibia/fibula, for example

☐☐☐☐☐☐☐☐☐☐☐☐☐☐

2. a. complete mobility — EBVMOLA — ☐ ☐ ☐ ☐ ☐ ☐ ☐
   b. articular discs — IMCESNI — ☐ _ ☐ _ _ _ ☐
   c. sutures and gomphosis — SSYENSAORRHT — ☐ ☐ ☐ ☐ _ _ _ ☐ _ _ ☐ _

Total: fluid secretor

☐☐☐☐☐☐☐☐ ☐☐☐☐☐☐☐☐

Crossword

Across

1 Direct current (abbr.)
3 Uric acid arthritis
6 CT of skin
11 Sodium (abbr.)
12 Unit of radiation
14 Acetabulum
17 Joint disease
20 California city (abbr.)
21 Flexion at Atlas
24 Practical Nurse (abbr.)
25 Type of amphiarthroses
27 A diarthritic joint
31 Alpha epsilon chi (Greek letters)
32 Cinderella's sister, for example
33 Articulation
35 Subjunctive preposition
36 ________ white, like synovial fluid
37 Vintner's apparatus
39 In regard to (abbr.)
40 Type of arthritis
42 Brings amino acids to site of protein synthesis
44 Southern college (abbr.)
46 Joint that allows rotation
47 Dislocation
49 Tough type of cartilage
52 Personality component
53 What marrow is for blood
54 What subs do
55 Knee joint type
57 Lambdoidal, for example
59 King
61 Prefix meaning immovable
62 Infusion
63 God
64 Joint
65 Disease carried by tick
67 National Organization of Women (abbr.)
68 Sudoriferous secretion
69 Articular cartilage

Down

1 Genetic material
2 Pliable support tissue
4 Acid involved in big toe inflammation
5 Sew
7 Prefix meaning outer
8 I am, you are, it ________
9 ________ vs. gel
10 Synovial fibrocartilage discs
13 Princess of Wales
15 Money
16 Toward, short form
18 DNA configuration
19 Slightly movable joint
22 This is made of fibrocartilage in a symphysis
23 Fluid-filled cavity in this type of joint
24 Fruit usually has one of these
26 One-half quart
28 Tibia and fibula are found here
29 Prefix meaning bone
30 Advanced Placement (abbr.)
34 Internal Affairs Division (abbr.)
36 Emergency Room (abbr.)
38 Abnormal tissue in some forms of arthritis
40 Type of movement at a synovial joint
41 Here and ________
42 Young Egyptian king
43 Shape of distal end of radius
44 Sixty-one to Caesar
45 "Horsey" diarthritic joint
46 Turn palms down
48 Turn sole of foot inward
50 Tiny piece
51 Egg (abbr.)
56 How a bone moves in a saddle joint
57 What you are if you know all the bones
58 Small child
60 Total number of metacarpals in Roman times
63 Injection into gluteus muscle (abbr.)
66 Time period (abbr.)

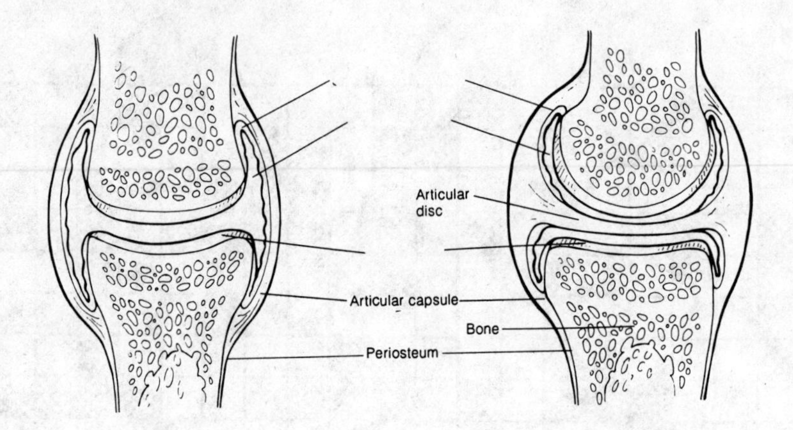

Anatomic Artwork

1. Label the area where synovial fluid is found.
2. Shade in the structure that secretes this fluid and label it.
3. Label the areas where cartilage is found.
4. What kind of joint is this? ________________
5. What kinds of movement are possible at this kind of joint? ________________
6. Give two examples of locations where this type of joint is found. ________________ ________________

Chapter 8

The Muscular System

Introduction

It has been said that were it not for our muscles, we would not be able to interact with our world. We are able to respond as an organism—to move toward a pleasing stimulus or flee from a noxious one—thanks to the muscles that move our skeleton. This type of muscle is appropriately called skeletal muscle.

In addition, the great muscle of the heart (cardiac muscle) pumps life sustaining blood throughout our bodies, and the muscles of our organs (smooth muscle) are active in moving substances like food, air, and urine through our bodies.

Anatomically there are two types of muscle: *striated* and *nonstriated*. The difference is due to the organization of the muscle proteins, *actin* and *myosin*. Functionally, there are also two types of muscle: *voluntary* and *involuntary*. This difference is based on the way in which the muscle is stimulated to contract. Voluntary muscles are totally dependent on nervous stimulation for the initiation of their action. The initiation of involuntary muscle contraction may or may not depend on nervous stimulation, as some are self-excitatory.

When anatomical and functional characteristics are combined, three types of muscle can be distinguished: *skeletal* (striated, voluntary), *smooth* (nonstriated, involuntary), and *cardiac* (striated, involuntary).

Muscle tissue has a special capacity to convert electrochemical energy (the stimulus) into mechanical energy (the contraction). After you have studied this process in the textbook, you should be able to do the following activities, as well as name the basic structural groups of muscles in the body.

Multiple Choice

___ 1. The "power stroke" of contraction is
   a. movement of the myosin cross bridge
   b. the thin myofilaments sliding past the thick
   c. actin moving past myosin
   d. all are correct descriptions

___ 2. In muscle cells, calcium is stored in the:
   a. T-tubules
   b. myofilaments
   c. sarcoplasmic reticulum
   d. none of these

___ 3. The time period after a stimulus during which a second stimulus cannot produce a response in the muscle cell is the:
a. lag period
b. refractory period
c. fusion period
d. subliminal period

___ 4. Which of the following statements is/are correct concerning treppe contractions?
a. They increase in strength with each subsequent contraction
b. The increased contractility of the muscle is related to increased availability of Ca^{2+} within the cell
c. both a and b are correct
d. neither a nor b is correct

___ 5. Which of these is found at the neuromuscular junction?
a. motor end plate
b. synaptic cleft
c. vesicles
d. all of these

___ 6. Which of the following characteristics is/are generally attributed to visceral smooth muscle?
a. self-excitable
b. plasticity
c. contracts as a whole sheet
d. all of these

___ 7. Which of the pairs below is/are correctly matched?
a. fibromyalgia–rheumatic disorders characterized by pain
b. spasm–abnormal contraction of muscle
c. myasthenia gravis–weakness of muscle caused by abnormality at neuromuscular junction
d. all are correct

___ 8. The purpose of myoglobin in skeletal muscle cells is:
a. assist anaerobic reactions
b. store oxygen for use during vigorous exercise
c. catabolize pyruvic acid
d. all of these

___ 9. Anatomic characteristics of skeletal muscle cells include all of the following *except*:
a. many nuclei
b. sytematic arrangement of myofilaments
c. tapered ends
d. T-tubules

___ 10. A high-energy molecule that can be used to produce more ATP during exercise is:
a. ADP
b. creatine
c. phosphocreatine
d. none of these

___ 11. Which of these statements is false with regard to tetanus?
a. The contractions are characterized by summation and fusion of twitch contractions
b. Tetanic contractions are related to frequency of stimuli
c. Tetanic contractions are responsible for body movements
d. The contractions result from decreased calcium levels in the sarcomeres

___ 12. A high-energy compound unique to muscle cells, which provides energy for regeneration of ATP, is:
a. lactic acid
b. tropomyosin
c. creatine phosphate
d. pyruvic acid

___ 13. A single brief contraction in response to a single threshold stimulus is:
a. twitch
b. incomplete tetany
c. contracture
d. tetany

___ 14. An impulse travels along the ________ and passes into the cell.
a. sarcoplasmic reticulum
b. sarcolemma
c. sarcoplasm
d. all of these

___ 15. Bundles of muscle fibers are enclosed by a sheath of connective tissue known as the:
a. endomysium
b. epimysium
c. perimysium
d. superficial fascia

___ 16. Which statement is/are true?
a. The insertion of a muscle is the more movable end
b. The origin and insertion of a skeletal muscle do not lie on the same bone
c. When a muscle contracts, it usually pulls toward the origin
d. All statements are correct

___ 17. Transmission of an impulse across the myoneural junction is mediated by:
a. acetylcholine
b. cholinesterase
c. calcium ions
d. sodium ions

___ 18. Most of the lactic acid produced in skeletal muscle during the oxygen debt:
a. is transported to the liver
b. is converted back to glucose or glycogen
c. both a and b
d. neither a nor b

___ 19. Aerobic respiration refers to:
a. glucose breaking down to lactic acid
b. glucose breaking down to pyruvic acid
c. pyruvic acid converting to lactic acid
d. pyruvic acid breaking down to carbon dioxide and water and releasing energy

___ 20. Which appropriately describes a sarcomere?
a. a unit enclosed by epimysium
b. a unit of the sarcoplasmic reticulum involved in calcium activation
c. a repeating unit within the fasciculus
d. none of these

___ 21. The importance of cholinesterase to the normal functioning of muscle lies in its role as:
a. an activator of troponin
b. depolarizing agent for the sarcolemma
c. a deactivator of ACh
d. none of these

___ 22. Muscles responsible for movement of food through the digestive tract from stomach to small intestine are:
a. voluntary
b. smooth
c. both a and b
d. neither a nor b

___ 23. Which of the following statements is/are correct?
a. The strength of muscle contraction is related to the number of fibers that are stimulated with a threshold stimulus
b. Striations in muscle results from the arrangement of actin and myosin filaments within the cell
c. Acetylcholine alters the permeability of skeletal muscle cells to sodium ion, causing the membrane to become more permeable to sodium
d. All of these are true

___ 24. Which muscle is not involved in moving the thigh?
a. rectus femoris
b. adductor longus
c. biceps brachii
d. gracilis

___ 25. A motor unit is the motor neuron and:
a. all other neurons stimulating the same fiber
b. all the fibers it stimulates
c. the simple twitch it causes
d. all of these

True/False

__________ 1. A *refractory* stimulus is the weakest stimulus that will cause a muscle fiber to contract.
__________ 2. The *sarcoplasmic reticulum* of muscle cells is comparable to the ER of other cells.
__________ 3. Since smooth muscle is nonstriated, there is an *absence* of actin and myosin.
__________ 4. ATP molecules that provide energy for contraction are situated on *myosin* cross bridges.
__________ 5. A *short* refractory period enables skeletal muscle to undergo tetanic contractions.
__________ 6. Muscle contraction *cannot* occur in the absence of oxygen.
__________ 7. Muscle *atrophy* is an increase in the diameter of the fiber.
__________ 8. Generation of tension with no shortening of a muscle is known as *isotonic* contraction.
__________ 9. Lactic acid accumulates when muscle cells undergo *aerobic* metabolism of glucose.
__________ 10. The *myofilaments* of muscle cells are made up of proteins.
__________ 11. The *perimysium* wraps around individual fibers and holds them together.
__________ 12. Muscles that contribute indirectly to movement are called *synergists*.

_____ 13. Stimuli above the threshold level produce *stronger* contractions in skeletal muscle cells than threshold-level stimuli.

_____ 14. The dark part of the A band is due to the presence of *myosin only.*

_____ 15. One way in which skeletal muscle differs from smooth and cardiac is that skeletal muscle is *self-excitatory.*

_____ 16. Cells of *smooth* muscle are often arranged in sheets or layers.

_____ 17. In *multi-unit* smooth muscle, every fiber has its own motor nerve endings.

_____ 18. The continued partial contraction of fibers within a skeletal muscle produces *muscle tone.*

_____ 19. Muscle contraction results from the sliding of *actin past myosin.*

_____ 20. Oxygen debt occurs along with increased production of *carbon dioxide.*

_____ 21. Muscle fatigue appears to result from buildup of *lactic acid* and decrease in availability of energy-rich compounds.

_____ 22. Transmission of action potentials from cell to cell in cardiac muscle occurs via *gap junctions.*

_____ 23. The *biceps brachii* muscle is named according to the number of heads and the location.

_____ 24. Shivering *increases* the rate of heat production by skeletal muscle fibers.

_____ 25. When a skeletal muscle maintains a sustained contraction it is termed *tetanus.*

Completion

1. The fibrous connective tissue covering that envelops muscle is the _____.
2. A motor neuron and the muscle fibers it innervates are collectively called a _____.
3. Transmission of an action potential from a motor neuron to muscle cell is mediated by a chemical compound. In skeletal muscle this compound is _____.
4. Muscles that relax while the prime mover (agonist) is contracting are called _____.
5. Uninucleate, nonstriated cells are characteristic of _____ muscle.
6. _____, similar to hemoglobin, is a protein found in muscle cells, which is responsible for giving the cells red color.
7. The gradual loss of contractility resulting from depressed energy sources and from waste accumulation is called _____.
8. _____ muscle is normally subject to our conscious control.
9. A stimulus that is strong enough to elicit an action potential is a _____ stimulus.
10. The continuation of the epimysium that attaches muscle to bone is called a _____.
11. Two different types of smooth muscle can be distinguished: _____ and _____.
12. The _____ period is the interval following stimulation during which a muscle cell is unresponsive to a second stimulus.
13. The area between two Z lines in a skeletal muscle is called a _____.
14. During exercise, the availability of oxygen to the muscle is increased by allowing the blood vessels to _____.
15. The _____ is the time in a simple twitch response when the calcium is being released from the ST and actin–myosin cross bridges are forming.

16. The ion that binds to troponin, thus exposing actin binding sites leading to muscle contraction, is _______________.
17. Fusion of skeletal muscle twitches into a single, prolonged contraction is called _______________.
18. Conversion of lactic acid to glycogen occurs in the _______________.
19. An action potential on the sarcolemma results in the release of _______________ ions from the sarcoplasmic reticulum.
20. The duration of contraction of smooth muscle lasts _______________ (longer, shorter) than that of skeletal muscle.
21. _______________ contractions are characterized by a constant length throughout the contraction with the development of tension.
22. The "head" of the _______________ molecule contains an actin-binding site and an ATP-binding site.
23. The light band, or I band, in a skeletal muscle cell contains _______________ filaments.
24. In cardiac muscle the boundary between cells is marked by _______________ discs.
25. Within a muscle, the fibers are bound together into groups or fasciculi by connective tissue called _______________.

Lost Sheep

Set 1

1. gastrocnemius, soleus, pectoralis, tibialis anterior
2. pectoralis major, infraspinatus, latissimus dorsi, trapezius
3. vastus medialis, rectus femoris, biceps femoris, gracilis
4. triceps brachii, biceps brachii, deltoid, latissimus dorsi
5. gluteus maximus, rectus abdominus, external oblique, linea alba
6. platysma, serratus, sternocleidomastoid, masseter
7. frontalis, temporalis, masseter, trapezius
8. intercostals, serratus, pectoralis major, diaphragm
9. superior rectus, inferior oblique, external oblique, lateral rectus
10. tendon, aponeurosis, bursa, origin

Set 2

1. lactic acid, anaerobic, oxygen, glycolysis
2. shortening, isometric, outside work, isotonic
3. single twitch, summation, treppe, tetanus
4. visceral, smooth, self-excitable, skeletal
5. ATP, CP, Ca^{2+}, creatine
6. collagen, actin, myosin, troponin
7. action potential, refractory period, latent period, fasciculation
8. acetylcholine, synaptic cleft, epimysium, sarcolemma
9. myasthenia gravis, spasm, atrophy, tonus
10. perimysium, sarcoplasmic reticulum, fasciculus, bundle
11. CP, ATP, ADP, glucose
12. heat production, acetylcholine production, body movement, posture maintenance
13. autonomic nervous system innervation, striations, tapered ends, long contraction
14. stretch, hormones, high oxygen, high acetyl

15. rhythmic contraction, dependency on nerve stimulation, functional unit, self-excitable
16. sarcomere, epimysium, tendon, perimysium
17. mitochondrion, sarcoplasmic reticulum, myofilament, epimysium
18. actin, tropomyosin, myosin, troponin
19. intercalated discs, voluntary control, functional unit, long refractory
20. motor neuron, acetylcholine, synaptic vesicle, ACh esterase

Matching

Set 1

1. voluntary, multinucleate
2. long refractory period
3. longest period of contraction
4. intercalated discs
5. shortest contraction period
6. nonstriated, involuntary
7. wide in middle, tapered at ends
8. striated, cells act as one functional unit

a. skeletal muscle
b. visceral (smooth) muscle
c. cardiac muscle

Set 2

1. flexes elbow joint
2. adducts thigh
3. depresses shoulder
4. turns head to one side
5. points the toe
6. extends leg
7. abducts upper arm

a. sternocleidomastoid
b. biceps brachii
c. quadriceps femoris
d. deltoid
e. pectoralis minor
f. gastrocnemius
g. gracilis

Set 3

1. actin molecules
2. no response to second stimulus
3. uncoordinated contraction, fasciculation
4. its movement exposes active sites on actin
5. myosin molecules
6. time from stimulation to contraction
7. accumulates in oxygen debt
8. high-energy molecules stored in muscle
9. muscle covering
10. sustained contraction

a. epimysium
b. lactic acid
c. creatine phosphate
d. latent period
e. tetanus
f. fibrillation
g. thin myofilaments
h. troponin
i. thick myofilaments
j. refractory period

Double Crosses

1. *Across:*
muscle that turns the palm upward

Down:
mild, constant muscle contraction

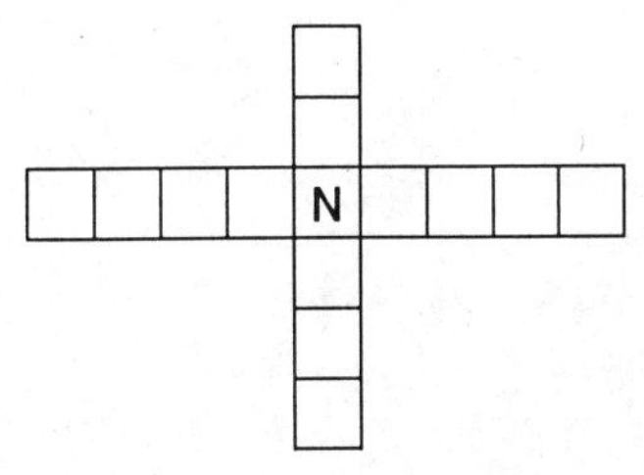

2. *Across*:
in-between the ribs

Down:
biceps that flex forearm

C

3. *Across*:
muscle that abducts arm

Down:
muscle that raises eyebrows

T

4. *Across*:
tailor's muscle

Down:
flexes and rotates thigh laterally

O

5. *Across*:
flexes abdomen (2 words)

Down:
extends thigh

B

Sleuthing

1. Jim is a marathon runner. He has been running eight miles a day for the last few years. Before a big race he eats a lot of starchy food and drinks more beer than usual as a means of "carbohydrate loading."
   a. Why ingest so much carbohydrate?

   b. What is happening to Jim's muscle metabolism as he engages in more and more strenuous running?

   c. When muscle fatigue develops in a long run, what could be a possible explanation?

   d. What is the explanation for the increase in size of Jim's calf muscles?

   e. Explain the effects of treppe and warmth on muscle performance.

Word Scrambles

1. a. frequent site of intramuscular injection	LUTGUSE	□ _ _ _ □ _ _
b. flexion's companion movement	SNEXEITON	_ □ _ _ _ _ _ □ □
c. draws scapula upward and medially	EOIMDRUBOSH	_ _ _ _ □ _ _ □ □ _ _
d. disease of wasting of muscle fibers	PYRSPYDTOH	_ □ _ □ _ _ _ _ _

Total: lactic acid accumulation in muscle is indicative of this

□□□□□□ □□□□

2. a. tissue specialized for contraction	UCEMSL	_ _ □ _ _ □
b. opposes agonist	NGOAINTATS	_ _ _ _ □ _ _ _ □ _
c. period during which additional stimuli produce no response	RTFAEORCYR	□ _ _ _ _ _ □ _ _ _
d. movable muscle attachment	SRITNIENO	_ □ _ _ _ _ □ _ _

Total: assists agonist

□□□□□□□□□

3. a. wide middle, tapered end	PSNILDE	□ _ □ _ _ _ _
b. connective tissue covering muscle	ISMIEYMPU	_ _ _ □ _ □ _ _ _
c. gives red color to some muscle cells	OLYBMOGNI	_ _ _ □ _ _ _ _ □
d. enzyme employed at myoneural junction	ECLETHISEONARS	_ _ _ _ _ _ _ _ □ _ _ □ _ _
e. results in muscles from "pumping iron"	HETYRORPYHP	□ _ _ _ □ _ _ _ _ _ _

Total: dorsal thigh muscles often injured by athletes

□□□□□□□□□□

Addagrams

1. a. from Z to Z — 1, 14, 4, 7, 17, 13, 3, 4, 9
   b. vessels that serve exercising muscles are . . . — 11, 18, 8, 14, 16, 9, 19
   c. type of contraction where constant tension is maintained — 10, 15, 12, 2, 6, 5, 18, 7

Total: this'll turn your head!

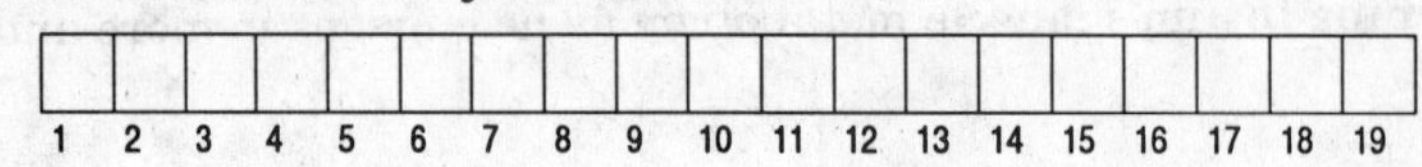

2. a. thin muscle filament — 4, 7, 1, 6, 5
   b. movement around an axis — 12, 17, 3, 13, 15, 16, 9, 18
   c. steady, partial contraction — 11, 9, 10, 2
   d. ion that releases binding site on actin — 8, 4
   e. animal often used in dissection — 14, 4, 1

Total: normal body movement

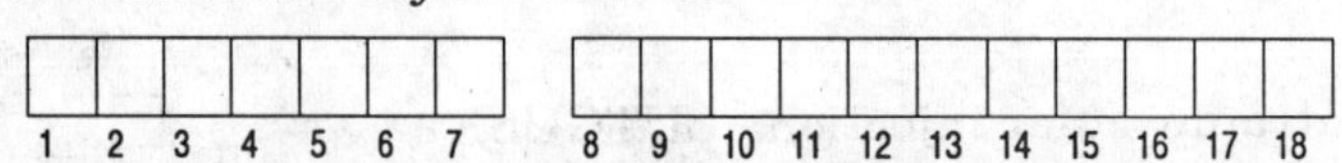

3. a. to turn palms down — 7, 9, 11, 5, 2, 3, 4
   b. shoulder cap — 12, 4, 1, 6, 11, 10, 12
   c. I band contains the thin filament . . . — 2, 1, 11, 5, 8

Total: time lag between stimulus and contraction

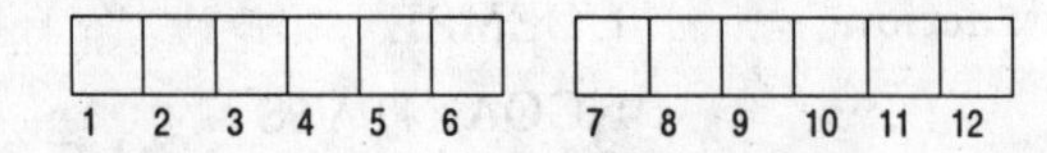

4. a. bound with actin — 16, 2, 11, 9, 11, 7, 6, 7
   b. enzyme that releases energy — 4, 5, 13, 15, 12, 17
   c. what sore muscles do — 15, 1, 10, 8
   d. generated by contraction — 14, 3, 4, 5

Total: on its own, it gives a loan, to ATP

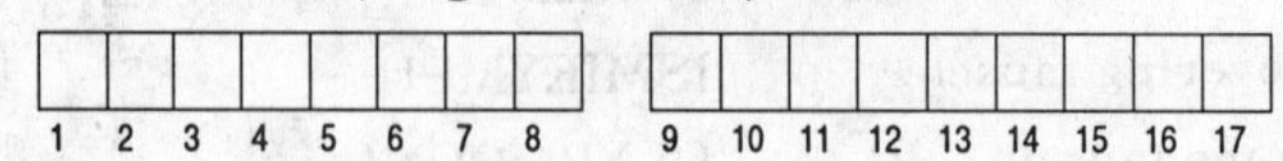

Crossword

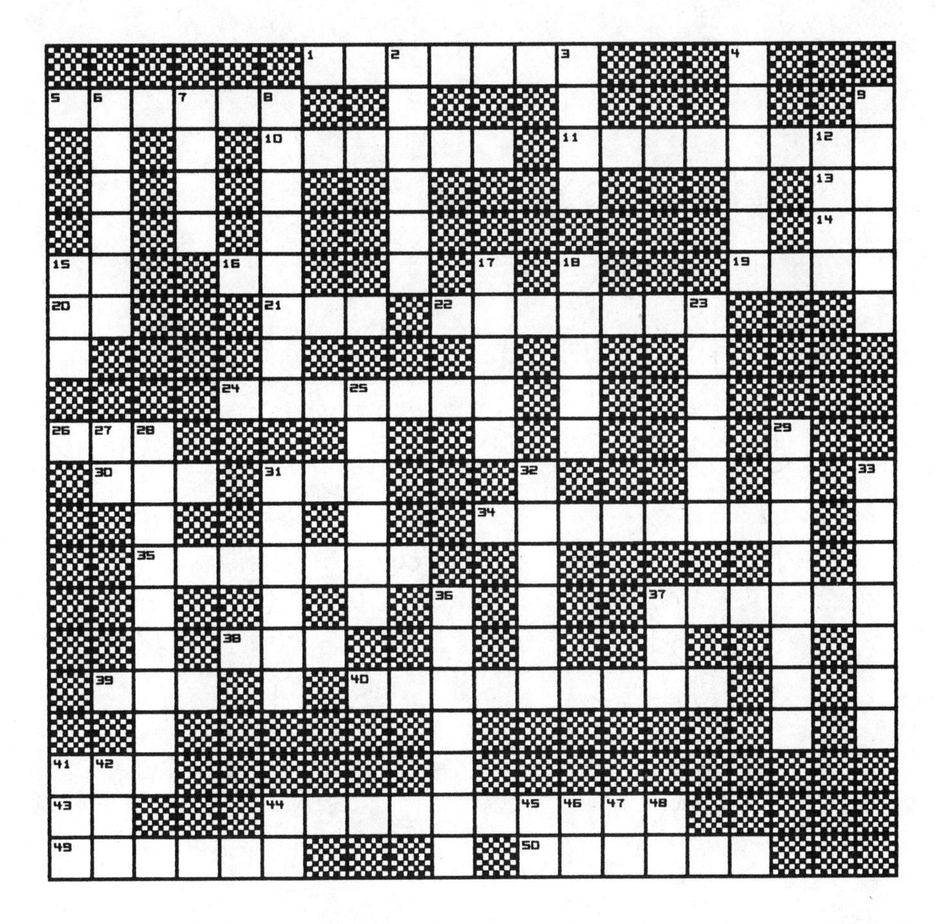

Across

1 Rapid repeated contractions
5 Increase in strength of contraction with stimulus of same strength
10 Ion to which muscle membrane becomes more permeable upon stimulation
11 Bundle of fibers
13 ________ and out
14 Indiana football team (abbr.)
15 Exclamation
16 All ________ nothing response of muscle
19 North Atlantic Treaty Organization
20 Afternoon and evening (abbr.)
21 Molecule containing energy
22 Ion that releases inhibition of actin by troponin
24 With oxygen
26 Type of fish
30 Muscle prefix
31 Negative vote
34 Protein associated with actin
35 Transverse ________ of myofiber
37 ________ choline
38 Space
39 Prefix meaning "on top of"
40 Strength of stimulus which will elicit a response
41 Retiring type
43 "And ________ says I"
44 Unresponsive
49 Attachment into movable bone
50 Respiratory reflex

Down

2 Single muscle contraction
3 Coffee house
4 Immovable attachment
6 What smooth muscle has
7 Prefix meaning "outer area"
8 chemical that breaks up ACh
9 Cordlike attachment of muscle
12 Dust from material
15 Mimic
17 Prefix meaning "muscle"
18 Protein myofilament
23 Thick protein myofilament
25 Used in aerobic reactions
27 Sound repeated in a mantra
28 Disease
29 Myoneural ________
31 Type of stimulation that skeletal muscle needs
32 Myasthenia ________
33 Acetyl ________
36 Type of muscle that beats constantly
37 ________-or-none response of muscle
41 A type of social security payment
42 Casual term of endearment
44 Right (abbr.)
45 First initials
46 Above
47 Regarding (abbr.)
48 "Oh ________ of little faith"

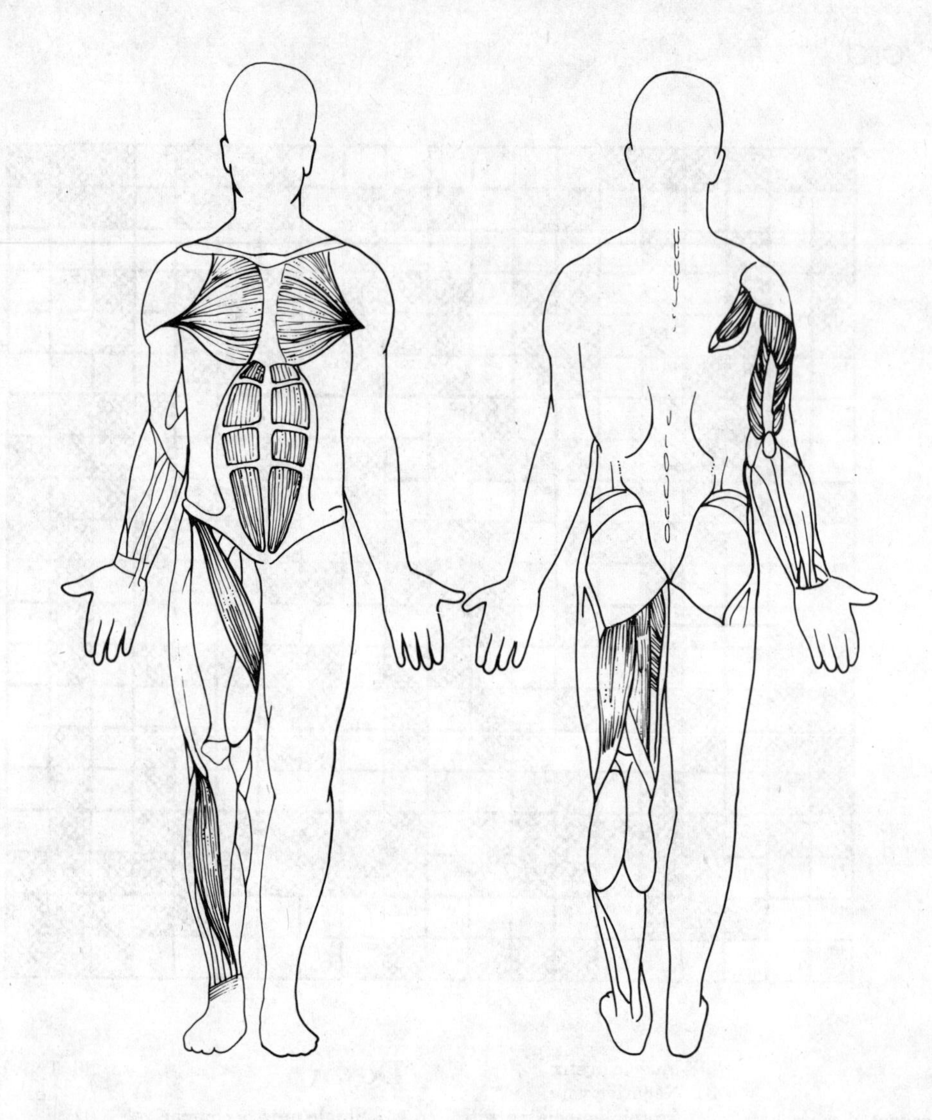

Anatomic Artwork

1. Indicate the direction of the fibers and label the muscle forming the “cap” on your shoulder.
2. Indicate the direction of fibers and label the muscle on the ventral/lateral/superficial aspect of the abdomen.
3. Indicate the direction of fibers and label the two-headed muscle of the calf.
4. Label the superficial muscle you sit on.
5. Indicate the direction of fibers and label the muscle that is the major flexor of the arm at the elbow.
6. Draw and label the muscle running across the upper back and neck which allows shrugging of the shoulder.
7. Indicate the direction of fibers and label the muscle pair that turns the head to the side opposite the one contracting.
8. Label the muscle of the lower extremity (which consists of four parts) that extends your leg.
9. Label the muscle whose border can be seen on the lateral posterior surface of the torso when the shoulder is drawn down.
10. Draw in and label the muscle that lifts your eyebrows in surprise.

Chapter 9

Nervous Tissue

Introduction

The human organism is continually responding to stimuli of almost infinite variety. This is vital for maintaining *homeostasis* and is accomplished, in large measure, by the activities of the nervous system. It is the nervous system that *senses* changes in the external and internal environments, *coordinates* and *integrates* these "data," and finally provides the "command signals" that allow appropriate *adjustments* to the changes by various effectors. The nervous system, as a coordinator, works in conjunction with the endocrine system, which is covered in a later chapter.

Although the activities of the nervous system are very complicated, they are carried out efficiently and rapidly by *neurons*, which are the conducting elements of nervous tissue. The conducting elements utilize the *membrane potential*, which exists in neurons to transmit nerve impulses along the cell membrane in the form of a *wave of depolarization*. The functioning of the nervous system is so vital that nervous tissue also has specialized "supportive" cells referred to as *neuroglia*.

After studying the information on nervous tissue in your text, you should have an understanding of these concepts and should be able to answer the following questions and complete the activities included.

Multiple Choice

___ 1. When the postsynaptic neuron exhibits a "near threshold excitation" and is prepared for subsequent stimuli, this is:
a. hyperpolarization
b. facilitation
c. summation
d. inhibition

___ 2. In which of the following would a nerve impulse travel fastest?
a. myelinated A fiber
b. myelinated C fiber
c. nonmyelinated B fiber
d. myelinated B fiber

___ 3. A PNS myelinated fiber can exhibit which of the following:
a. saltatory conduction
b. regeneration
c. both a and b
d. neither a nor b

___ 4. In a polarized neuron at rest, which ion is concentrated outside the cell?
a. calcium
b. potassium
c. sodium
d. none of these

___ 5. Neuroglial cells that line the ventricles of the brain are:
a. oligodendrocytes
b. microglia
c. Schwann cells
d. ependymocytes

___ 6. Which of the statements below concerning impulse conduction is/are true?
a. Speed of impulse transmission is influenced by the diameter of the fiber
b. Myelinated fibers conduct impulses at greater speed than nonmyelinated fibers
c. Repolarization of neurons occurs within milliseconds, thus permitting a rapid volley of impulses to be conducted
d. All of these statements are true

___ 7. Which of the following statements is false?
a. The transmission of an impulse across the synapse is chemically mediated
b. The synapse automatically transmits the impulse from one neuron to another
c. Neurotransmitter is contained within the presynaptic neuron's vesicles
d. There is one-way conduction at a chemical synapse

___ 8. Release of neurotransmitter from vesicles is directly related to:
a. calcium ions flooding into the end bulb
b. presence of passive ion channels
c. closure of voltage-gated channels
d. all of these

___ 9. It can be said that in a resting neuron, the:
a. membrane is polarized
b. inside of the cell is more electropositive than the outside
c. concentration of K^+ ions is greater outside the cell
d. all of these

___ 10. If a neuron receives a threshold level stimulus, it transmits an impulse based on the:
a. amount of neurotransmitter released
b. all-or-none principle
c. concept of continuous conduction
d. all of these

___ 11. Synaptic conduction of a nerve impulse can be affected by:
a. certain diseases
b. drugs
c. changes in pH
d. all of these

___ 12. The ability of neurons to respond to stimuli and convert them into electrical impulses is called:
a. depolarization
b. facilitation
c. excitability
d. summation

True/False

______ 1. Phagocytic cells of nervous tissue are called *microglia.*

______ 2. *Afferent* neurons convey impulses away from the central nervous system.

______ 3. Neurons, like muscle fibers, conduct impulses according to the *all-or-none law.*

______ 4. A hyperpolarized neuron "fires" *more easily* than a facilitated neuron.

______ 5. Association neurons are found only in the *peripheral* nervous system.

______ 6. *Neurofibrils* help transport nutrients within the neuron.

______ 7. Side branches of an axon are called *end bulbs.*

______ 8. The sodium–potassium pump contributes to the *polarization* of the membranes.

______ 9. A neuron will *respond maximally* to a subthreshold–level stimulus.

______ 10. Transmission of an action potential from neuron to neuron is mediated by chemicals called *transmitters.*

______ 11. The presence of myelin and nodes of Ranvier are responsible for *saltatory* conduction.

______ 12. In a postsynaptic neuron when the exictatory effect is greater than the inhibitory effect, but less than the threshold level, *facilitation* results.

Completion

1. The supportive and protective cells of the nervous system are collectively called ______.
2. The chromophilic ______ in a neuron is a form of protein-synthesizing endoplasmic reticulum.
3. The efferent components of the peripheral nervous system are referred to as ______ neurons.
4. Chemicals called ______, located in the synaptic vesicles, determine whether or not an impulse passes from one neuron to another.
5. Small unmyelinated gaps in the myelin sheath of the neurons in the PNS are called ______.
6. The term "bipolar" refers to a ______ classification while "association" refers to a ______ classification of neurons.
7. Integral proteins in the plasma membrane are associated with ______, which are passages for the movement of materials through the membrane often in response to voltage or chemical changes.
8. A membrane that at rest has a net positive charge externally, relative to a net negative charge internally, is referred to as a ______ membrane.
9. Restoring the resting membrane potential after an action potential is called ______.
10. The area where two neurons "communicate" and transmit an impulse is called a ______.

11. Localized changes in membrane potential which are of short duration and of decreasing intensity as they travel are called ________________.
12. If transmission of an impulse at the synapse is inhibited, the postsynaptic membrane has been ______________.

Lost Sheep

1. myelin, C fibers, pain impulses, 0.5 m/sec
2. synaptic end bulbs, neurotransmitter, dendrite, collateral
3. astrocyte, microglia, ependymal, Schwann
4. depolarization, inhibitory transmission, postsynaptic neuron more negative, increased permeability to K^+ and Cl^-
5. Nissl bodies, nodes of Ranvier, neurofibril, nucleus
6. saltatory conduction, myelin sheath, energy expenditure elevated, node of Ranvier
7. calcium ions, synaptic transmission, vesicles, dendrite
8. simultaneous firing of numerous presynaptic end bulbs, summation, hyperpolarization, increased Na^+ flow into postsynaptic neuron
9. myelin sheath, oligodendrocyte, Schwann cell, microglia
10. sensory, afferent, receptor, efferent

Matching

Set 1

___ 1. additional stimuli fail to generate a response
___ 2. process of making the outside of the membrane relatively negative to inside
___ 3. near threshold excitation level
___ 4. cumulative addition of simultaneous release of neurotransmitter by several end bulbs
___ 5. inhibition of transmission
___ 6. role of postsynaptic neuron

a. depolarization
b. facilitation
c. refractory period
d. integration
e. hyperpolarization
f. summation

Set 2

___ 1. forms myelin sheath within the CNS
___ 2. phagocytic
___ 3. attaches neurons to blood vessels
___ 4. lines ventricles of brain
___ 5. produces myelin sheath in the PNS
___ 6. assists in circulation of cerebrospinal fluid

a. oligodendrocyte
b. astrocyte
c. microglia
d. Schwann cell
e. ependymal cell

Double Crosses

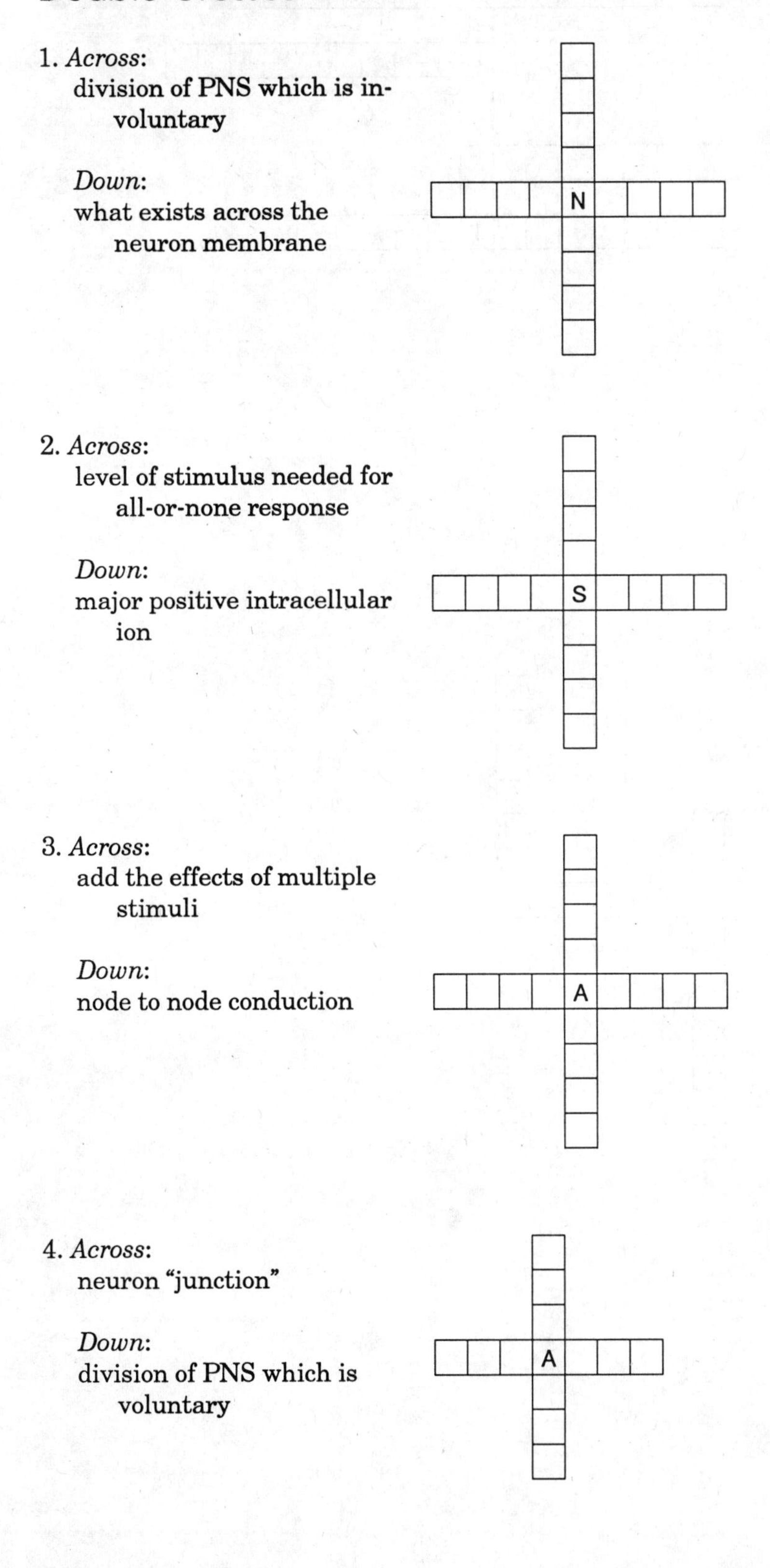

1. *Across*:
   division of PNS which is involuntary

   Down:
   what exists across the neuron membrane

2. *Across*:
   level of stimulus needed for all-or-none response

   Down:
   major positive intracellular ion

3. *Across*:
   add the effects of multiple stimuli

   Down:
   node to node conduction

4. *Across*:
   neuron "junction"

   Down:
   division of PNS which is voluntary

5. *Across*:
   a. a neuron may have many of these
   b. a capacity to do this is greatly reduced in neurons
   c. neurons that transmit toward the CNS
   d. ________ line ventricles
   e. this occurs when a near-threshold level is achieved in a neuron
   f. what neurons do with an impulse

   Down:
   neuron that transmits away from the brain and spinal cord to the effector

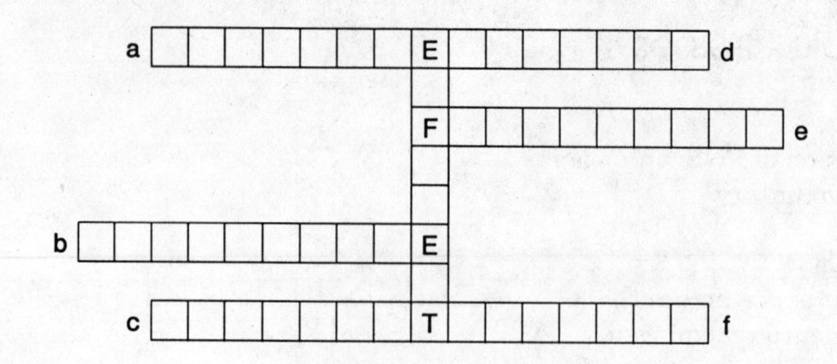

6. *Across*:
   period when additional stimuli cannot elicit a response

   Down:
   a. node name
   b. division of nervous system consisting of nerves
   c. star-shaped glial cells
   d. name for neuron that transmits from receptors

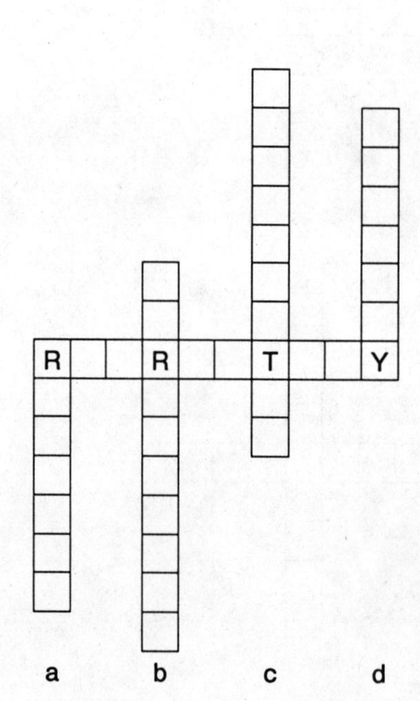

Word Cage

A D F S U B T H R E S H O L D
L T Y U E M K N B D T B K E M
O L I G O D E N D R O C Y T E
R T A S F G L P W A C T J S F
L A C T R O L O A Z F R E Y F
A X O N D S A L T A T O R Y E
R B N H I P U A D E N D N O R
E X I D E B A R B A T M O T E
H E D M U L T I P O L A R H N
P O L A R D I Z F G N L U M T
I N E U R O G E B R A I E Y I
R O L I G O D D E N D R N Z E
E F G H N I T R S Y N A P S E
P N O I T A T I L I C A F A C

1. Division of the nervous system which consists of cranial and spinal nerves.
2. Produces myelin sheath in CNS.
3. Its ends contain neurotransmitter in vesicles.
4. Neurons with many dendrites.
5. Type of neuron that conveys motor impulses from brain and spinal cord to effectors.
6. Neuron membrane at rest is ________.
7. A stimulus that is too weak to elicit a response.
8. Conduction that occurs in myelinated fibers.
9. The "junction" of neurons.
10. Conduction unit of nervous tissue.

Word Scrambles

1. a. electrical "first step" in a nerve impulse — EORADAIZLPTNOI — _ _ ☐ _ _ _ _ _ _ _ ☐ _ _ ☐

   b. exists across a polarized membrane — TILTPNEOA — ☐ ☐ _ _ ☐ ☐ _ ☐ _

   c. characteristic of nervous tissue — BITIXAILYTCE — ☐ _ ☐ _ _ _ _ _ _ ☐ _ ☐

   d. stimulus level needed to elicit a response — RSLTDOEHH — _ _ ☐ _ ☐ _ _ _ _

   e. Schwann cells — EUOLMORNEMYCTSE — ☐ _ ☐ _ _ _ _ _ _ ☐ _ _ _ _ ☐

Total: the "integrator" cell

2. a. period of no additional action potentials — ERCTAFRROY — □ _ _ _ □ _ □ □ _ □
   b. part of nervous system with brain and spinal cord — ERLNATC — □ _ □ □ _ _ □
   c. ion concentrated inside the neuron — OTSIMSAPU — _ □ _ _ _ □ _ □ _
   d. cell process — DNRTEIDE — □ _ □ _ _ _ □ _
   e. phagocytic cell — IROLIGCMA — _ □ □ □□ _ _ _ □

Total: "node to node"

□□□□□□□□ □□□□□□□□□□

Addagrams

1. a. sheath of Schwann — 6, 3, 13, 14, 15, 4, 9, 1, 1, 7
   b. receives impulse and sends impulse to cell body — 10, 12, 16, 10, 14, 5, 8, 9
   c. afferent neuron — 17, 3, 11, 17, 15, 14, 2

Total: quickest conductors of impulses

□□□□□□□□□□ □□□□□□□
1 2 3 4 5 6 7 8 9 10 11 12 13 14 15 16 17

2. a. ________ way conduction at the synapse — 7, 8, 4
   b. depolarization, an ________ potential — 11, 9, 3, 6, 7, 12
   c. cells that serve to support and protect neurons — 1, 15, 6, 2, 15
   d. bulb where vesicles are located — 14, 13, 5
   e. what Schwann cells form — 16, 10, 4, 2, 3, 10

Total: sometimes open, sometimes closed

□□□□□ □□□ □□□□□□□□
1 2 3 4 5 6 7 8 9 10 11 12 13 14 15 16

Crossword

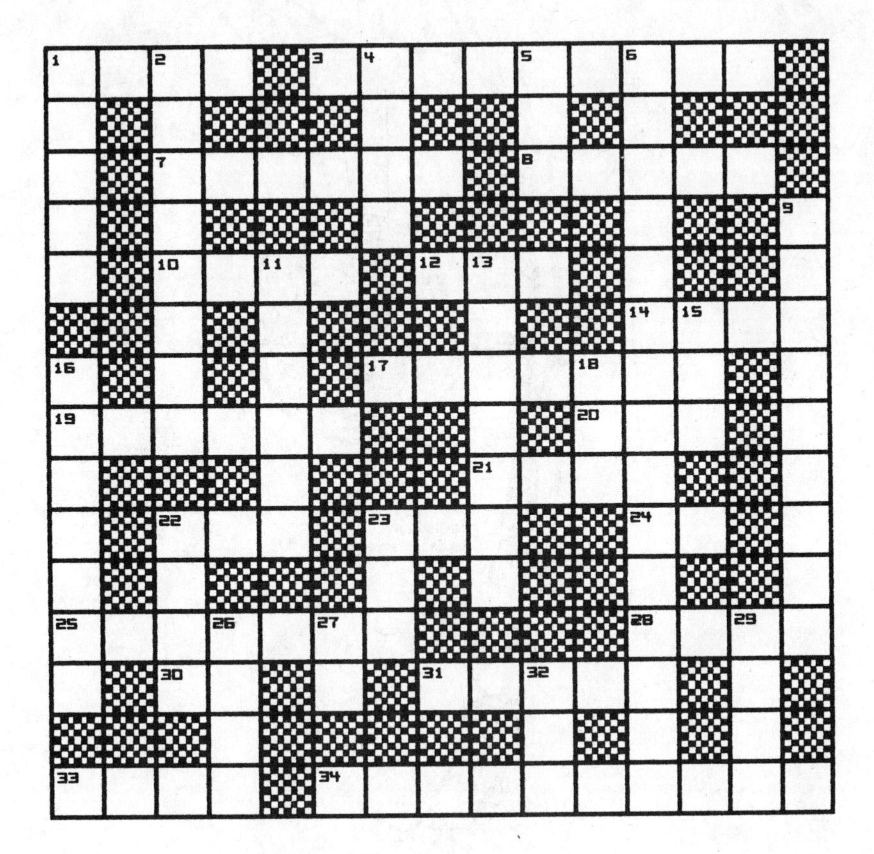

Across

1 Space between Schwann cells
3 Conduction along myelinated fiber
7 Functional unit of nervous system
8 Cost
10 When sodium is out and potassium is in, the neuron is at ________
12 Brain's bath
14 Attempt to lose weight
17 Node of ________
19 Covering that speeds conduction
20 Nervous system that is involuntary (abbr.)
21 Where ducks can be found
22 With honors: ________ laude
23 Brain and spinal cord (abbr.)
24 Egyptian sun god
25 Voluntary nervous system
28 What ribs form
30 Indefinite article (before vowel)
31 Myelin is this type of material
33 ________ atrial node: the heart's pacemaker
34 Branch of an axon

Down

1 Modified ER in the neuron body
2 Conducts impulse toward cell
4 Neurons have only one
5 Is conserved by saltatory conduction
6 Forms myelin sheath in CNS
9 Part of the blood–brain barrier
11 Extracellular ion that moves across membrane to start an action potential
13 Where neurons "communicate" with each other
15 Tax people
16 What a neuron conducts
18 Scottish man's name
22 What can result when neurons shut down
23 Three hundred to a Roman
26 Latin year
27 Preposition
29 Collective name for ependyma, microglia, astrocytes, etc.
32 This can result from a decrease in the blood circulation to the brain

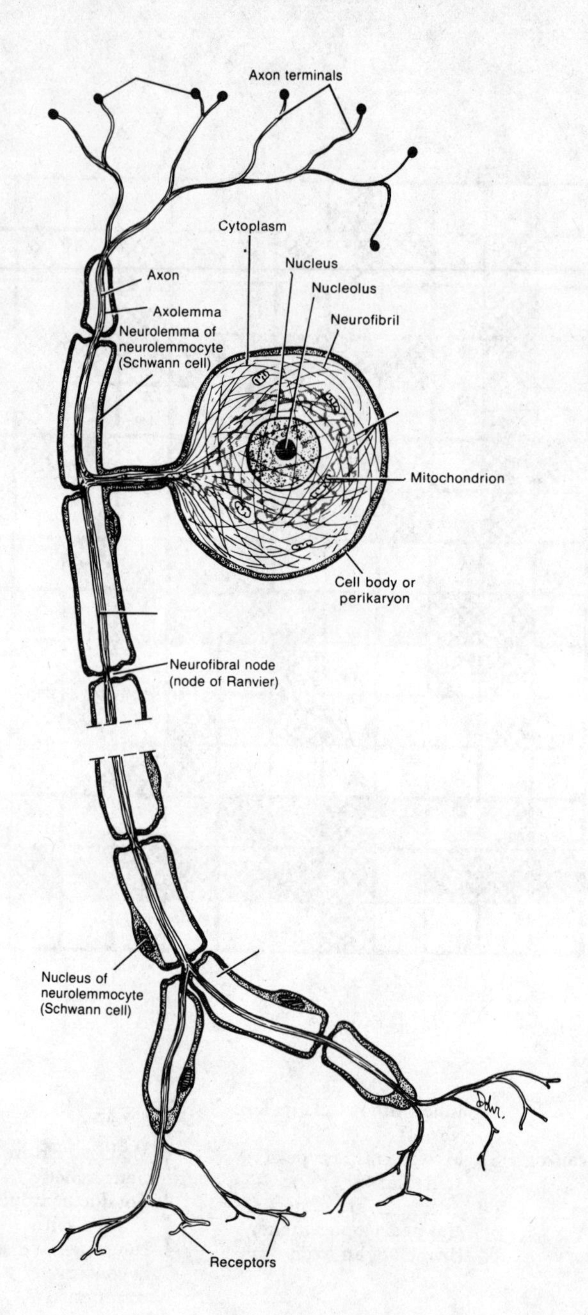

Anatomic Artwork

1. Label the organelle that is modified endoplasmic reticulum.
2. Shade in and label the substance that, when present, results in saltatory conduction.
3. Identify and label the cell that is responsible for forming number 2 above.
4. Label the structures that would be responsible for transmission of an impulse to a succeeding neuron.
5. Label the structure that transmits the impulse toward the cell body.
6. Indicate by arrows the direction a nerve impulse would travel in this neuron.
7. This illustration would represent which type of neuron based on function?

Chapter 10

Central and Somatic Nervous System

Introduction

In the course of maintaining homeostasis, the nervous system is a very important component since it is one of the body's control centers. *Integrating, coordinating,* and *controlling* many of the body's activities is the *brain,* which houses a myriad of control centers for vital functioning as well as thinking and emotions. It is the *spinal cord* and *peripheral nerves* that transmit impulses from the receptors to the brain and from the brain back to effectors so that adaptation to changes in the environment can be made.

After completing the work in your text, you should have an understanding of these concepts and be able to answer the following questions and complete these activities.

Multiple Choice

___ 1. The spinal cord extends through the vertebral canal from the foramen magnum to the:
   a. 7th cervical vertebra
   b. 12th thoracic vertebra
   c. 2nd lumbar vertebra
   d. 1st sacral vertebra

___ 2. Which of the following spinal cord tracts ascends?
   a. ventral spinothalamic
   b. lateral corticospinal
   c. lateral reticulospinal
   d. ventral corticospinal

___ 3. Which of the following statements concerning the meninges is/are correct?
   a. The outer dura mater has several folds into the cranial cavity between the cerebral lobes
   b. The subarachnoid space, located between the arachnoid and pia mater, contains cerebrospinal fluid
   c. The pia mater is a thin vascular membrane
   d. All statements are true

___ 4. Which of the following pairs is/are mismatched?
a. spinal nerve dorsal root–sensory root
b. spinal nerve ventral root–motor root
c. thoracic spinal nerves–cauda equina
d. cervical spinal nerves–8 pairs

___ 5. A reflex arc would have which of the following?
a. receptor
b. afferent pathway
c. effector
d. all of these

___ 6. Coordination of muscular activity is a function of the:
a. cerebral peduncles
b. cerebellum
c. thalamus
d. medulla oblongata

___ 7. Which of the following statements is correct?
a. Appetite control is a function of the thalamus
b. The midbrain is a conduction pathway and reflex center
c. The cerebellum has sensory centers for thirst and speech
d. Respiratory centers are located in the basal ganglia

___ 8. The structure that connects the cerebral hemispheres and facilitates the sharing of cerebral information is the:
a. corpus callosum
b. cingulate sulcus
c. fourth ventricle
d. aqueduct of Sylvius

___ 9. Somasthetic association areas:
a. store memories of past sensory experiences
b. permit determination of exact shape and texture
c. integrate and interpret sensations
d. all of these

___ 10. Inability to chew might result from damage to which cranial nerve?
a. facial
b. trigeminal
c. abducens
d. vagus

___ 11. A memory trace in the brain is called an
a. association
b. engram
c. nucleus
d. none of these

___ 12. Which of the following is not found in the gray matter of the nervous system?
a. neuron cell bodies
b. myelinated tracts of neurons
c. nonmyelinated dendrites
d. nonmyelinated axons

___ 13. Which of the following statements is false concerning the spinal cord?
a. It consists of 31 segments, each giving rise to a pair of spinal nerves
b. It has a central canal that runs the length of the cord
c. It has two enlargements, the cervical and thoracic enlargements
d. There is an H-shaped central core of gray matter

___ 14. Which of the following is a descending motor pathway in the spinal cord, which relays information for precise movements of skeletal muscles:
a. posterior column pathways
b. pyramidal pathways
c. anterior spinothalamic tracts
d. none of these

___ 15. The major functions of the spinal cord are to:
a. relay sensory and motor impulses into and out of the brain
b. serve as a center for reflexes
c. both a and b
d. neither a nor b

___ 16. Which of the following statements concerning reflexes is correct?
a. Reflexes that cause secretion of glands or smooth muscle contractions are called visceral reflexes
b. Reflex arcs are a form of conduction pathway
c. A reflex is a rapid response to changes in the internal and external environment which helps maintain homeostasis
d. All statements are correct

___ 17. The branches of a spinal nerve are referred to as:
a. tracts
b. pathways
c. rami
d. plexus

___ 18. Which of the following is not part of the diencephalon?
a. thalamus
b. fourth ventricle
c. hypothalamus
d. all are part of the diencephalon

___ 19. Inactivation of the RAS produces:
a. hunger
b. increased sensory activity
c. sleep
d. elevated blood pressure

___ 20. In most people, the left cerebral hemisphere is more important for which of the following?
a. spoken and written language
b. space and pattern perception
c. musical and artistic awareness
d. imagination

___ 21. Which of the following is true concerning the limbic system?
a. It is comprised of parts of the cerebrum and diencephalon
b. It assumes a primary function in emotions
c. both a and b
d. neither a nor b

___ 22. The degree of representation of a particular body part on the motor cortex is:
a. proportional to the precision of the movement required
b. sensitivity of the body part
c. size of the body part
d. all of these

___ 23. Lack of muscle coordination is called:
a. monoplegia
b. ataxia
c. TIA
d. none of these

___ 24. A bundle of neuron fibers outside the central nervous system is called a:
a. tract
b. plexus
c. nerve
d. nucleus

___ 25. Which of the following occurs in the pyramids of the medulla?
a. dysplasia
b. decussation
c. plexation
d. segmentation

True/False

__________ 1. The covering of the brain and spinal cord is called the *epineurium*.

__________ 2. CSF passes from the *lateral* ventricles into the fourth ventricle through the cerebral aqueduct.

__________ 3. A muscle stretch reflex is an example of a *visceral* reflex.

__________ 4. Cell bodies of neurons outside the CNS are concentrated in structures called *plexi*.

__________ 5. The center that controls the cardiovascular system functions, such as heart rate, is located in the *pons*.

__________ 6. *Extrapyramidal* pathways help coordinate head movements with visual stimuli and play a major role in equilibrium.

__________ 7. The cell bodies of sensory neurons are located in the *dorsal* root ganglion.

__________ 8. The projections of the H-shaped area of gray matter in the spinal cord are referred to as *horns*.

__________ 9. The inner layer of the meninges is the *pia mater*.

__________ 10. In the meninges CSF circulates in the *subdural* space.

__________ 11. A bundle of fibers in the central nervous system is called a *ganglion*.

__________ 12. Spinal nerves have several branches known as *rami*.

__________ 13. The brachial plexus constitutes the nerve supply for the *pelvic girdle and lower extremity*.

__________ 14. The cavities of the brain are called *ventricles*.

__________ 15. Cerebrospinal fluid is formed by filtration of fluid from blood through the *circle of Willis*.

______________ 16. The *blood–brain barrier* includes the tighter cell junctions and thicker basement membrane of the capillaries, as well as the astrocytes that surround the capillaries.

______________ 17. The cerebral peduncles are found primarily in the *pons.*

______________ 18. The *midbrain,* a part of the diencephalon consisting of gray matter organized into nuclei, is a relay center for sensory impulses.

______________ 19. The principal connection between the pituitary and nervous system in the brain is found in the *medulla.*

______________ 20. REM and NREM sleep *alternate* throughout the night.

______________ 21. *Projection* fibers transmit impulses from the cerebrum to other parts of the brain and spinal cord.

______________ 22. Impulses concerning texture of a substance, traveling from the hand to the brain, would terminate in the *parietal* lobe of the cerebral cortex.

______________ 23. Dopamine, norepinephrine, and serotonin are examples of biogenic amines that are *neurotransmitters.*

______________ 24. Cranial nerves I, II, and VIII are *mixed* nerves.

______________ 25. Interruption of the blood supply to the brain can result in *TIA* or a CVA.

Completion

1. The coverings of the brain and spinal cord are known as the ______________.
2. There are ______________ pairs of spinal and ______________ pairs of cranial nerves.
3. The part of the brain that houses the control centers for food intake and regulation of body temperature is the ______________.
4. The special blood vessel at the base of the brain which supplies the all important nutrients to the brain is the circle of ______________.
5. External hydrocephalus results from an obstruction, which causes CSF to accumulate in the ______________.
6. Cardiac, vasomotor, and respiratory centers are located in the ______________.
7. The ______________ contain motor fibers that connect the cerebral cortex to the pons and spinal cord.
8. Muscle tone is depressed and respiration and pulse rate are increased and irregular in ______________ sleep.
9. Folds in the cerebral cortex are known as ______________.
10. ______________ transmit impulses from the gyrus of one cerebral hemisphere to the corresponding gyrus in the other cerebral hemisphere.
11. The ______________ is a large area of the brain with white matter arranged much like a "tree."
12. Biogenic amines, neuropeptides, and some amino acids which establish communication between neurons are called ______________.
13. Nerves that have both sensory and motor fibers are referred to as ______________ nerves.
14. Herpes zoster causes an acute infection in the peripheral nervous system commonly referred to as ______________.

15. Degeneration of dopamine-producing neurons in the brain results in a disorder known as ________________ .
16. Branches of spinal nerves which are components of the autonomic nervous system are called rami ________________ .
17. ________________ are networks of the ventral rami of the spinal nerves.
18. Each cerebral hemisphere has ________________ lobes.
19. ________________ are networks of capillaries in the ventricles which form CSF.
20. The third and fourth ventricles of the brain are connected by the ________________ .
21. The brain stem consists of the ________________, ________________, and ________________ .
22. The diffuse network of neurons found throughout the brain stem and involved in arousal is called the ________________ .
23. Loss of the sense of smell could result if the ________________ cranial nerve were damaged.
24. A mixed cranial nerve that travels throughout the thoracic and abdominal cavities is the ________________ nerve.
25. Abnormal and irregular discharges of electricity from millions of neurons is common in ________________ .

Lost Sheep

1. phrenic, vagus, glossopharyngeal, trigeminal
2. skeletal muscle coordination, temperature regulation, water balance, food intake
3. corpus callosum, spinothalamic, corticospinal, fasciculus gracilis
4. CSF, subarachnoid space, ventricles, hypothalamus
5. cerebral aqueduct, epidural space, central canal, fourth ventricle
6. unmyelinated, horns, nucleus, tract
7. cauda equina, central canal, cervical enlargement, postganglionic neuron
8. dorsal root ganglion, ventral root, sensory nerve, receptor
9. cervical, brachial, plexus, rami
10. diencephalon, thalamus, hypothalamus, fourth ventricle
11. blood–brain barrier, tight junctions, thin basement membrane, astrocytes
12. REM sleep, regular respiration, dreaming, depressed muscle tone
13. right cerebral hemisphere, spoken language, numerical skill, reasoning
14. corpus striatum, cerebral peduncles, putamen, basal ganglia
15. somasthetic sense, cerebral cortex, thalamus, frontal lobe

Matching

Set 1

___ 1. helps return CSF to the bloodstream
___ 2. regulates body temperature
___ 3. white matter shaped like a tree
___ 4. connects the cerebral hemispheres
___ 5. functions in emotions
___ 6. area where sensory input to the CNS terminates
___ 7. contains the hunger and thirst centers
___ 8. cerebral nuclei that help regulate large subconscious movements of skeletal muscles

a. limbic system
b. arachnoid villi
c. hypothalamus
d. corpus callosum
e. cerebellum
f. association areas
g. cerebral cortex
h. basal ganglia
i. pons

___ 9. serves as a relay station between medulla and higher centers
___ 10. provides an integrative function for the cerebrum

Set 2

___ 1. commonly called a stroke
___ 2. loss of motor function in both legs
___ 3. a demyelinating disease
___ 4. temporary condition due to an interference of brain's blood supply
___ 5. genetic disease due to a lack of lysosomal enzyme
___ 6. common in young people and often follows a viral infection
___ 7. abnormal electrical discharges often resulting in seizures
___ 8. degeneration of dopamine-producing neurons
___ 9. caused by the herpes zoster virus
___ 10. inflammation of a nerve

a. shingles
b. TIA
c. CVA
d. neuritis
e. Parkinson's disease
f. Tay-Sachs disease
g. Reye's syndrome
h. multiple sclerosis
i. diplegia
j. epilepsy

Double Crosses

1. *Across:*
middle layer of meninges

Down:
fourth cranial nerve

H

2. *Across:*
functioning unit in PNS

Down:
cerebral cortex single elevation

R

3. *Across:*
ventral root of spinal nerve carries this kind of information

Down:
second cranial nerve

T

Sleuthing

1. Stephanie has been diagnosed as having Parkinson's disease. Her treatment includes injection of a fat-soluble medication called L-dopa, which is the immediate precursor of the neurotransmitter dopamine.
   a. What are the symptoms of this disease?

   b. What causes these symptoms?

   c. What part of the brain is affected in Parkinson's disease?

   d. Why is injection of L-dopa an effective treatment?

Word Scrambles

1. a. circular brain artery — SWLLII — ☐ _ _ _ _ ☐
   b. inner meningeal layer — AIP TMARE — ☐ ☐ ☐ ☐ ☐ _ ☐ ☐
   c. consistent, predictable arc — RLXFEE — ☐ _ _ _ _ ☐
   d. controls and integrates ANS and many endocrine functions — SUMAOPYHLATH — _ ☐ ☐ _ ☐ ☐ ☐ ☐ ☐ _ _ _
   e. previous name of the VIII cranial nerve — YOTRAUDI — ☐ _ ☐ _ ☐ _ _ ☐

   Total: rubrospinal, tectospinal, and vestibulospinal, for example

   ☐☐☐☐☐☐☐☐☐☐☐☐☐☐ ☐☐☐☐☐☐☐

2. a. largest part of brain — MCUERRBE — ☐ _ _ _ ☐ _ _ ☐
   b. a subdivision of the lentiform nucleus — PNUETAM — _ _ _ _ ☐ ☐ _
   c. cranial nerve dealing with sense of smell — YORLOFCAT — _ ☐ _ _ _ _ _ _ ☐
   d. a state of wakefulness — SCSONNSSUOCIE — _ _ _ ☐ _ ☐ _ _ ☐ _ _ _ _
   e. results when brain has reduced blood supply — AIT — ☐ ☐ _

   Total: wishbone shape; functions in emotions

   ☐☐☐☐☐☐ ☐☐☐☐☐☐

Addagrams

1. a. bundle of sensory and motor fibers in the midbrain — 9, 2, 18, 16, 11, 1, 13, 4
   b. cuneatus and gracilis in the spinal cord — 14, 12, 8, 1, 10, 1, 16, 15, 17
   c. where cranial nerves originate or terminate — 5, 3, 12, 10, 11
   d. ________ callosum — 1, 7, 6, 9, 16, 8

Total: the brain's shock absorber

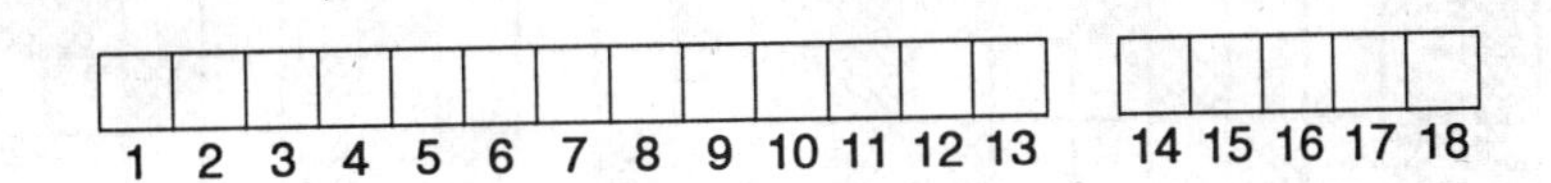

2. a. coverings in CNS — 25, 2, 18, 4, 18, 19, 24, 20
   b. stroke — 5, 14, 8
   c. groove on cerebral surface — 20, 6, 7, 11, 6, 22
   d. bundle of myelinated fibers — 3, 9, 15, 11, 16
   e. syndrome that may follow the flu or chickenpox — 1, 24, 21, 2, 20
   f. neuritis of nerve to the leg — 20, 5, 13, 10, 23, 17, 11, 8
   g. transient ischemic attack — 12, 4, 10

Total: controls consciousness and sleep

1 2 3 4 5 6 7 8 9   10 11 12 13 14 15 16 17 18 19   20 21 22 23 24 25

Crossword

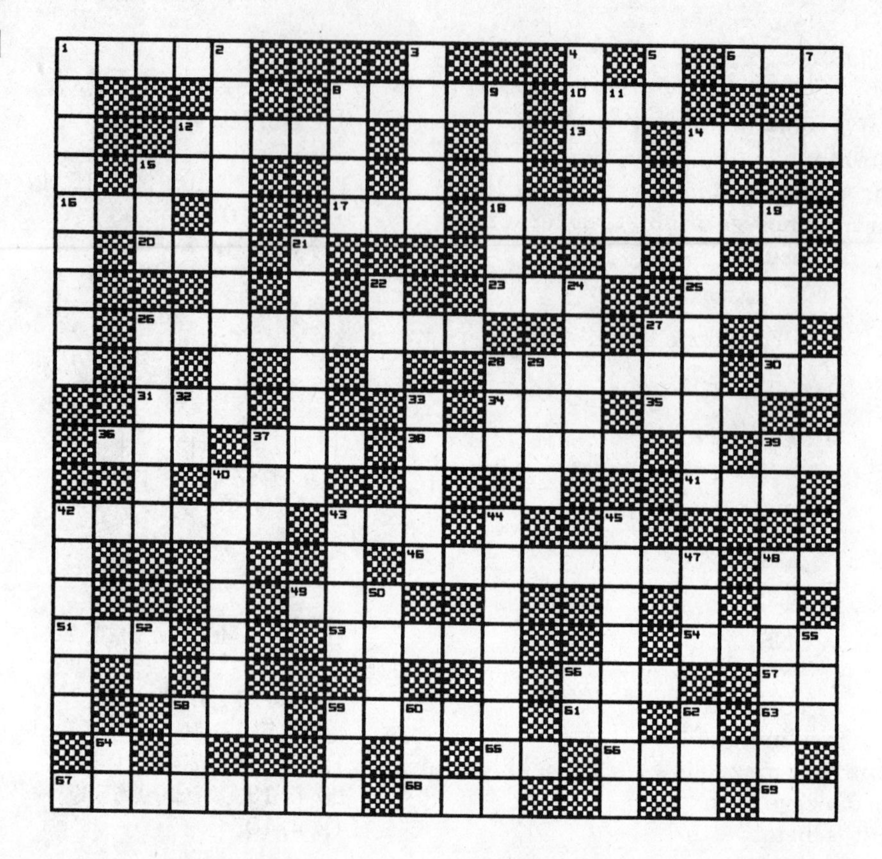

Across

1. Cranial Nerve II
6. Found in ventricles (abbr.)
8. Cerebral 'mound'
10. Lou Gehrig's Disease (abbr.)
12. Cerebellum's ________ vitae
13. Marriage promise: "I ______"
14. Spinal ________
15. "Honest ________" (president)
16. Has 31 pairs of spinal and 12 pairs of cranial nerves (abbr.)
17. Japanese currency
18. Where emotions and reasoning are housed
20. Dessert
23. Tree cutting instrument
25. Blind animals
26. The kind of motor activity controlled by the cerebellum
27. To exist
28. ________ tap
30. Care giver (abbr.)
31. Famous uncle of U.S.
34. Piece of land
35. National League of Nurses (abbr.)
36. Inner meningeal covering
37. Means of transport
38. This kind of center is found in the spinal cord
39. Amino acid (abbr.)
40. "My ________ Sal"
41. Mother (nickname)
42. Small gland within brain
43. Medullary area to control heart rate (abbr.)
46. A relay center to the cerebrum
48. Afternoon (abbr.)
49. Fleshy tissue surrounding mouth
51. Hypoglossal nerve symbol
53. Ganglia burried within the cerebrum
54. Singing voice
56. Fetal form often dissected
57. Indefinite article
58. Laboratory animal
59. Extensions of gray matter in the cord
61. Toward
63. Pronoun
65. Not down
66. Unit of the PNS, CNS
67. Sylvius' Canal
68. Girl's nickname
69. Negative response

Down

1. Cerebral lobe which interpretes visual stimuli
2. Has white and gray matter in a tree arrangement
3. Part of central nervous system
4. Writing tablet
5. Advanced science degree (abbr.)
7. A "here today, gone tomorrow" style
8. Made mostly of cell bodies: ________ matter
9. Cerebral depression
11. Each cerebral hemisphere has four
12. Prefix meaning away
14. The fourth ventricle is below this part of the brain
15. Small snake
19. The ventral root of a spinal nerve carries this kind of information
21. This spinal cord root lacks a ganglion
22. Father (nickname)
24. Tracts are made of this kind of matter
26. This sense is interpreted in the occipital lobe
27. Prevent
28. State legal fund (abbr.)
29. North or south magnetic center
32. Drinkers reform group
33. Organization of white matter in spinal cord
37. Coolidge's nickname
39. Morning (abbr.)
40. Clusters of cell bodies outside the CNS
42. Nerve "network"
43. Where a baby sleeps
44. Tract that connects the cerebral hemispheres: corpus ________
45. What the Limbic System controls
47. A place to get in shape
48. Part of the basal ganglia
50. Texas town: El ________
52. Personality component
55. Number of pineal glands
56. Physical therapy (abbr.)
58. Fish eggs
59. A thermoreceptor can detect when something is ________
60. Tear, or tombstone marking (abbr.)
62. Time periods (abbr.)
64. Intelligent Quotient (abbr.)

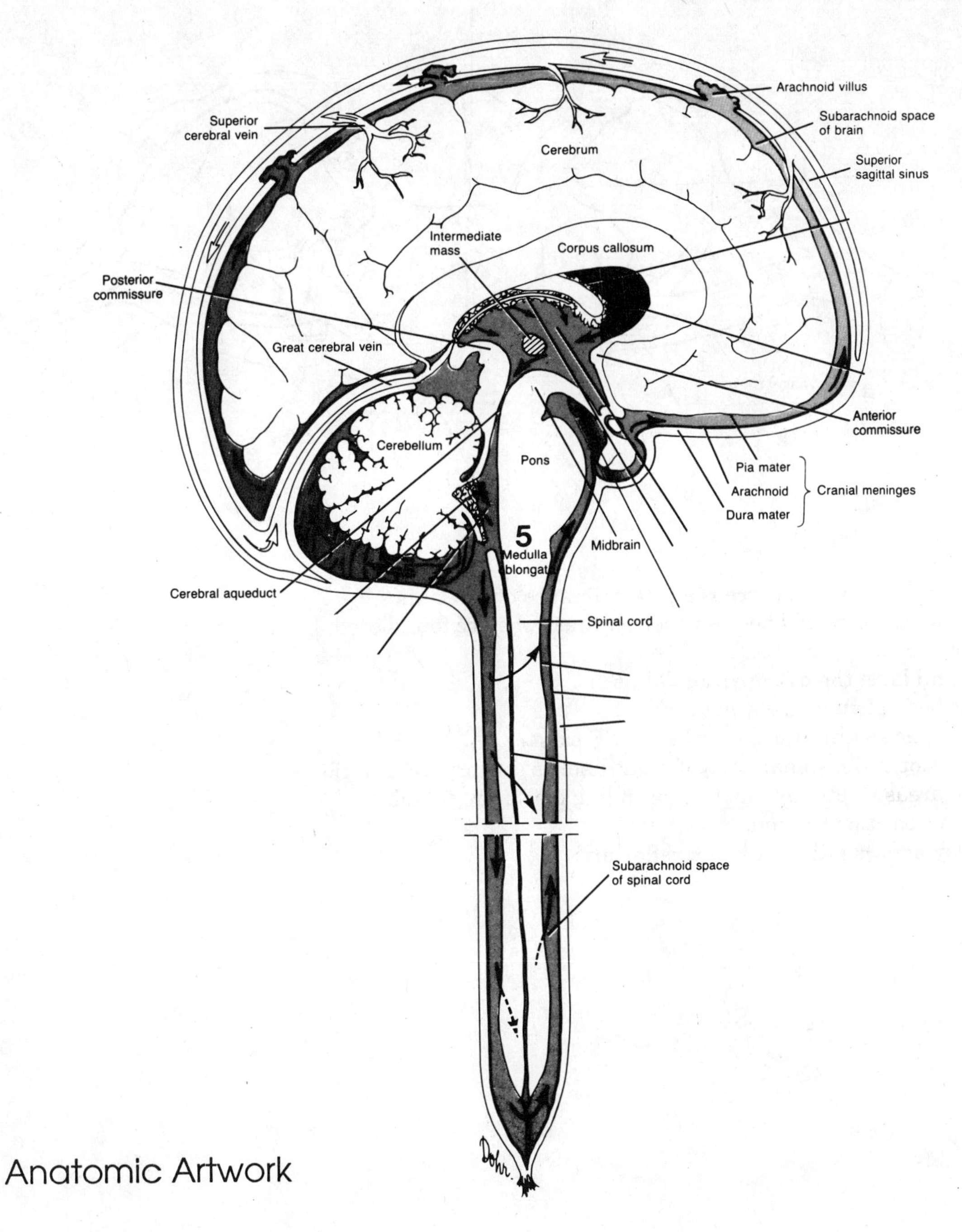

Anatomic Artwork

Figure 1

1. Shade in and label the areas within the brain where CSF is found.
2. Indicatewhere CSF flows within the spinal cord.
3. Label the areas where CSF is formed.
4. Indicate the primary area that allows you to interpret these written instructions.
5. Indicate the area where heart rate is controlled.
6. Indicate the large white tract that connects the cerebral hemispheres.
7. Indicate the area of the brain where cranial nerves V, VI, and VII originate.
8. Label the three layers of the cerebral meninges.
9. Indicate the area of the brain that, when damaged, can cause ataxia.
10. Label the area of the brain that has control centers for hunger, thirst, and body temperature regulation.

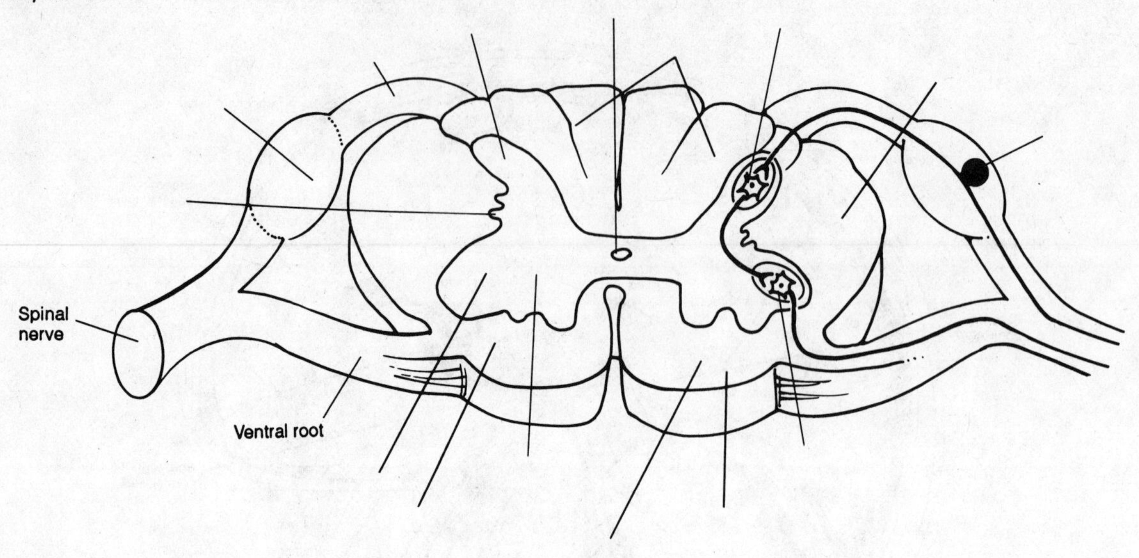

Figure 2

1. Shade in the area where there is a predominance of cell bodies.
2. Label the area where cell bodies of sensory neurons are found.
3. Label the horns.
4. Draw in and label the extrapyramidal paths.
5. Label the body of an efferent neuron.
6. Label the space within the cord where CSF exists.
7. Label the root of the spinal nerve that transmits sensory information.
8. Label the areas where myelinated ascending tracts are found.
9. Label an association neuron.
10. Indicate by arrows a three neuron reflex arc.

Chapter 11

Autonomic Nervous System

Introduction

The autonomic nervous system (ANS) is considered to be an *efferent system* that regulates *visceral* activities *automatically, involuntarily,* and *without conscious control.* The tissue that comprises this part of the overall nervous system is organized into two divisions: *sympathetic* and *parasympathetic divisions.* Most organs have innervation from both divisions and the two divisions have antagonistic effects on any given structure. The autonomic nervous system is not a separate system but rather is integrated with the rest of the nervous system, particularly at the level of the hypothalamus. This CNS structure exerts a great deal of influence and control over autonomic functions.

After completing the material in the textbook, you should have an understanding of the organization and operation of both sympathetic and parasympathetic divisions of the ANS. In addition, you should be able to answer the following questions and complete the activities that follow.

Multiple Choice

___ 1. Which of the following statements is true concerning autonomic fibers?
   a. Autonomic fibers are regarded as visceral efferent fibers
   b. Autonomic fibers are organized as preganglionic and postganglionic neurons with the ganglion between the two
   c. Autonomic fibers regulate visceral activities such as gastrointestinal movement
   d. All statements are true

___ 2. In the ANS as compared with the somatic nervous system, effectors would not include the following:
   a. cardiac muscle
   b. glands
   c. skeletal muscle
   d. smooth muscle

___ 3. Which of the following would be found in an autonomic efferent pathway?
   a. sensory receptor
   b. afferent neuron
   c. prevertebral ganglion
   d. none of these

___ 4. Acetylcholine is released from all of the following *except*:
a. sympathetic preganglionic neurons
b. parasympathetic preganglionic neurons
c. most sympathetic postganglionic neurons
d. parasympathetic postganglionic neurons

___ 5. Which statement concerning the autonomic nervous system is false?
a. The sympathetic division is a mixed division having both cholinergic and adrenergic fibers
b. The parasympathetic division is totally adrenergic
c. Most visceral effectors have dual innervation with the two divisions producing antagonistic effects
d. In stress, the sympathetic division tends to dominate

___ 6. The sympathetic nervous system:
a. in most cases releases ACh from the postganglionic neuron
b. stimulates digestion
c. is mimicked by adrenalin
d. emanates from the sacral spinal cord

___ 7. Which of the following ganglia are found in the parasympathetic division?
a. preganglionic ganglia
b. terminal ganglia
c. prevertebral ganglia
d. trunk ganglia

___ 8. Which of the following effects would be the result of sympathetic activity?
a. dilation of pupils of the eye
b. constriction of blood vessels of the viscera
c. increased heart rate
d. all of these

___ 9. The parasympathetic fibers emerge from the CNS in which area?
a. sacral region of spinal cord
b. thoracic region of spinal cord
c. both a and b
d. neither a nor b

___ 10. The sympathetic division is primarily concerned with which activity?
a. conserving the body's energy
b. restoring the body's energy levels
c. expending energy and preparing to combat stress
d. lessening the body's level of arousal

True/False

__________ 1. Adrenergic postganglionic fibers produce *norepinephrine*.

__________ 2. Most structures of the body have *either* sympathetic or parasympathetic innervation.

__________ 3. The ANS can be referred to as a visceral, *afferent* system.

__________ 4. The *hypothalamus* exerts a controlling and integrating effect on the autonomic system.

__________ 5. Parasympathetic *preganglionic* fibers tend to be quite long.

_______________ 6. The effects of *sympathetic* activity tend to have a widespread effect in the body.

_______________ 7. The effects of *cholinergic* fibers are short-lived and local.

_______________ 8. *Sympathetic* ganglia are usually near or within the visceral effector.

_______________ 9. The parasympathetic is a *stimulatory* division.

_______________ 10. The ANS is generally considered to be entirely *motor*.

Completion

1. Sacral parasympathetic preganglionic fibers synapse with postganglionic neurons in _______________ ganglia.
2. The division of the ANS that is totally cholinergic is the _______________ division.
3. Postganglionic adrenergic fibers utilize the neurotransmitter _______________ .
4. Prevertebral ganglia receive preganglionic fibers from the _______________ division of the ANS.
5. Sympathetic stimulation would _______________ (increase/decrease) cardiac activity.
6. Preganglionic neurons conduct efferent impulses from the CNS to autonomic _______________ .
7. Increased blood levels of glucose and epinephrine would occur as a result of _______________ stimulation.
8. The _______________ division is primarily concerned with restoring the body's energy.
9. Cardiac muscle, smooth muscle, and glands are examples of visceral _______________ .
10. The effects of both the sympathetic and parasympathetic divisions on any given structure or organ are _______________ .

Lost Sheep

1. increased kidney function, parasympathetic stimulation, elevation of salivary gland secretion, contraction of gallbladder
2. promotes glycogenolysis, promotes gluconeogenesis, increased sympathetic stimulation, parasympathetic
3. visceral effectors, efferent pathway, somatic, autonomic
4. hunger, nausea, conscious recognition, somatic nervous system
5. acetylcholine, autonomic, norepinephrine, dopamine
6. sympathetic trunk ganglion, prevertebral ganglion, dorsal root ganglion, terminal ganglion
7. sympathetic, terminal ganglion, fight-or-flight, bronchiole dilation
8. postganglionic axon, norepinephrine, parasympathetic, long fiber
9. fight-or-flight response, increased heart rate, increased digestive activity, increased epinephrine and norepinephrine secretion
10. sympathetic, cholinergic, postganglionic fibers, parasympathetic

Matching

___ 1. efferent path has one neuron
___ 2. regulates effectors such as skeletal muscles
___ 3. stimulates secretion of digestive juices
___ 4. has two divisions that function in an involuntary manner
___ 5. stimulation decreases rate and strength of heart beat
___ 6. has prevertebral ganglia in the conduction pathway
___ 7. effects of stimulation are short-lived and basically local
___ 8. most of its postganglionic fibers are adrenergic
___ 9. distribution of fibers is limited to the head and viscera primarily
___ 10. usually has very short preganglionic neuron
___ 11. totally cholinergic
___ 12. its action results in large energy expenditure
___ 13. has one neuron efferent pathway
___ 14. stimulation of skeletal muscle
___ 15. regulates the size of pupils of the eye

a. sympathetic
b. parasympathetic
c. autonomic
d. somatic

Double Cross

Across:

a. where neurons "talk" to each other
b. number of different types of ganglia in ANS
c. where preneurons and postneurons meet
d. uses lots of ATP
e. what the parasympathetic division is

Down:

what most sympathetic postganglionic fibers are

A a
R b
G c
d C e

Word Cage

V A D E F F E R E N T H G I C

I H C D A G E R I C L M K I I

S A H D S B C D U A L Y T L G

C D O Y A Y J S K E M E E D R

E A L G A S M P A T H A R I E

R N I R A M P P I T L E M A N

A Y N E D A T Y A F H U I L E

L B E N D A R P A T E A N M R

A V R E D G M X W T H I A L D

V I G A S Y J L O P D E L M A

F I I U S Y P M A T H E T O L

D E C A G A N G L I A F G I D

N O R E P I N E P H R I N E C

S A N O D E V E R T D E B R A

P E R I M I N C R E A S E B M

1. Sympathetic stimulation causes heart rate to ________ .
2. Two neurons make up this pathway in the ANS.
3. Postganglionic neurons relay impulses from autonomic ganglia to these effectors.
4. What the sympathetic system is famous for expending.
5. These form the "sympathetic chain."
6. Parasympathetic neurons are totally ________ .
7. This division produces fight-or-flight response.
8. These are the ganglia of the parasympathetic division.
9. When both divisions are present in an organ, it is said to have this kind of innervation.
10. Promotes restoration of the body's nutrients.
11. Neurotransmitter secreted by most sympathetic postganglionic fibers.
12. The type of neuron in #11 above is called ________ .

Sleuthing

1. Laura was nervously pacing the room the night before her anatomy final exam. Although she had studied, she was "nervous" and was having an anxiety attack.
   a. Were her symptoms due to sympathetic or parasympathetic stimulation?

   b. If the ANS is subconscious, how was a "cerebral" sensation causing this effect?

   c. How is the anxiety produced?

Word Scrambles

1.	a. a type of autonomic ganglion	PLREAVRBTEER
	b. structures that carry out responses	SERFOFTEC
	c. the level at which the ANS operates	SSUUBCCOISON
	d. totally cholinergic division	CPIATREASHYTMPA
	e. parasympathetic cranial and sacral	WOOUFTL

Total: what sympathetic and parasympathetic neurons are

2.	a. a cluster of cell bodies outside CNS	NGOANIGL
	b. name for junctions of autonomic fibers with glands, for example	RNOETUCREOEFFS
	c. ANS division with very wide distribution, even the skin	CSIYTMEPHAT
	d. adrenergic fibers secrete this transmitter	ENNORRHEPPIENI
	e. area where parasympathetic neurons emerge from CNS	LSAARC

Total: a sympathetic effect

Addagram

1. a. this liver activity is increased by sympathetic activity — 5, 9, 17, 14, 2, 7, 16, 11, 8, 16, 12, 16, 3, 10,21
   b. parasympathetic stimulation of this organ increases insulin secretion — 1, 6, 15, 14, 18, 16, 6, 21
   c. narrowing of blood vessels and airway — 14, 19, 20, 21, 4, 18, 13, 14, 4, 10, 19, 20

Total: sympathetic are long, parasympathetic are short

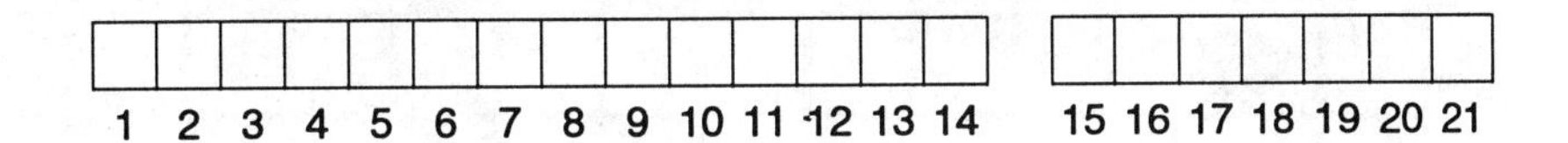

Crossword

Across

1 Division that provides for the fight-or-flight response
6 Accomplish
7 What the head does when you sleep in a chair
9 Droop
12 Fish eggs
13 Egyptian sun god
15 Most sympathetic postganglionic fibers are of this type
18 To say something untrue
19 Not young
20 Association of doctors (abbr.)
22 Land parcel
23 Small bed
24 These ganglionic fibers are very short in the parasympathetic ANS
26 Exclamation of surprise
27 Man's name (abbr.)
29 Short sleep
30 Personality component
33 Highest card in a 52–card deck
34 What the parasympathetic division does with energy

Down

1 Place where neurons communicate?
2 Involuntary Nervous System (abbr.)
3 Editor (abbr.)
4 Toward (short form)
5 Utilizes acetylcholine
8 How most organs are innervated by ANS
10 I am, you ________
11 Where pre and post synapse
14 The ________ ganglionic neuron can be short in sympathetic stimulation of
16 What sympathetic stimulation of bronchioles causes them to do
17 In regard to (abbr.)
20 Parasympathetic Neurotransmitter (abbr.)
21 Modus Operandi (abbr.)
22 Large amounts are expended with sympathetic stimulation
25 Not different
28 Change color
31 A "girl" that "comes out" (abbr.)
32 Not off
33 Morning (abbr.)

Chapter 12

Sensations

Introduction

The body has complex specialized *receptors* that are capable of detecting changes in the environment, both external and internal. These receptors effect a *generator potential,* which turns a *stimulus* into a *nerve impulse.* This travels to the brain for interpretation or translation into a sensation. Most receptors are triggered by specific kinds of stimuli. All the body's senses operate together to provide an integrated appraisal of the environment and thus help maintain homeostasis.

After studying the material in your text, you should have an understanding of the many types of receptor and their classification, as well as knowledge about the major *special senses,* such as vision, hearing, and balance. You should then be able to answer the following questions and complete the activities included.

Multiple Choice

___ 1. Which of the following is necessary to "experience" a sensation?
 a. An adequate stimulus must be present
 b. A receptor must convert the stimulus into a nerve impulse
 c. The impulse must be conducted to the brain where it is translated into a sensation
 d. All of these are necessary

___ 2. The function of a generator potential is to:
 a. trigger an environmental change
 b. convert the stimulus into an impulse
 c. receive the impulse in the brain
 d. translate the impulse into a sensation

___ 3. The process by which the brain refers sensations to their point of stimulation is called:
 a. adaptation
 b. projection
 c. proprioception
 d. modality distinction

___ 4. Which of the following provides information about the internal environment?
 a. enteroceptors
 b. exteroceptors
 c. both a and b
 d. neither a nor b

___ 5. Which of the following receptors would respond to any kind of stimulus?
a. thermoreceptors
b. mechanoreceptors
c. nociceptors
d. chemoreceptors

___ 6. The ability to know the body's position in space is called:
a. gustatory sense
b. vibratory sense
c. tactile sense
d. kinesthetic sense

___ 7. Which of the following is necessary for a substance to be "smelled"?
a. It must be capable of becoming gaseous
b. It must be water soluble
c. It must be lipid soluble
d. All of these are necessary

___ 8. Which of the following pairs is mismatched?
a. gustatory sense–taste buds
b. visual sense–macula lutea
c. auditory sense–organ of Ruffini
d. proprioception–muscle spindles

___ 9. The cornea of the eye is part of the:
a. fibrous tunic
b. choroid
c. vascular tunic
d. retina

___ 10. Which of the following statements is/are true with regard to the eye?
a. When the radial muscles contract, the pupils dilate
b. The circular and radial muscles are controlled by the ANS
c. The pupil is constricted when the circular muscles contract
d. All statements are true

___ 11. Which of the following pairs is mismatched?
a. myopia–nearsightedness
b. presbyopia–loss of lens elasticity
c. astigmatism–surface irregularities on lens/cornea
d. emmetropia–farsightedness

___ 12. The ability of the lens to change shape for clearer focusing is called:
a. refraction
b. accommodation
c. convergence
d. none of these

___ 13. Which of the following procedures is/are required for image formation in the retina of humans:
a. refraction
b. accommodation
c. convergence
d. all of these

___ 14. Which of the following is *not* found in the internal ear?
a. saccule
b. ampulla
c. tectorial membrane
d. pinna

___ 15. The vestibule is associated with the:
a. auditory canal
b. bony labyrinth
c. tympanic membrane
d. ossicles

___ 16. Which of the following "covers" the hair cells of the organ of Corti?
a. basilar membrane
b. membranous labyrinth
c. vestibular membrane
d. tectorial membrane

___ 17. The bony labyrinth:
a. is a series of cavities in the temporal bone
b. contains perilymph
c. contains a series of fluid-filled sacs and tubes called the membranous labyrinth
d. all of these

___ 18. Which of the following is/are correctly paired?
a. capula–crista
b. macula–otoliths
c. membranous labyrinth–endolymph
d. all of these

___ 19. Which cranial nerve is responsible for transmitting information regarding equilibrium?
a. I
b. II
c. VIII
d. none of these

___ 20. The organ of Corti is located in the:
a. cochlear duct
b. macula lutea
c. Eustachian tube
d. none of these

___ 21. Which statement is true concerning sound?
a. Sound waves are produced from alternate compressions and decompressions of air molecules
b. Pitch is related to the frequency of vibrations
c. The greater the force of vibration, the louder the sound
d. All statements are true

___ 22. Which of the following is *not* part of the "hearing" apparatus?
a. fovea centralis
b. organ of Corti
c. ossicles
d. tympanic membrane

___ 23. Which of the following absorbs excess light rays within the eye?
a. ciliary body
b. choroid
c. tectorial membrane
d. lacrimal apparatus

___ 24. The production of tears is a function of the:
a. nasolacrimal sac
b. lacrimal gland
c. cornea
d. none of these

___ 25. The inability to "taste" food when a person has a cold is due to:
a. nonfunctioning taste receptors
b. blocked olfactory receptors
c. both a and b
d. neither a nor b

True/False

__________ 1. The Eustachian tube serves to equalize the pressure on the *tympanic membrane.*

__________ 2. The *utricle* in the ear plays a role in maintainin static equilibrium.

__________ 3. The semicircular canals are concerned with *static* equilibrium.

__________ 4. The tympanic antrum is located in the *external* ear.

__________ 5. The *incus* "fits" against the oval window.

__________ 6. The *basilar* membrane is located between the scala media and the scala tympani.

__________ 7. The *vestibule* contains the utricle and saccule.

__________ 8. Changes in light-sensitive pigments within the *photoreceptors* are responsible for initiating impulses that produce visual sensations.

__________ 9. The eye contains *two* refractive media.

__________ 10. *Scotopsin* is found in cones, which are responsible for vision in dim light.

__________ 11. The *vitreous* humor is constantly produced by filtration from blood.

__________ 12. The *iris* of the eye consists of radial and circular muscle fibers under the control of the ANS.

__________ 13. Single binocular vision is, in part, the result of *convergence.*

__________ 14. An increase or decrease in the curvature of the lens is called *accommodation.*

__________ 15. Excessive intraocular pressure results in *glaucoma.*

__________ 16. One method to control pain is to interfere with nerve impulse transmission to the *cerebrum* through the use of analgesics.

__________ 17. Free nerve endings are the branching ends of *axons* from sensory neurons.

__________ 18. *Cutaneous* sensations include tactile sensations, thermosensations, and pain.

__________ 19. There are four types of taste bud, which are *uniformly* distributed over the tongue.

__________ 20. *Olfactory* receptors respond to substances that are gaseous, water soluble, and lipid soluble.

Completion

1. The greatest concentration of cones is massed in an area of the retina known as the _______________.
2. Night blindness results from a deficiency of vitamin _______________, which is needed for production of _______________.
3. The posterior cavity of the eye is filled with a gelatinous material called the _______________.
4. The size of the _______________ regulates the amount of light that enters the eye.
5. The area on the retina where sensory neurons exit as the optic nerve is referred to as the _______________.
6. The shape of the lens can be adjusted by contraction/relaxation of the _______________ muscle.
7. The membranous labyrinth is filled with a fluid called _______________.
8. The spiral organ of Corti is located on the _______________ membrane.
9. The tympanic membrane is the boundary between the _______________ ear and _______________ ear.
10. The phenomenon in which pain is felt in an area removed from the actual pained structure is called _______________.
11. _______________ refers to the specific characteristics of a sensation which allows it to be distinguished from other types.
12. Exteroceptors are located near the body's _______________.
13. Pacinian corpuscles are receptors for _______________.
14. The area where some fibers from the eye cross over to the opposite side is called the _______________.
15. The receptor organs for equilibrium are located in the internal ear in the _______________, _______________, and _______________.

Lost Sheep

1. Merkel's discs, Ruffini's end organs, Pacinian corpuscles, organ of Corti
2. pain sensation, thermoreceptor, 8° C, 75° F
3. olfaction, superior conchae, cranial nerve II, olfactory bulbs
4. cones, fovea centralis, dim light, retina
5. cornea, aqueous humor, vitreous humor, ciliary body
6. night blindness, vitamin A deficiency, cones, rhodopsin
7. Eustachian tube, tympanic antrum, ossicles, inner ear
8. static equilibrium, otoliths, utricle and saccule, crista
9. stapes, round window, middle ear, incus
10. vision, proprioception, smell, hearing
11. pain, free nerve endings, quick adaptation, cerebral recognition
12. hair cells, photoreceptors, organ of Corti, dynamic equilibrium
13. otitis media, eustachian tube, bacteria, trachoma
14. modality, olfaction, taste, mechanoreceptor
15. vascular tunic, choroid, cornea, ciliary body

Matching

Set 1

1. blind spot
2. lines eyelid
3. crystalline protein
4. radial and circular muscles
5. vascular, pigmented layer

a. conjunctiva
b. choroid
c. iris
d. optic disc
e. lens

Set 2

1. utricle
2. perilymph
3. scala vestibuli and tympani
4. "lid" on organ of Corti
5. semicircular canals

a. scala vestibuli
b. cochlea
c. otoliths
d. ampulla
e. tectorial membrane

Set 3

1. clouding of the lens
2. contagious conjunctivitis
3. Meniere's syndrome
4. vertigo
5. motion sickness

a. sensation of spinning
b. excessive stimulation of vestibular apparatus
c. trachoma
d. increased amounts of endolymph
e. cataract

Sleuthing

1. Mrs. Burns is traveling by plane from New York to California. She has a bad head cold and sinusitis. As a result, she is experiencing a great amount of discomfort in her ears. It was particularly painful on takeoff when the cabin was being pressurized. She also found the meal that was served to be basically tasteless.
   a. Explain why she was experiencing trouble with her ears.

   b. Physiologically, why was there no taste to the meal.

2. When individuals are emotionally upset, they often cry a lot. This can result in a "runny nose."
   a. Explain why lots of tears are produced.

   b. Why does this give the crier a runny nose?

Word Scrambles

1. a. characterized by increased ocular pressure — ACMGOULA — _ _ ☐ _ _ ☐ ☐ _
   b. pigment in rods — HOIOPSDRN — _ _ _ ☐ _ _ _ _ ☐
   c. found in posterior chamber (2 words) — IESVURTO MOHRU — _ _ ☐ _ _ ☐ _ _   _ _ ☐ _ _
   d. vascular layer — ODHIRCO — ☐ _ _ _ ☐ ☐ _
   e. crossing of optic nerve fibers — AHSCIMA — ☐ _ ☐ _ _ _ ☐

   Total: lens curvature adjustments for varying distances

   ☐☐☐☐☐☐☐☐☐☐☐☐☐

2. a. unique characteristic of each sensation — YMOTIDLA — ☐ _ _ ☐ _ _ ☐ _
   b. depression on retina — AFEVO — ☐ _ _ ☐ ☐
   c. photoreceptive layer — ATENIR — ☐ _ _ ☐ _ _
   d. potential — RGOETNEAR — ☐ _ _ ☐ _ _ _ _ _

   Total: when a visual sensation persists after removal of stimulus

   ☐☐☐☐☐☐☐☐☐☐

Addagrams

1. a. innermost layer of eye — 6, 2, 1, 7, 16, 8
   b. spiral organs of inner ear — 3, 5, 14, 4, 7
   c. tear-producing apparatus — 9, 15, 3, 6, 7, 10, 15, 9
   d. symmetry exhibited by humans — 13, 7, 9, 8, 4, 17, 14, 8, 9
   e. mathematical average — 12, 11, 15, 16

   Total: canopy for sound receptors

☐	☐	☐	☐	☐	☐	☐	☐	☐		☐	☐	☐	☐	☐	☐	☐	☐
1	2	3	4	5	6	7	8	9		10	11	12	13	14	15	16	17

2. a. white of the eye — 1, 5, 10, 2, 7, 11
   b. "hammer" — 3, 14, 10, 17, 2, 9, 18
   c. little sac in bony vestibule — 1, 16, 8, 13, 9, 10, 2
   d. "anvil" — 4, 15, 13, 9, 18
   e. regulates amount of light entering the eye — 6, 12, 6, 1

   Total: organs of dynamic equilibrium

☐	☐	☐	☐	☐	☐	☐	☐	☐	☐	☐	☐		☐	☐	☐	☐	☐	☐
1	2	3	4	5	6	7	8	9	10	11	12		13	14	15	16	17	18

3. a. membrane that forms floor of cochlear duct — 1, 8, 12, 2, 7, 8, 9
   b. parasympathetic nervous system's area of origin — 5, 9, 8, 3, 11, 14, 12, 8, 5, 9, 8, 7
   c. window covered by stapes — 4, 10, 8, 7
   d. canal of Schlemm: scleral venous ________ — 12, 13, 15, 6, 12

Total: needs two eyes

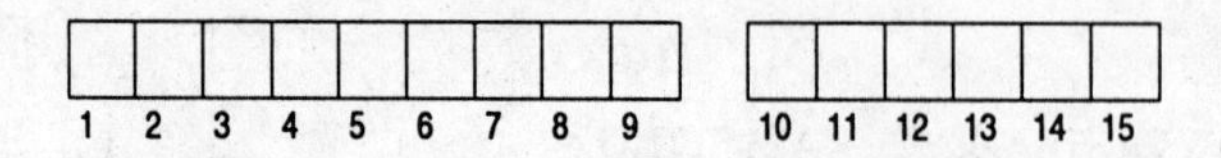

Crossword

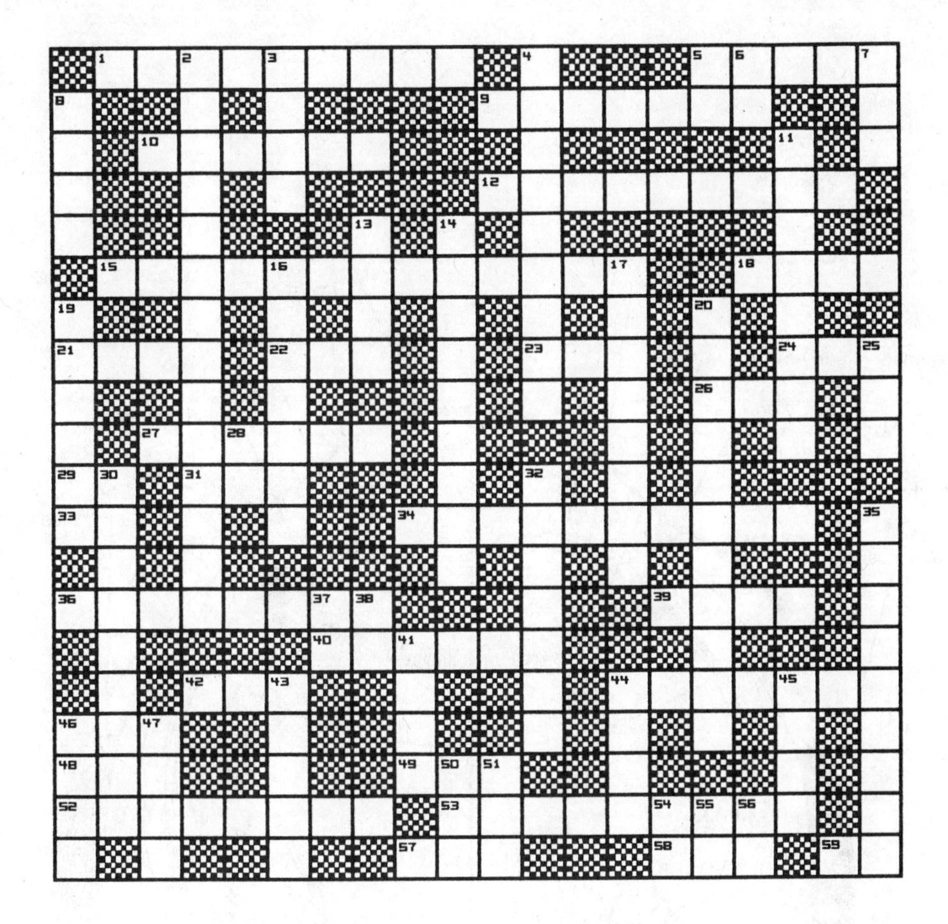

Across

1 Types of sensations that include thermal, tactile, and pain sensations
5 Ten cent pieces
9 What neurons do to impulses
10 Tactile disc's other name
12 A potential that changes the resting membrane potential
15 Provides information about the body's position
18 These cells and supporting cells form the organ of Corti
21 A plot of land
22 Enemy
23 TV creature of the 1980s
24 This goes up in glaucoma (abbr.)
26 Used by fisherman
27 Contains photoreceptors
29 Not hi but ________
31 Needle and ________
33 Man's nickname
34 Receptors for pain
36 Tear-producing apparatus
39 Functions in accommodation
40 Rods are absent from this area
42 High-energy compound
44 Proprioceptive receptor between skeletal muscle fibers
46 Specialized for vision in dim light
48 Anger
49 Eastern Standard Time (abbr.)
52 Membrane found in ear
53 Similar to intracellular fluid and found in the membranous labyrinth
57 Shape of sphenoid bone
58 Seen by people when they are angry
59 ________You Like It"

Down

2 Sensitive to changes in temperature
3 Brand name sneaker
4 What exists across the neuron's membrane
5 Nation's capital (abbr.)
6 He, she, or ________
7 Speak
8 Window associated with the stapes
11 The specific kind of sensation
13 Functions in bright light
14 Touch, pressure, and vibration are examples of tactile ________
16 End organs of ________
17 This pain occurs when the organ involved is innervated by the same section of spinal cord
19 Area on retina
20 This sense tells us the degree to which muscles are contracted
25 Apple, cherry, or peach dessert
28 7th musical note on the scale
30 Refers to sense of smell
32 Body, process, and muscle found in the vascular tunic
35 Gustatory receptors
37 Morning (abbr.)
38 City of Angels (abbr.)
41 Concentrated in central fovea
43 Outer ear
44 Type of boat
45 Foot race for short distances
46 Ceremony
47 A "short" presentation (abbr.)
50 Body of water
51 Explosive material
54 Time period (abbr.)
55 "________ and My Gal" Broadway show
56 Police Department (abbr.)

Anatomic Artwork

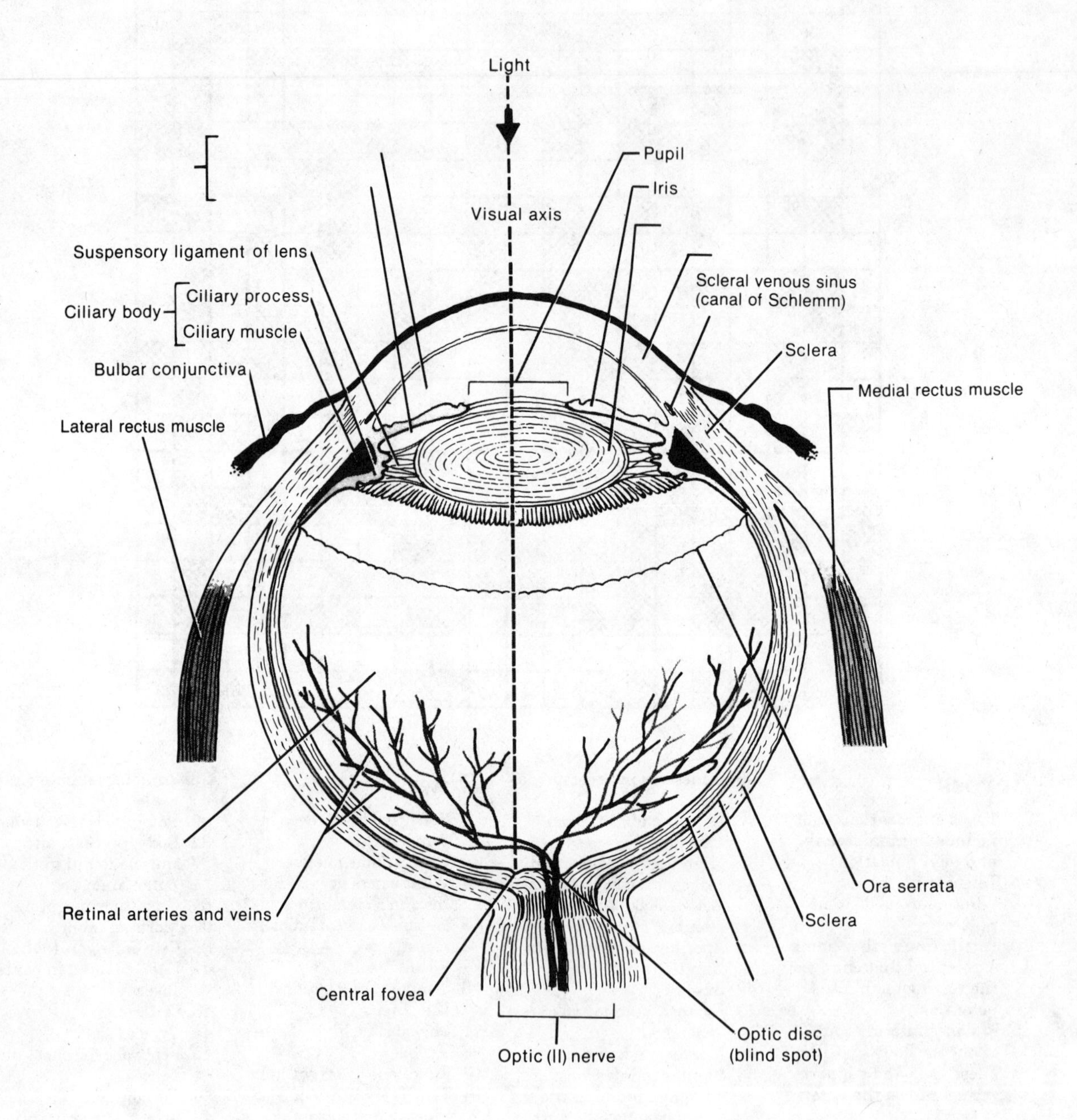

Figure 1

1. Label the vascular tunic.
2. Shade in and label the area that has the vitreous humor.
3. Label the photosensitive layer.
4. Label the structures that can often become cloudy in old age.
5. Indicate where aqueous humor is located by labeling the area and shading it in.

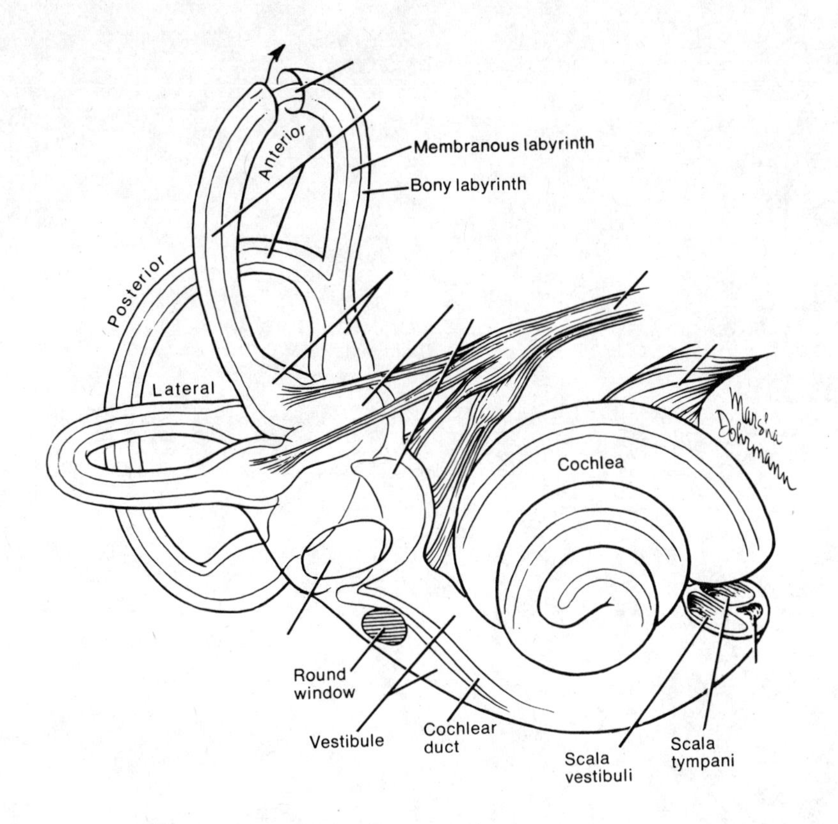

Figure 2

1. Label the semicircular canals.
2. Label the areas that have receptors for static equilibrium.
3. Label the structure associated with the stapes.
4. Label the branches of the VIII cranial nerve.
5. Shade in all areas that are filled with endolymph.
6. Label the bony labyrinth that is concerned with hearing.
7. Label the specific area where the receptors for hearing are found.

Chapter 13

The Endocrine System

Introduction

Distributed in various areas of the body are the *ductless glands,* which comprise the endocrine system. These glands secrete chemicals called *hormones* into the blood and body fluids. Hormones travel through the body in blood to receptors in *target structures* where they function in the *regulation* of a wide variety of physiological processes.

Control and *coordination* of the body's development, growth, and maintenance are functions of two complementary systems: the nervous and endocrine systems. In general, the action of the nervous system is rapid and short term. The endocrine system, on the other hand, operates more slowly, and its effects are of longer duration. Many of the major endocrine glands are controlled through *feedback* mechanisms by the hypothalamic–pituitary system. Others are influenced by the blood concentration of specific substances like glucose or calcium. In addition, there are hormonelike substances that are released and are active in local tissues. These are referred to as prostaglandins.

After completing the reading in your text, you should have an understanding of how these distant and diverse glands function with the same basic "modus operandi" and thus linked together as a system. You should also be able to answer the questions that follow as well as complete the activities.

Multiple Choice

___ 1. Which of the following is secreted by the anterior pituitary?
a. oxytocin
b. cortisol
c. prolactin
d. antidiuretic hormone

___ 2. A decrease in the osmotic pressure of plasma would inhibit the release of which hormone?
a. STH
b. ADH
c. LTH
d. TSH

___ 3. If the blood levels of sodium are decreased, the production of which hormone would be increased?
a. cortisol
b. thyroxine
c. calcitonin
d. none of these

___ 4. Increased release of ACTH would result in which of the following?
a. decreased blood levels of growth hormone
b. decreased blood levels of parathyroid hormone
c. increased blood levels of norepinephrine
d. increased blood levels of cortisol

___ 5. Which of the statements below is true?
a. Thyroxine stimulates oxidative metabolism
b. Cortisol is active in both carbohydrate and protein metabolism
c. Insulin plays a role in the uptake of amino acids by tissues
d. All statements are true

___ 6. Which of the following endocrine secretions is/are *not* antagonistic?
a. parathyroid hormone–calcitonin
b. insulin–glucagon
c. epinephrine–norepinephrine
d. more than one of these

___ 7. Destruction of the alpha cells of the pancreas might result in:
a. hypoglycemia
b. hypercalcemia
c. Addison's disease
d. acromegaly

___ 8. Thymosin is produced by the:
a. thyroid gland
b. parathyroid glands
c. both a and b
d. neither a nor b

___ 9. Growth hormone:
a. increases production of antibody-forming cells
b. regulates increase in the length of bones
c. aids in regulating blood levels of sodium and potassium
d. stimulates the production of insulin

___ 10. Abnormal skin pigmentation may result from excess secretions of which gland?
a. thyroid
b. pituitary
c. pineal
d. thymus

___ 11. Normal thyroxine production is dependent on which of the following ions?
a. sodium
b. calcium
c. potassium
d. iodine

___ 12. Diabetes insipidus results from lack of:
a. glucocorticoids
b. insulin
c. glucagon
d. ADH

___ 13. Which of the following statements is false?
a. Some endocrine glands are under "neural" control
b. Most endocrine secretions are regulated by negative feedback mechanisms
c. Hormone production is local and constant
d. Endocrine secretions tend to modify cellular metabolism

___ 14. Which of the following is a gonadotropic hormone?
a. ADH
b. STH
c. FSH
d. MSH

___ 15. Which of the following is closely associated with the ANS?
a. thyroid gland
b. pancreas
c. adrenal glands
d. pineal gland

___ 16. Which of the following statements is false?
a. Growth hormone is produced in children and in adults
b. FSH is functional in both males and females
c. Both cortisol and aldosterone raise blood sugar levels
d. Cushing's syndrome results from hypersecretion of adrenocortical hormone

___ 17. Which structure releases thyrotropin releasing factor?
a. thryoid gland
b. anterior pituitary
c. posterior pituitary
d. none of these

___ 18. Which of the following statements is/are true?
a. FSH maintains sperm production
b. LH stimulates ovulation
c. Prolactin is found in both males and females
d. All the statements are true

___ 19. Which of the following thyroid secretions is "active"?
a. diiodotyrosine
b. triiodothyronine
c. thyroglobulin
d. none of these

___ 20. Hormones are:
a. organic compounds
b. secreted by endocrine glands
c. generally controlled by negative feedback
d. all of these

___ 21. The second messenger for protein hormones is:
a. adenyl cyclase
b. cyclic AMP
c. phosphodiesterase
d. kinases

___ 22. Which of the following pairs is mismatched?
a. calcitonin–parathyroid
b. aldosterone–adrenal cortex
c. glucagon–pancreas
d. oxytocin–posterior pituitary

___ 23. Deficiency of aldosterone secretion is likely to result in:
a. edema and slow heart rate
b. lower blood pressure and diminished blood volume
c. increased FSH production
d. acromegaly

___ 24. Which of the following can be affected by growth hormone?
a. protein metabolism
b. carbohydrate metabolism
c. lipid metabolism
d. all of these

___ 25. Which of the following pairs is/are correctly matched?
a. TSH–thyroxine
b. adrenal gland–cortisol
c. insulin–pancreas
d. all of these

True/False

__________ 1. The *thymus* gland plays a role in the body's immune system.

__________ 2. Addison's disease, characterized by weight loss, low blood pressure, and general weakness, results from a hyposecretion of the *adrenal cortex*.

__________ 3. *Insulin* promotes an increase in blood glucose levels.

__________ 4. The *anterior pituitary*, derived from nervous tissue, houses the termination point of neurons that begin in the hypothalamus.

__________ 5. Chemically, hormones may be amines, protein, and *carbohydrates*.

__________ 6. *Mineralocorticoids* are secreted by the adrenal glands located on top of the kidneys.

__________ 7. Secretions of endocrine glands are most often controlled by *positive* feedback.

__________ 8. Angiotensin II stimulates the release of *aldosterone* by the adrenal cortex.

__________ 9. The primary effect of *ADH* is to aid uterine muscle contraction during childbirth.

__________ 10. *Thyroxine* secretion by the thyroid gland is regulated by TSH.

__________ 11. Stimulation of the *adrenal cortex* can cause sympathomimetic effects.

__________ 12. *Protein* hormones require a second messenger, which activates protein kinases within the cell.

__________ 13. Sufficient levels of *insulin* are required for normal protein synthesis.

__________ 14. All hormones require a *receptor* in order to create the desired effect.

__________ 15. The production of *mineralocorticoids and glucocorticoids* is regulated by ACTH.

__________ 16. Increasing the blood level of calcium is the primary effect of *calcitonin*.

__________ 17. Secretion of most anterior pituitary hormones is under the influence of regulating factors from the *hypothalamus*.

_____________ 18. Nervousness, increased heart rate, and elevated metabolic rate are symptomatic of *hyperparathyroidism.*

_____________ 19. Excessive excretion of potassium can result from increased secretion of *aldosterone.*

_____________ 20. ADH stimulates the reabsorption of *sodium* ions by the kidney.

_____________ 21. The function of glucagon is to *decrease* blood sugar levels.

_____________ 22. Somatomedins released by the *pituitary* regulate the growth of the skeleton.

_____________ 23. The *pineal gland* is suspended from the base of the brain and secretes several hormones that regulate other endocrine glands.

_____________ 24. The thyroid and parathyroid glands are situated in the *neck.*

_____________ 25. Inability of the beta cells to produce insulin results in *diabetes insipidus.*

Completion

1. The blood levels of _______________ control the release of parathyroid hormone.
2. Secretion of hormones is most often regulated by _______________ .
3. _______________ are small proteins synthesized by the liver; they mediate the effects of growth hormone.
4. Diabetes insipidus results from lack of _______________, while diabetes mellitus results from a deficiency of _______________ .
5. Cushing's syndrome results when the cortex of the _______________ gland is overactive.
6. Secretion of glucocorticoids by the adrenal cortex is stimulated by _______________, which is secreted by the _______________ .
7. Beta cells of the islets of Langerhans secrete the hormone _______________, which _______________ (increases/decreases) blood glucose levels.
8. Kidney tubular reabsorption of _______________ is stimulated by ADH.
9. The function of _______________ is to lower blood calcium levels.
10. The presence of a particular receptor for a hormone in the cell membrane makes the cell a specific _______________ for that hormone.
11. The reabsorption of sodium and excretion of potassium is promoted by the hormone _______________ .
12. Secretion of hormones of the pituitary gland is largely regulated by factors from the _______________ of the brain.
13. The enzyme _______________ catalyzes the increased production of cyclic AMP.
14. The _______________ has both endocrine and exocrine functions.
15. _______________ is secreted by the anterior pituitary; it regulates the release of thyroxine.
16. Myxedema, which results from dysfunction of the pituitary gland, is due to the lack of _______________, which regulates secretion of thyroid hormones.
17. Secretory activity of the parathyroid glands is controlled directly by blood levels of _______________ .
18. Hyposecretion of the _______________ glands is often characterized by tetany.
19. The hormones produced by the _______________ are often referred to as sympathomimetic compounds.

20. Substances produced by the hypothalamus, which regulate the release of pituitary hormones, are referred to as ______________ .
21. The second stage of the general adaptation syndrome is the ______________ .
22. Releasing factors travel between the hypothalamus and pituitary by means of ______________ .
23. Skin pigmentation is under the control of ______________ hormone.
24. Deficiency in ______________ can ultimately cause circulatory failure resulting from loss of sodium salts.
25. ______________ are secreted in response to stress.

Lost Sheep

1. cyclic AMP, adenyl cyclase, second messenger, AChase
2. thyroid, thymosin, calcitonin, serum Ca^{2+}
3. glucagon, beta cells, glycogenolysis, rise in blood sugar
4. posterior pituitary, releasing factors, oxytocin, ADH
5. anterior pituitary, growth hormone, erythropoietin, ACTH
6. FSH, LH, TSH, gonadotropic hormone
7. protein, amine, hormone, carbohydrate
8. epinephrine, aldosterone, cortisone, androgens
9. pineal, brain sand, pituitary, melatonin
10. giantism, acromegaly, Addison's disease, dwarfism
11. oxytocin, vasopressin, antidiuretic hormone, thyrotropin
12. Cushing's, primary aldosteronism, goiter, Addison's
13. acromegaly, hypothyroidism, giantism, excess GH
14. tyrosine, monoiodotyrosine, thyroxine, DIT
15. glucocorticoid, reabsorption of sodium, gluconeogenesis, mobilization of fatty acids
16. beta cells, insulin, pancreas, exocrine
17. alpha cells, lipogenesis, glucagon, glycogenolysis
18. PTH, calcium, vitamin C, phosphate
19. ACTH, growth hormone, PTH, testosterone
20. prostaglandins, local hormone, PG, PTH

Matching

Set 1

___ 1. adrenal gland
___ 2. pineal gland
___ 3. pancreas
___ 4. pituitary gland
___ 5. parathyroid glands

a. oxyphil and principal cells
b. alpha and beta cells
c. regulating factors
d. medulla and cortex
e. brain sand

Set 2

___ 1. thyroid gland
___ 2. anterior pituitary
___ 3. parathyroid gland
___ 4. posterior pituitary
___ 5. adrenal cortex

a. aldosterone
b. calcitonin
c. FSH
d. PTH
e. ADH

Set 3

___	1. myxedema	a.	stimulates release of growth hormone
___	2. melatonin	b.	part of thyroid hormone
___	3. GRF	c.	hypersecretion of parathyroid hormone
___	4. MSH	d.	stimulates thyroxine production
___	5. FSH	e.	hyposecretion of thyroid hormone
___	6. tetany	f.	stimulates secretion of cortisol
___	7. osteoporosis	g.	pineal gland hormone
___	8. epinephrine	h.	stimulates sperm production
___	9. ACTH	i.	hyperthyroidism
___	10. iodine	j.	secretion of adrenal medulla
___	11. goiter	k.	influences skin pigmentation
___	12. TSH	l.	glucocorticoid
___	13. cortisone/cortisol	m.	characteristic of hypoparathyroidism
___	14. oxytocin	n.	stimulates release of ACTH
___	15. CRF	o.	positive feedback control

Double Crosses

1. *Across:*
   pancreatic hormone
   Down:
   secretes epinephrine:
   adrenal ________

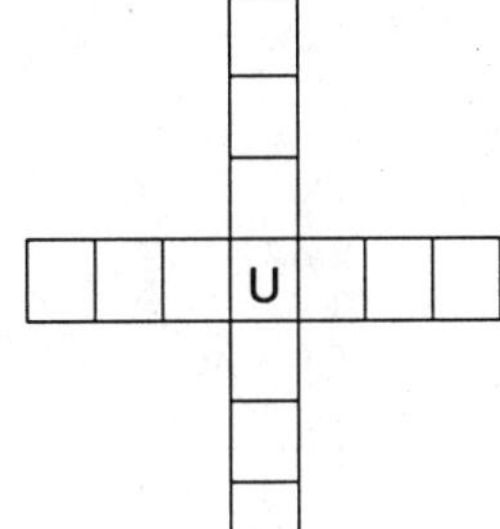

2. *Across:*
   influenced by TSH
   Down:
   chemical class of the sex
   hormones

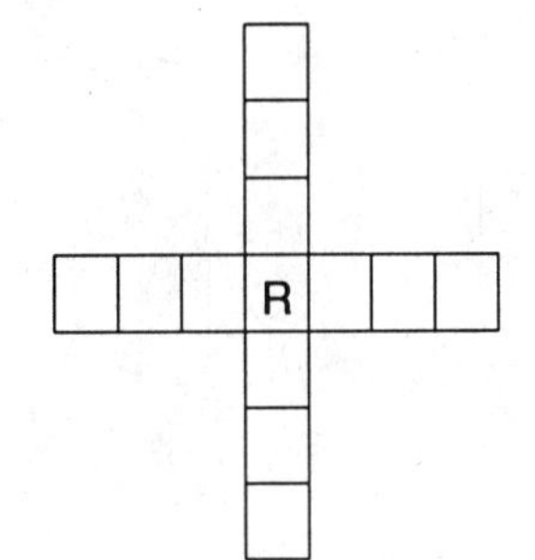

3. *Across*:
   ductless gland
   Down:
   cells that produce PTH

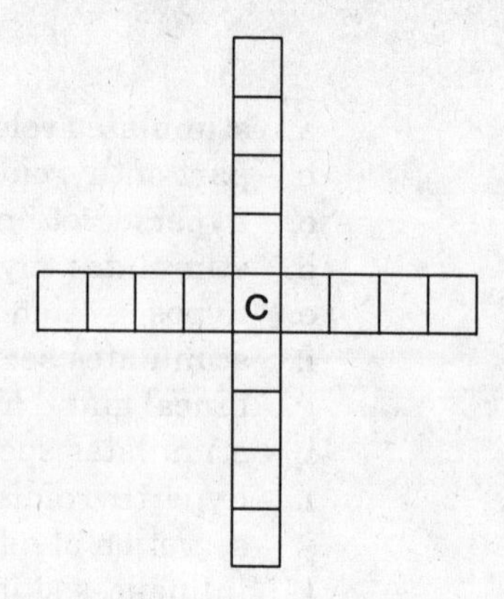

4. *Across*:
   collective term for male sex hormones
   Down:
   exerts negative feedback on TSH

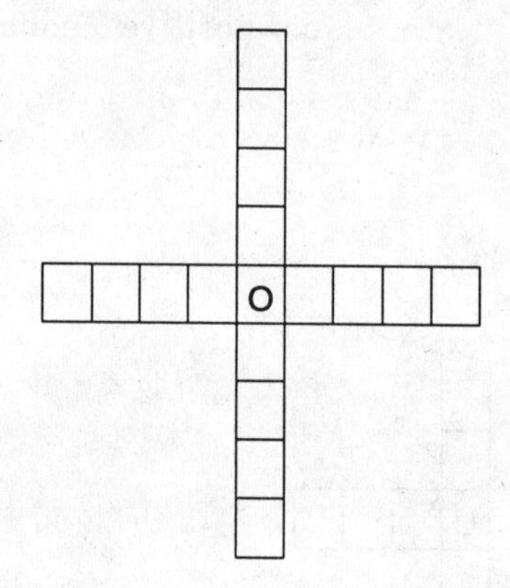

5. *Across*:
   powerful hormone "mimickers"
   Down:
   a. insulin belongs to this class of hormone
   b. has follicles but no eggs
   c. these run between hypothalamus and posterior pituitary
   d. react with intracellular receptor

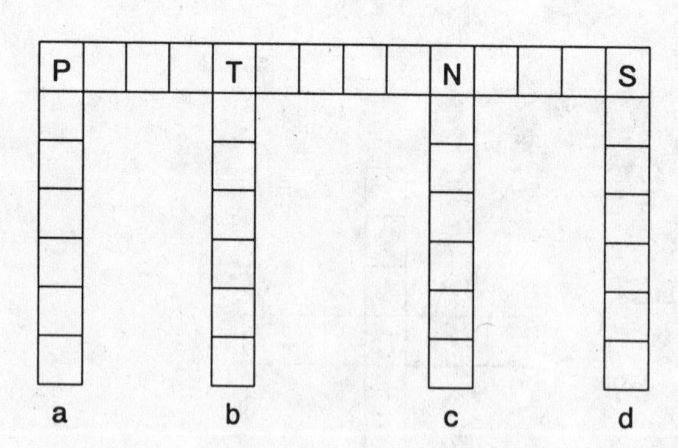

Sleuthing

1. Debbie has been suffering from rheumatoid arthritis. Her therapy has included treatment with cortisol.
   a. Why is cortisol used in such cases?

   b. To what undesirable effects can this therapy lead?

   c. In normal adrenocortical function, how is the release of glucocorticoids regulated?

2. Joe is 28 and diagnosed as having a dysfunctional posterior pituitary, resulting in negligible hormone release.
   a. What is another name for this gland?

   b. What specific hormone levels are affected in Joe?

   c. What is the principal abnormality associated with this type of dysfunction?

   d. What typical symptoms would Joe exhibit without treatment?

Word Scrambles

1. a. pineal hormone	NLTIMEANO	☐ _ _ _ _ _ _ _ _
b. stimulates uterine contraction	YTOIXNOC	_ ☐ _ _ _ _ _ _
c. calcitonin secretor	DRYOIHT	_ _ ☐ _ _ _ _
d. converts immature T lymphocytes to mature ones	MYSOHNIT	_ _ _ ☐ _ _ _ _
e. enlarged thyroid	ROIGET	_ _ _ _ ☐ _
f. male sex hormone	GOANEDRN	☐ _ ☐ _ _ _ _ _
g. part of adrenal	LAMULDE	_ ☐ _ _ _ _ _

Total: cold intolerance and "puffy" face

☐☐☐☐☐☐☐☐

2. a. released by nerve cells EOHNUURMSOR _ _ ☐ _ _ _ _ _ _ _ _
   b. mechanism of control for some endocrine glands (2 words) EVENGAIT EFBEDKAC _ _ _ _ ☐ _ _ _ _ ☐ _ _ _ _ _ _
   c. lipids that mimic hormone action POTGADNSINLASR _ ☐ _ _ ☐ ☐ ☐ _ _ _ _ ☐ _ _
   d. chemical class of estrogen and testosterone ERSOTDIS ☐ ☐ ☐ _ _ _ _ ☐

Total: what hormones act on

☐☐☐☐☐☐ ☐☐☐☐☐☐

Addagrams

1. a. pituitary gland's other name 7, 2, 3, 10, 3, 1, 8, 14, 11, 14
   b. organic class for estrogen 14, 6, 4, 9, 10, 13, 12
   c. amount of GH secreted in giantism: ________than normal 15, 10, 5, 4

Total: exophthalmos results from this

☐☐☐☐☐☐☐☐☐☐☐☐☐☐☐

1 2 3 4 5 6 7 8 9 10 11 12 13 14 15

2. a. glucocorticoid produced in greatest quantities 9, 15, 5, 12, 13, 18, 10, 7
   b. effect on osteoclastic activity is opposite PTH 14, 6, 7, 9, 2, 12, 8, 3, 16, 3
   c. common name for epinephrine 6, 17, 11, 4, 3, 6, 7, 16, 3
   d. secreted by pineal gland 1, 4, 7, 6, 12, 10, 3, 13, 3

Total: primarily affect electrolyte metabolism

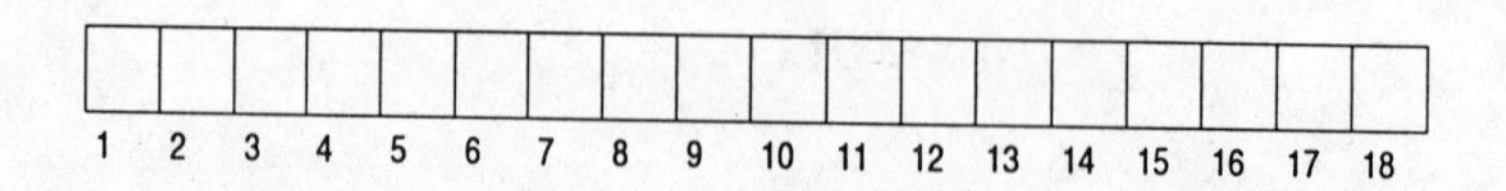

Crossword

Across

1 What you do to blood calcium levels when parathyroid hormone is secreted
3 In regard to (abbr.)
5 Small snake
6 Type of play needed for success on the golf course
10 Insipidus or mellitus
13 Gland that has brain sand
15 Where thyroid gland is located
16 Computer command system
17 Indefinite article
18 Chart
19 Increases osteoclastic activity
22 What thyroid gland does to iodine and amino acids
25 Ductless gland
27 Has a medulla and cortex
30 Correct manuscript
31 Alternating current (abbr.)
32 Frightens
35 Nine men in baseball
37 What can occur during withdrawal from alcohol or drugs
39 Preposition
40 Hormone that surges just before ovulation (abbr.)
41 Organic secretion with a target
44 Hyposecretion of this can result in low blood calcium
48 Chinese or British drink
49 Second messenger (2 words)
53 4th note on musical scale
54 A hormone that affects other endocrine glands
55 Milligrams (abbr.)
56 Stairs
57 Where humerus is located
59 Exaggerated cry
60 Needed to transport glucose across cell membranes
63 Fat soluble or a type of lipid
64 Negative reply
65 Number of parathyroid glands
66 Controls the secretion of glucocorticoids
67 Type of bread
68 Respiratory therapist (abbr.)
69 Street (abbr.)
71 Often used in the summer
75 Basic unit of life
76 A hormone secreted by the anterior pituitary
79 Where protein hormone receptors are found
80 Where milk is processed
81 Casper might say this

Down

1 Rapid ambulation
2 Man's name
3 Needed after a hard day's work
4 Extraterrestrial (abbr.)
5 Atop a horse
6 Type of tree
7 Article
8 Sleep happening (abbr.)
9 Cells that secrete glucagon
11 Sympatomimetic secretion of adrenal medulla
12 Big snakes
13 Has endocrine and exocrine functions
14 Amino acid (abbr.)
19 Mathematical constant
20 See 7 down
21 Farm implement
22 Ion regulated by PTH (abbr.)
23 This is affected by thyroxine levels
24 Secretes sympatomimetic hormones
26 Adrenalin's "partner" (abbr.)
27 Secreted by the pituitary to regulate the adrenal cortex
28 Sodium (abbr.)
29 California city (abbr.)
31 Classified
33 Egyptian sun god
34 Runs through some trees
35 This stimulates release of thyroxine (abbr.)
36 Surrounds a castle
38 Preposition
42 These interact with hormones
43 Time period
45 Secretes calcitonin
46 Protein hormone is the first of these
47 Not down
49 Laboratory dissection animal
50 Near the shore
51 Steroids pass right through this
52 Has two lobes
53 Gonadotropin secreted by pituitary (abbr.)
55 The level of influence of MSH in humans
58 Fish eggs
61 Not out
62 Decay
65 Sinatra movie "______ Here to Eternity"
66 Alcoholics Anonymous (abbr.)
70 Used to surface roads or roofs
72 Highest card in the deck
73 Not young
74 Afternoon (abbr.)
76 Gram (abbr.)
77 Lung disease (abbr.)
78 Seven Dwarfs tune: "Hi ________"

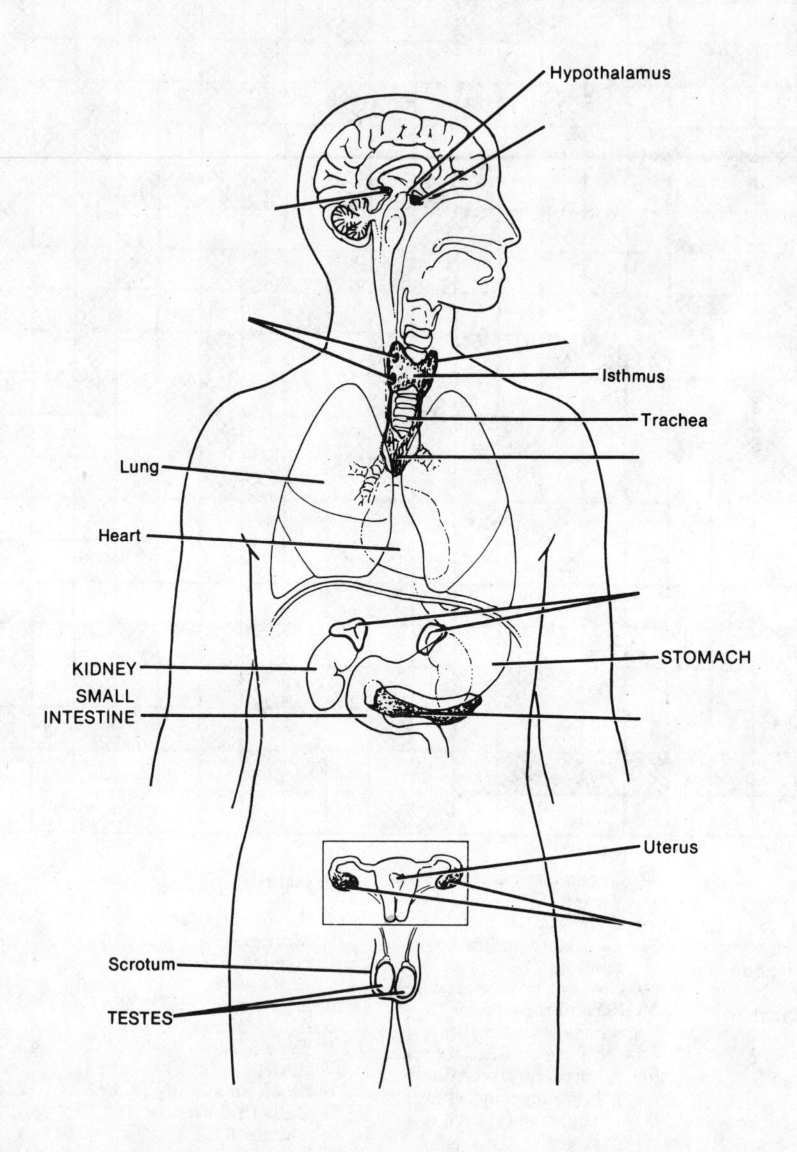

Anatomic Artwork

1. Label the structure that has alpha, beta and delta cells.
2. Label the gland that secretes hormones that are significant factors in the retention of sodium ions.
3. Label the gland that secretes calcitonin.
4. Label the gland that is greatly influenced by secretions of the hypothalamus.
5. Label the glands that secrete a hormone that has an opposite effect to calcitonin.
6. Label the gland that plays an important role in the development and maintenance of our immune defenses.
7. Label the glands that secrete hormones that are responsible for the development and maintenance of the mammary glands.
8. Label the gland that secretes a hormone that is produced in darkness.

Chapter 14

The Cardiovascular System: Blood

Introduction

Like a river that flows through the fields, the coursing of blood through the body serves a similar purpose: the *delivery* of the necessities for life and the *removal* of *metabolic* wastes. Blood can be thought of as a means of transportation within the body, for it carries gases (O_2 and CO_2), hormones, ions, nutrients, and wastes. Ordinarily, no cell is more than about 25 μm from a tiny branch of the vascular system, and it is at this level that *exchange* occurs between blood, cells, and their surrounding fluids.

Because of this exchange, blood loses certain substances and gains others as it flows. Yet its composition remains quite constant. For example, the temperature and the pH, Na^+, glucose, and urea levels in the blood are fairly steady in their values. To effect this, blood is cleansed by the kidneys, refreshed by the lungs, intestines, and liver, and buffered by its protein and ion buffer systems.

The fluid portion, or *plasma,* carries all soluble substances, as well as large protein molecules. The *formed elements,* erythrocytes, leucocytes, and platelets, perform their own various functions. For example, red cells carry oxygen and carbon dioxide; white cells represent a mobile defense system against invading microorganisms; platelets help prevent loss of this vital fluid, blood. It is as though the river has, in addition to all it carries, supplementary mechanisms to keep the fields (body) in a constant, homeostatic, and healthy condition.

Now that you have come to appreciate the various roles and aspects of the blood by studying the textbook, test your knowledge by working at the questions and activities in this chapter.

Multiple Choice

___ 1. When platelets come into contact with a damaged blood vessel:
   a. they adhere to the collagen of the vessel
   b. they become large and sticky
   c. they activate more platelets
   d. all of these

___ 2. Which of these develop from a hemocytoblast?
   a. erythrocytes
   b. neutrophils
   c. agranulocytes
   d. all of these

___ 3. Which is/are correctly paired?
a. B cell–antibodies
b. T cells–carry carbon dioxide
c. eosinophils–active against transplanted tissue
d. all of these are correctly paired

___ 4. The blood's first defense against bacterial infection is the:
a. lymphocyte
b. monocyte
c. erythrocyte
d. neutrophil

___ 5. Small tears in a vessel are repaired by:
a. collagen plugs
b. neutrophil plugs
c. platelet plugs
d. all of these

___ 6. Which is/are involved in extrinsic clotting?
a. tissue factor
b. thromboplastin
c. calcium
d. all of these

___ 7. Which of these blood types can theoretically donate blood to a type AB person?
a. A
b. B
c. O
d. all of these

___ 8. Agglutination refers to:
a. the clotting of blood
b. the clumping of platelets
c. the clumping of RBCs
d. all of these

___ 9. Which of these cells would be counted in a differential count?
a. reticulocytes
b. eosinophils
c. erythrocytes
d. platelets

___ 10. Which is/are true concerning antibodies?
a. They are made by B cells
b. They inactivate antigens
c. They are proteins
d. All of these are true

___ 11. Which of these may occur in erythroblastosis fetalis?
a. hemolysis of RBCs
b. anemia
c. the baby may be transfused in utero
d. all of these are true

___ 12. In the final step of clotting:
a. damaged endothelium releases thromboplastin
b. platelets begin to aggregate
c. fibrinogen is converted into fibrin
d. none of these

___ 13. The rate of RBC production can be gauged by the percentage in circulating blood of:
a. reticulocytes
b. platelets
c. eosinophils
d. all of these

___ 14. A floating clot is a(n):
a. thrombus
b. fiber
c. agglutinate
d. embolus

___ 15. The fraction of plasma proteins primarily responsible for creating osmotic pressure is the:
a. globulin fraction
b. albumin fraction
c. lipoprotein fraction
d. all of these

___ 16. Which of these is not found in plasma?
a. enzymes and hormones
b. gamma globulin
c. fibrin
d. bicarbonate

___ 17. Which of these is a function of blood?
a. production of hormones
b. it carries gamma globulin
c. pH buffering
d. manufacture of nutrients

___ 18. Which of these areas is/are involved in RBC breakdown?
a. liver
b. spleen
c. bone marrow
d. all of these

___ 19. Inability to absorb vitamin B^{12} results in:
a. hypochromic anemia
b. aplastic anemia
c. pernicious anemia
d. inherited spherocytic anemia

___ 20. Control over the rate of RBC production is effected by:
a. thombopoietin
b. severity of infection
c. erythropoietin
d. none of these

___ 21. Blood treated with heparin will:
a. not agglutinate
b. hemolyze
c. have a reduced number of platelets
d. not clot

___ 22. Which is/are true about WBCs?
a. They are larger than RBCs
b. They are larger than platelets
c. They are less numerous than platelets
d. all of these are true

___ 23. Which is true of the RBC?
a. small nucleus
b. largest blood cell
c. most numerous blood cell
d. none of these

___ 24. Where are the agglutinogens of the ABO system located?
a. on the surface of the WBC
b. on the surface of the RBC
c. in the plasma
d. all of these places

___ 25. When an agglutination of red cells occurs:
a. They swell and rupture
b. They become lodged in small vessels
c. They undergo hemolysis
d. All of these

True/False

___________ 1. The blood consists of approximately 55 percent *plasma.*

___________ 2. Normally, most of the red marrow is taken up by *red cell* production.

___________ 3. *All blood cells* arise from the same type of primitive stem cell.

___________ 4. In general, white blood cells are *smaller* and more numerous than red blood cells.

___________ 5. Antibodies are carried in the *plasma.*

___________ 6. Basophils develop in the *same type* of marrow as red blood cells do.

___________ 7. When red blood cells are destroyed in marrow, the iron is stored, the amino acids are recirculated, and the heme is excreted *via bilirubin.*

___________ 8. Lymphocytes are involved in the *immune response.*

___________ 9. The reticulocyte is *anucleate* but is not a mature cell.

___________ 10. *Pinocytosis* is the process whereby white cells squeeze out of capillaries.

___________ 11. White blood cells migrate to areas of inflammation by an ameboid movement called *diapedesis.*

___________ 12. Hematocrit is a relative ratio of red cells to *white cells.*

___________ 13. A hematocrit of 72 is indicative of *polycythemia.*

___________ 14. A red blood cell count of *5000/mm*^{3} is considered to be about normal.

___________ 15. The cytotoxic cell is considered to be a type of *B lymphocyte.*

_____________ 16. Generally, a 1 percent level of reticulocytes in circulating blood is considered to be a *high level.*

_____________ 17. *Interstitial* fluid bathes cells.

_____________ 18. The largest portion of the protein content of blood is made up by the *albumin* fraction.

_____________ 19. Blood is *less* viscous than water.

_____________ 20. A person who is type AB blood carries A and B *agglutinins.*

_____________ 21. A *thrombosis* might be treated with a clot preventive like heparin or coumadin.

_____________ 22. When a mother is *Rh^+ and the baby is Rh^-*, the subsequent baby may suffer from erythroblastosis fetalis.

_____________ 23. The fate of a monocyte when it enters the tissues is to become a *tissue macrophage.*

_____________ 24. *Lymphocytes give rise to plasma cells* when triggered by an antigen.

_____________ 25. Clotting of the blood can be initiated both by factors within blood and by factors from *damaged tissue.*

Completion

1. Bilirubin is a breakdown product of the _____________ portion of hemoglobin.
2. A substance involved in the dissolution of a clot is _____________.
3. Thromboplastin is involved in the initiation of the _____________.
4. Heparin, histamine, and serotonin are released by the _____________.
5. When hemoglobin carries carbon dioxide it is called _____________.
6. The normal pH of plasma is _____________.
7. Platelets develop from a giant cell known as a _____________.
8. The blood type that carries no agglutinins of the ABO system is type _____________.
9. Incompatibility in a blood transfusion leads to _____________ of the red blood cells.
10. The average amount of hemoglobin is about _____________ in men and _____________ in women.
11. When blood clots, the fluid expressed by the clot is known as _____________.
12. Sealing off a blood vessel, followed by plugging and clotting, is part of a process known as _____________.
13. Proteins involved in clot formation are produced in the _____________.
14. Plasma proteins carry out functions such as _____________.
15. The tightening of the fibrin clot is called _____________.
16. An individual with type AB blood will have _____________ type agglutinins in his or her plasma.
17. The function of blood is _____________ and _____________.
18. Leukemia prevents the normal production of _____________ in the marrow.
19. The most abundant white cell is the _____________.
20. A low level of white cells is called _____________.
21. An abnormal increase in red blood cells is characteristic of a condition called _____________.
22. _____________ is the process of blood cell formation.

23. The ion necessary for catalyzing many reactions of blood clotting is ______________.
24. Sickle cell anemia is characterized by an abnormality in the molecule ________________.
25. The way neutrophils operate is to ________________ the foreign microorganism.

Lost Sheep

1. neutrophil, lymphocyte, basophil, eosinophil
2. 120 days, biconcave, 15μm, hemoglobin
3. polycythemia, anemia, leukemia, uremia
4. shrink, sticky, become irregular in shape, adhere to collagen
5. phagocytosis, hemolysis, ameboid movement, diapedesis
6. WBC, leucocyte, nucleus, biconcave
7. parasitic infection, basophil, heparin, histamine
8. agglutination, clumping, clotting, blood typing
9. endothelium, intrinsic, blood vessel lining, occurs in seconds
10. prothrombin, fibrinogen, thromboplastin, fibrinolysin
11. carbon monoxide, carbon dioxide, oxygen, nitrogen
12. iron, biliverdin, B^{12}, intrinsic factor
13. leukemia, polycythemia, hemorrhagic anemia, sickle cell trait
14. type A blood cells, type B blood cells, type AB blood cells, type O blood cells
15. pernicious, hemolytic, hemophilia, sickle cell
16. tissue factor, thromboplastin, Rh factor, prothrombin
17. circulating antibodies, albumins, fibrinogen, small size
18. fibrinogen, glucose, amino acids, electrolytes
19. thrombocyte, platelet, fragment, hemoglobin
20. globin, antibody, protein, oxygen

Matching

Set 1

___ 1. has A agglutinogens
___ 2. cells agglutinate in type A anti-serum (carrying A agglutinins)
___ 3. least common blood type
___ 4. represents 85 percent of white, North American population
___ 5. carries no agglutinogens of the ABO system

a. type A
b. type B
c. type AB
d. type O
e. Rh^+

SET 2

___ 1. hemolysis leads to this
___ 2. platelet count less than $70{,}000/mm^3$
___ 3. hematocrit of 20
___ 4. hemostasis disorder
___ 5. low level of WBCs
___ 6. low O_2 delivery to body cells
___ 7. WBC count $500{,}000/mm^3$
___ 8. hematocrit of 60
___ 9. low hemoglobin per RBC
___ 10. uncontrolled production and accumulation of WBCs

a. thrombocytopenia
b. polycythemia
c. anemia
d. leukopenia
e. leukemia
f. hypoxia

SET 3

___ 1. agranular, phagocytic
___ 2. secretes heparin
___ 3. combats allergic reactions
___ 4. develops into antibody producer
___ 5. causes symptoms of allergies
___ 6. second responder to bacterial invasion
___ 7. 3–5 lobed nucleus
___ 8. normally first line of defense in inflammation
___ 9. becomes a wandering macrophage in the tissues
___ 10. known as a mast cell in tissues

a. neutrophil
b. lymphocyte
c. basophil
d. monocyte
e. eosinophil

Double Crosses

1. *Across*:
   red blood cell
   Down:
   breakdown product of hemoglobin

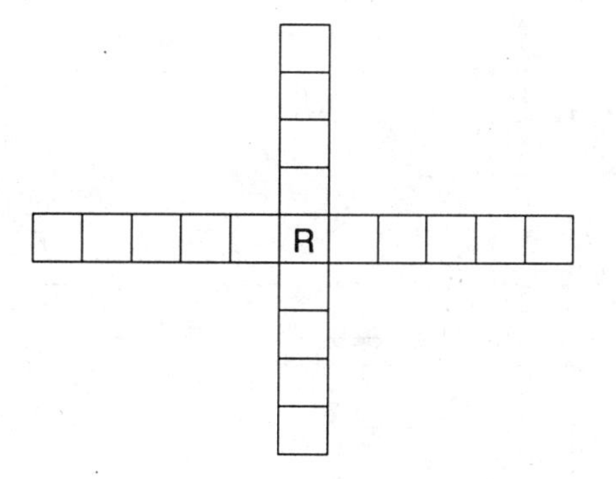

2. *Across*:
   white blood cell
   Down:
   process of blood cell formation

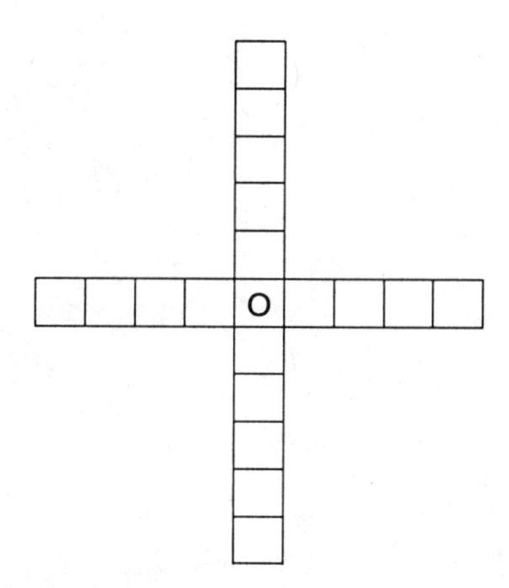

3. *Across:*
   a. cell fragments involved in clotting
   b. basophils release histamine and can cause an ________ reaction

   Down:
   immature red blood cells

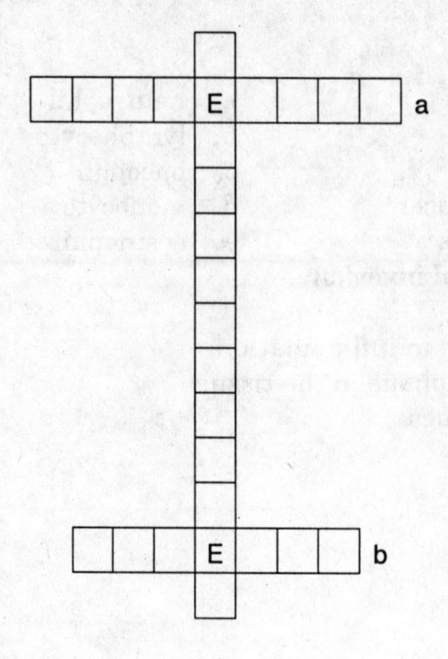

4. All words start from "A" in center
   a. plasma proteins
   b. type of anemia
   c. environmental factor affecting RBC production
   d. condition of abnormal immune response
   e. response to antigen
   f. foreign bodies
   g. tissue factor substance will ________ prothrombin
   h. blood at pH 7.5

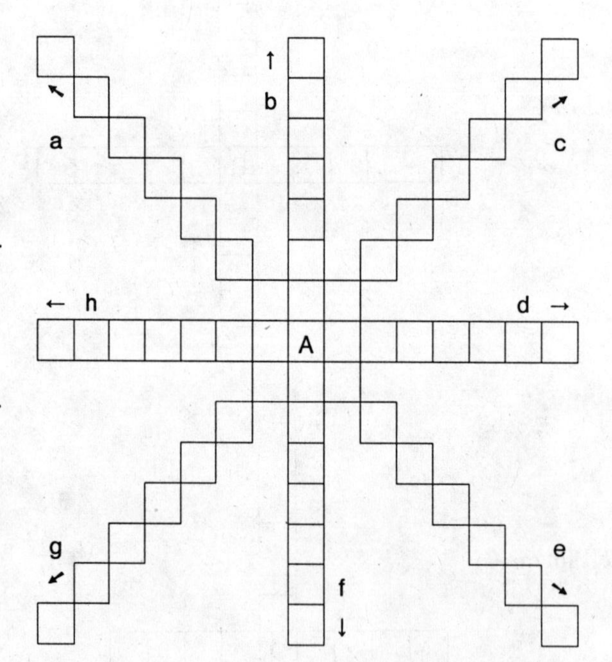

Sleuthing

1. Maryjo is Type O^-. She marries a B^+ man and they have a baby with type B^+ blood.
   a. Will there be a problem with Maryjo's first pregnancy? Why or why not?

b. Will there be a problem with Maryjo's second pregnancy? Why or why not?

c. What will happen in her blood upon exposure to baby's blood during the first delivery?

d. What will this (c) do to the second baby? What will the clinical symptoms be?

e. How could this (c) be prevented?

f. Would there be a problem if Maryjo were O^+ and her baby were B^-? Explain.

Word Scrambles

1. a. foreign body — NTEAIGN — ☐ _ ☐ ☐ ☐ _ ☐
   b. iron-containing pigment — IMOGHOLEBN — _ _ _ ☐ ☐ _ _ _ _ _
   c. enzyme involved in thrombin formation — HBOLATOSPMRNIT — ☐ _ _ _ _ _ _ _ _ _ _ _ ☐ ☐
   d. maintains osmotic pressure in blood — MINUBLA — ☐ ☐ _ ☐ _ _ _

   Total: results from incompatibility

   ☐☐☐☐☐☐☐☐☐☐☐☐☐

2. a. helps to "put a finger in the dike" — LAEPTLTE — ☐ _ _ ☐ ☐ _ _ _
   b. percentage of red cells — CHARTOMIET — ☐ ☐ _ _ _ _ _ ☐ ☐ ☐
   c. first one out — TERPNUOLHI — ☐ _ _ _ ☐ ☐ _ _ ☐ _
   d. fighter — TIOBYDNA — _ _ _ _ _ ☐ _ ☐

   Total: red rate controller

   ☐☐☐☐☐☐☐☐☐☐☐☐☐☐

Addagrams

1. a. fluid squeezed out of clot — 4, 6, 10, 5, 7
   b. ________ poiesis — 13, 16, 7, 11
   c. symbol for chemical silver — 14, 15
   d. what macrophages eat — 1, 10, 8, 3, 13
   e. ion needed in clotting — 9, 8
   f. blood cell production: hemo ________ — 12, 11, 2, 16, 17, 2, 17

Total: big scavengers

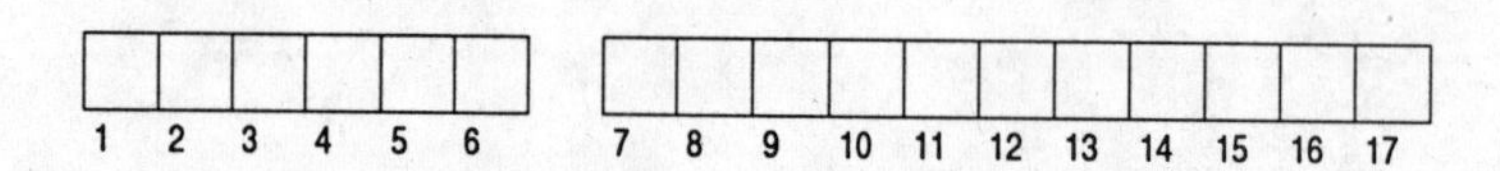

2. a. blood disorder due to lack of hemoglobin — 3, 13, 11, 5, 12, 6
   b. fibrinolysin — 7, 2, 3, 4, 5, 12, 13
   c. plug tidbit — 7, 2, 3, 10, 11, 2, 11, 10
   d. ________ thrombin — 1, 8, 9

Total: busy passenger

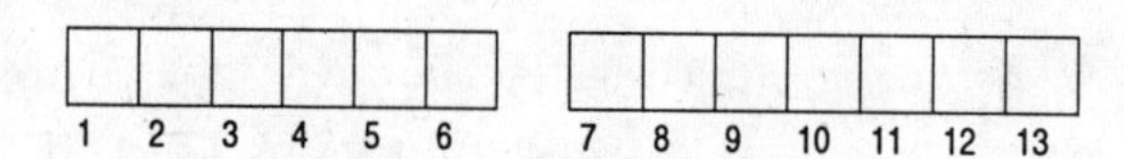

Crossword

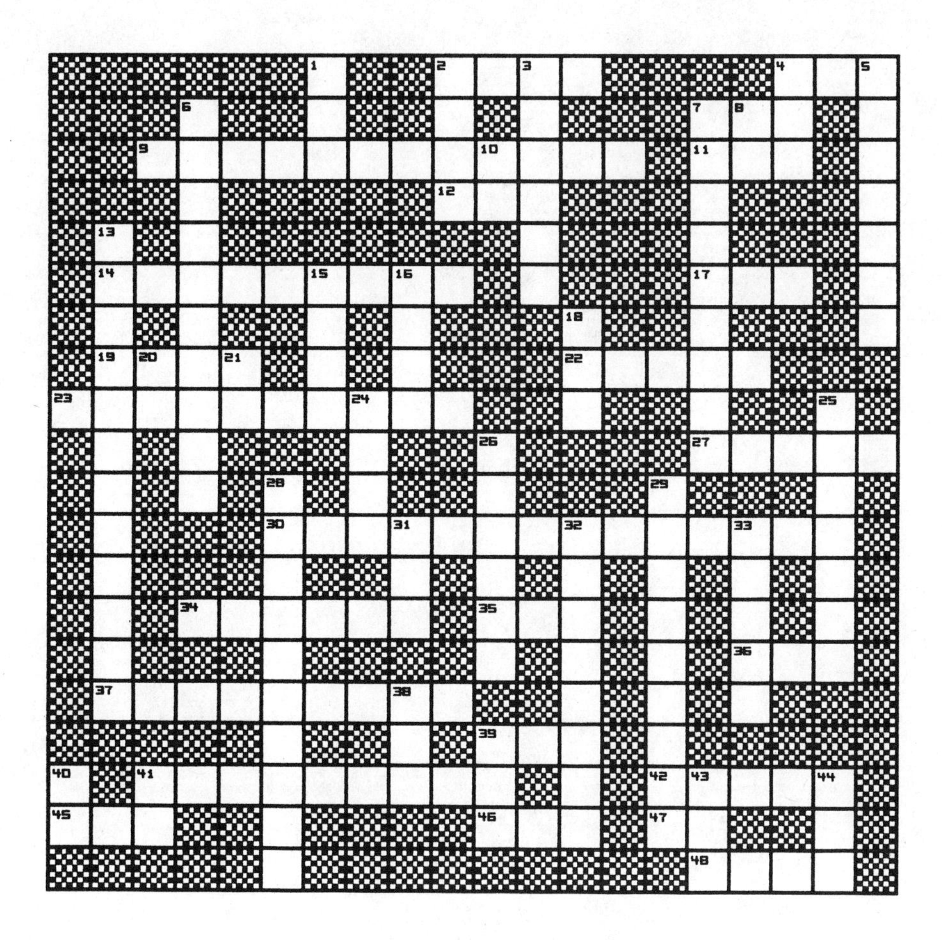

Across

2 Element held in hemoglobin
4 Anatomy friend
7 Marker
9 Immature red cell
11 Red cell (abbr.)
12 Egg drink
14 Type of clotting
17 Inherited abnormality of hemoglobin (abbr.)
19 Recognition feature
22 Liquid of life
23 What a clot undergoes
27 Type of lymphocyte
30 Process of red cell formation
34 Liquid of blood
35 "Hi" to Juan
36 Vessel
37 White cell
39 "Sun" to Ricardo
41 Ability to kill
42 In-between
45 Type of antigen on white cell (abbr.)
46 Little guy
47 Negative
48 Goner

Down

1 White cell (abbr.)
2 A god
3 Carried by hemoglobin
4 Count 'em all
5 ________cyte—platelet
6 Percentage of reds in whole
7 Function of blood
8 Antibody (abbr.)
10 Carbon monoxide (chemical symbols)
13 All the many types of white
15 Tidy
16 Going in
18 Blood grouping
20 "Where'er you________?"
21 Each (abbr.)
24 ________-bitty
25 Digests clot
26 Where reds and whites and platelets get started
28 Study of blood
29 Breakdown product of heme
31 ________juana—border town of Mexico
32 Functional tidbit
33 Plasma minus fibrinogen
38 Pal from Dallas
39 Hemolytic anemia (abbr.)
40 Factor in erythroblastosis fetalis
41 Elements used in clotting (chemical symbol)
43 Bow head in affirmation
44 Type of cell that is most prevalent in blood

Chapter 15

The Cardiovascular System: Heart

Introduction

The delivery of nutrients and the removal of tissue and cellular wastes are impossible without the constant movement of blood through the body. Blood is pumped at a rate of about 5 liters per minute by the heart (the *cardiac output)* and travels through a system of arteries, capillaries, and veins. In an average human adult, the heart muscle spontaneously depolarizes and contracts in the range of about 100,000 times a day. The heart muscle is *self-excitable.* While the initiation of contraction is independent of nervous control, the rate and strength of its action can be altered by sympathetic and parasympathetic stimulation. The strength of contraction is also dependent on venous return to the heart, for within limits, the greater the fill, the stronger the contraction.

After you have become familiar with the material on the heart in the textbook, you should be able to complete the following activities and answer the following questions.

Multiple Choice

___ 1. Major parasympathetic innervation for the heart is carried by the:
  a. glossopharyngeal nerve
  b. accelerans nerve
  c. vagus nerve
  d. hypoglossal nerve

___ 2. Sympathetic stimulation of the heart affects:
  a. the conduction system
  b. the atria
  c. the ventricular myocardium
  d. all of these

___ 3. Cardiac tamponade refers to:
  a. increased fluid in the pericardial cavity
  b. endocardial inflammation
  c. myocardial pathology
  d. all of these

___ 4. Which controls the heart rate to increase speed?
a. sympathetic stimulation
b. blocking parasympathetic stimulation
c. both a and b
d. neither a nor b

___ 5. When baroreceptors sense a high blood pressure, there follows:
a. a decrease in heart rate
b. a decreased force of contraction
c. a decrease in cardiac output
d. all of these

___ 6. Pacemaker tissue:
a. is located in the SA node
b. needs sympathetic nerves to start the heart
c. utilizes parasympathetic nerves to excite the heart
d. all of these

___ 7. Cardiac output equals:
a. stroke volume x rate
b. about 70 ml x 70 bpm
c. is about 5 liters per minute
d. all of these

___ 8. Which is the correct flow pattern of blood through the heart?
a. right atrium, left atrium, right ventricle, left ventricle
b. right atrium, right ventricle, left atrium, left ventricle
c. left atrium, left ventricle, right ventricle, right atrium
d. none of these is correct

___ 9. Which of these is/are a risk factor for a myocardial infarction?
a. high cholesterol (CH) level
b. cigarette smoking
c. obesity and lack of exercise
d. all of these

___ 10. Which of these is/are correctly paired?
a. QRS–diastasis
b. T–ventricular diastole
c. P–atrial depolarization
d. none of these is correct

___ 11. Which is true concerning the description of the heart?
a. located in the mediastinum
b. 2/3 to left of midline
c. apex pointed toward left foot
d. all of these are true

___ 12. Inflammation of the pericardium could involve the:
a. fibrous layer
b. serous layer
c. epicardium
d. all of these

___ 13. What pressure must be overcome in the left heart, in order to open the semilunar valves?
a. 120 mm Hg
b. 80 mm Hg
c. 22 mm Hg
d. 8 mm Hg

___ 14. Which has the greatest duration?
a. atrial diastole
b. atrial systole
c. ventricular diastole
d. ventricular systole

___ 15. Which best describes the volume change in the atrium during ventricular systole?
a. slow decrease
b. slow steady increase
c. rapid increase
d. no change

___ 16. What valve closes when the left ventricle begins diastole?
a. mitral
b. aortic semilunar
c. tricuspid
d. pulmonary semilunar

___ 17. Which valve opens just before rapid filling of the right ventricle?
a. pulmonary semilunar
b. mitral
c. tricuspid
d. aortic semilunar

___ 18. Which is correctly paired for fetal circulation?
a. ductus arteriosus–pulmonary vein to aorta
b. foramen ovale–connects right atrium to left atrium
c. interventricular septum–allows flow between ventricles
d. none of these is correct

___ 19. Which of these is/are found in the tetralogy of Fallot?
a. overriding aorta and incomplete ventricular septum
b. pulmonary artery stenosis
c. right ventricular hypertrophy
d. all of these

___ 20. The coronary (transverse) sulcus on the outside of the heart:
a. separates the atria from the ventricles
b. separates the auricles from the ventricles
c. contains fat and coronary blood vessels
d. all of these

___ 21. Which of these is correctly paired?
a. ventricular fibrillation–heart failure
b. heart block–Purkinje fibers
c. atrial flutter–life threatening
d. none of these is correct

___ 22. Which of these affects the rate of the heart?
a. potassium level changes in blood
b. body temperature
c. sex and age
d. all of these

___ 23. Which of these aids in the conduction of action potentials throughout heart muscle?
a. cardiac skeleton
b. gap junctions
c. coronary blood supply
d. all of these

___ 24. Intercalated discs of the heart:
a. cause fibrillation
b. receive blood from the inferior vena cava
c. aid in the conduction of APs throughout the myocardium
d. all of these

___ 25. Which of these is not an abnormality associated with the heart?
a. pulmonary embolism
b. cardiac tamponade
c. mitral stenosis
d. coronary artery disease

True/False

__________ 1. Stroke volume is about *150 ml.*

__________ 2. *Cardiac output* is measured as approximately 70 ml times 70 bpm.

__________ 3. The "slow down" in the conduction system occurs at the *AV node.*

__________ 4. When both ventricles and atria are at rest, it is known as *diastasis.*

__________ 5. *Left* heart failure can lead to pulmonary edema.

__________ 6. A sphygmomanometer measures *apical pulse.*

__________ 7. A narrowing of the semilunar valves is known as *stenosis.*

__________ 8. The *semilunar* valve prevents blood from backflowing down the pulmonary artery into the ventricle.

__________ 9. Ventricular contraction pushes the semilunar valve *open.*

__________ 10. When the heart rate increases, the *relaxation period* shortens.

__________ 11. A ballooning of the arterial wall is known as a *varicosity.*

__________ 12. A patent (open) ductus arteriosus in the adult leads to blood flowing from the *pulmonary artery to the aorta.*

__________ 13. The usual way *rate* is altered is by way of autonomic control, circulating chemicals, and temperature.

__________ 14. *Cardiac output* depends on the rate and the amount of blood ejected with each beat.

__________ 15. The left heart pumps into the high-pressure *pulmonary* circuit.

__________ 16. The bicuspid (mitral) valve is on the *right* side of the heart.

__________ 17. The heart *depends on nervous stimulation.*

__________ 18. Stimulation of the vagus nerve *speeds up* the heart beat.

__________ 19. Epinephrine increases rate *but not* strength of heart contraction.

__________ 20. Sympathetic nerves innervate the conduction system *and* the atria and ventricles.

__________ 21. An increase in blood pressure stimulates the *cardioinhibitory* center in the medulla.

__________ 22. Blood is pumped out to the body from the *ventricles* of the heart.

__________ 23. The *pons* is the area of the brain which contains the cardioregulatory centers.

__________ 24. The thickest wall of the heart is the *myocardium.*

__________ 25. The T wave of an ECG (EKG) represents ventricular *relaxation.*

Completion

1. Average stroke volume for the heart is about __________ .
2. Deposits of cholesterol and fatty plaques in the walls of the arteries are a condition known as __________ .
3. Excess fluid production from the inflammatory process in pericarditis can restrict the filling of the heart. This condition is known as __________ .
4. The ion that aids in the contractile process for the myocardium is __________ .
5. The __________ valve prevents the backflow of blood into the ventricle from the aorta.
6. Blood is drained from the heart muscle and flows into the right atrium by way of a vessel known as the __________ .
7. An interventricular septal defect in the adult would allow blood to flow from the __________ to the __________ because of the pressure difference.
8. The blood flow to the heart could be obstructed when the __________ circulation experiences atherosclerotic plaques building up and spasms in the vessels.
9. Systolic pressure in the left ventricle reaches __________ mm Hg.
10. In the adult, the only artery to carry poorly oxygenated blood is the __________ .
11. What brain center in the medulla has control to slow down the heart rate? __________
12. Impulses pass readily across cell boundaries by way of specialized areas known as __________ .
13. The amount of blood ejected by the heart in a minute is __________ liters.
14. Diastolic blood pressure is associated with __________ of the heart.
15. A reduced oxyhemoglobin content in arterial blood, accompanied by dark blue nail beds and mucous membranes, is a condition known as __________ .
16. Blood pressure is sensed in regions of the artery known as __________ .
17. When the heart muscle is damaged and unable to pump all the blood in it, it is known as __________ heart failure.
18. An enlarged right ventricle as a result of hypertension in the pulmonary circuit is known as __________ .
19. The mitral valve has __________ (number of) cusps.
20. The AV valves are prevented from prolapsing by tendonlike cords called __________ .
21. Impulses are slowed in the conduction system of the heart at the __________ .
22. Atrial depolarization is represented in the __________ wave of the ECG (EKG).

23. What kind of valve is in the pulmonary artery? ________________
24. Blood returns to the heart into the ________________.
25. How would cardiac muscle be characterized? ________________

Lost Sheep

1. pulmonary artery, vena cava, right atrium, left atrium
2. SA node, mitral valve, bundle of His, Purkinje fibers
3. foramen ovale, ductus arteriosus, vena cava, umbilical vessels
4. pulmonary artery, 120 mm Hg, deoxygenated blood, right heart
5. Starling's Law, increased stretch, increased contraction, decreased venous return
6. high BP, shock, thready pulse, rapid pulse
7. epinephrine, vagus, parasympathetic, slow heart beat
8. ventricular systole, AV valve closes, semilunar valve opens, blood enters atrium
9. atrial depolarization, P wave, precedes QRS wave, atrial repolarization
10. norepinephrine, epinephrine, sympathetic stimulation, vagal control
11. auricle, interventricular septum, coronary artery, pulmonary artery
12. fibrous pericardial sac, myocardium, papillary muscle, right ventricular wall
13. pulmonary artery, superior vena cava, coronary sinus, aorta
14. cardiac skeleton, trabeculae carneae (muscle ridges), left atrial wall, papillary muscles
15. cardiac vein, coronary artery, coronary sinus, carotid artery
16. gap junctions, inherent rhythmicity, self-initiating (self-excitatory), nonstriated
17. pressure drop, "milking of veins," negative chest pressure, edema
18. female, neonate, well-trained athlete, person in a sympathetic response
19. increased peripheral resistance, decreased blood pressure, increased heart rate, baroreceptors
20. pulmonary artery, mitral, superior vena cava, tricuspid

Matching

Set 1

___ 1. shows a long refractory period
___ 2. secretes fluid
___ 3. differentiates into excitatory tissue
___ 4. lines heart and root of aorta
___ 5. responsible for pumping
___ 6. covers heart
___ 7. involuntary, striated, and branched
___ 8. self-excitatory
___ 9. contractile tissue
___ 10. is continuous with capillary lining

a. epicardium
b. myocardium
c. endocardium

Set 2

___ 1. atrial depolarization
___ 2. ventricular depolarization
___ 3. ventricular repolarization
___ 4. largest wave
___ 5. atrial repolarization

a. T wave
b. QRS wave
c. P wave
d. none of the above

Set 3

Use all letters from right-hand column that apply to left-hand column.

___	1. aortic	a.	AV valve
___	2. mitral	b.	tricuspid valve
___	3. pulmonary	c.	bicuspid
___	4. between two chambers on right side	d.	semilunar valve
___	5. between two chambers on left side		

Set 4

___	1. begins impulse	a.	conducting (Purkinje) fibers
___	2. last part of atrium to depolarize	b.	AV node
___	3. delays	c.	bundle of His
___	4. fastest area to self-excite	d.	bundle branches
___	5. pacemaker	e.	SA node
___	6. affected by autonomic system for rate-setting		
___	7. runs down either side of interventricular septum		
___	8. at top of interventricular septum		
___	9. carries impulse into ventricular myocardium		
___	10. sets rate		

Double Crosses

1. *Across:*
   a. quivers ineffectively
   b. _________ tendineae

   Down:
   circuit out of right ventricle

L a

R b

2. *Across:*
   a. _________ of Fallot
   b. relating to lower chamber

   Down:
   space between lungs

A a

I b

3. *Across*
   a. 70 ml x 70 bpm (2 words)
   b. small muscle projection in ventricle
   c. ________ manometer

   Down:
   exchange blood vessel

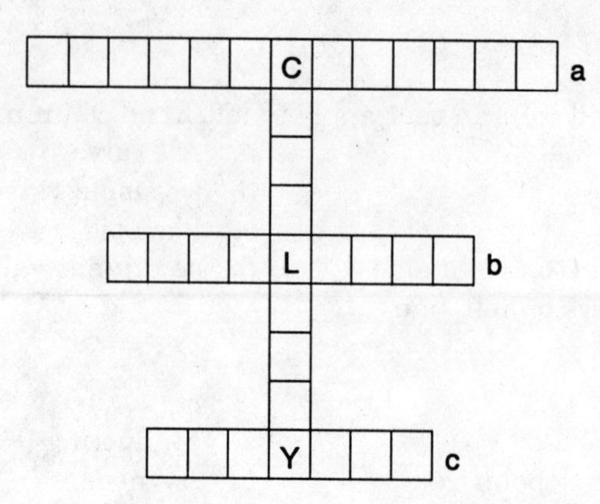

4. All words start from "A" in center
   a. pain associated with blocked coronary artery
   b. arch
   c. upper chamber
   d. vessel carrying blood away from heart

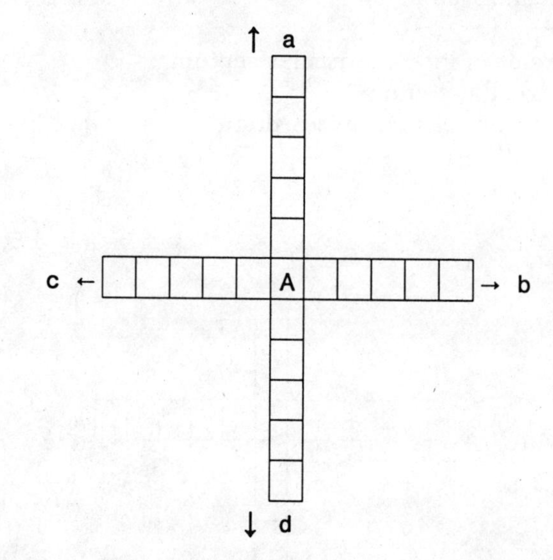

Sleuthing

1. Ellen has had angina attacks with increasing frequency over the past six months. The doctor decided to do an arteriogram, which showed a narrowing in the coronary vessel running down the anterior portion of the heart.
   a. What is the consequence if untreated?

   b. What is involved in surgical intervention?

   c. What is the major cause of this type of narrowing?

Word Scrambles

1. a. major blood vessel — ORATA — _ _ □ □ _
   b. rapid involuntary response to low blood pressure — FLEREX — _ _ □ □ □ _
   c. base of vessel — KURTN — □ _ □ _ _

Total: cardiac arrhythmia

□□□□□□□

2. a. gland affecting WBC development — SHYMUT — □ _ _ _ _ □
   b. having to do with the heart — DACRICA — _ _ _ □ _ _ □
   c. abnormal heart sound — RUMRUM — _ □ _ _ □ _

Total: fetal shunt

□□□□□□

Addagrams

1. a. pulse measures the ________ of heart beat — 4, 6, 8, 7
   b. the heart beats about 12 times every ________ seconds — 8, 9, 10
   c. ________ regulatory center in the medulla — 1, 6, 4, 5, 12, 3
   d. what occurs after prolonged asystole — 11, 15, 14, 8, 2
   e. neurotransmitter that stimulates heart — 13, 15

Total: heart strings

1	2	3	4	5	6	7

8	9	10	11	12	13	14	15

2. a. type of blood that is well oxygenated — 1, 3, 9, 7, 10, 11, 15, 19
   b. narrowing of a valve or vessel — 22, 9, 21, 8, 5, 22, 4, 22
   c. foramen ________ (in embryo) — 5, 6, 18, 19, 21
   d. blocks a coronary vessel — 12, 14, 5, 2
   e. heart rate — 16, 1, 9, 21
   f. one of the factors affecting blood pressure — 17, 5, 19, 13, 0, 21
   g. amount of blood in heart after it beats (abbr.) — 21, 22, 20

Total: "flaps"

1	2	3	4	5	6	7	8	9	10	11	12	13	14	15	16

17	18	19	20	21	22

Crossword

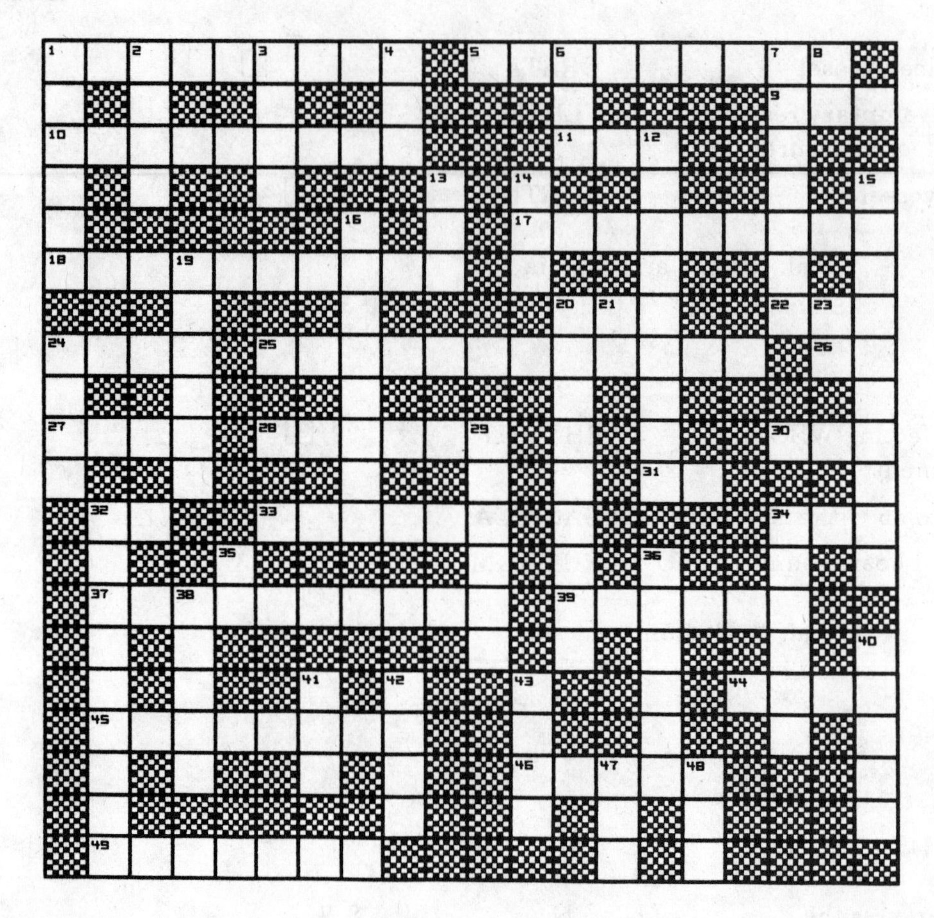

Across

1 Valve in aorta
5 Chordae ________
9 Grecian jug
10 Vessel serving lungs
11 "Hi" to Juanita
17 The pumper
18 Heart muscle
20 Writing implement
22 Tiny creature
24 Party men
25 Speed center
26 Neurotransmitter that accelerates rate
27 ________ cardium—cover
28 Receiver
30 Center that slows (abbr.)
31 Volume/min (abbr.)
32 Volume/beat (abbr.)
33 Tetralogy name
34 Animal doctor
37 Abnormal rhythm
39 Venous ________—what comes back
44 Heart's job
45 Lack of oxygen to tissues
46 Fluid that gets pumped
49 Contractile phase

Down

1 Wall
2 Concentration unit to chemist
3 Forearm bone
4 Medical picture: x-________
6 New (combining form)
7 Outer flap
8 Emergency Room (abbr.)
12 Involuntary control
13 Rate units (abbr.)
14 Upper great vein (abbr.)
15 The heart can ________ and therefore beat continually (2 words)
16 Layer near the organ
19 Bi________ or tri________ valve
20 Muscle anchor
21 Extraterrestrial (abbr.)
23 Lane (abbr.)
24 Second heart sound
29 Bicuspid
32 Heart man's law
34 Circulation that returns blood to heart
35 Substance of plaques (abbr.)
36 ________sclerosis
38 Stretch and grab
40 When a vessel constricts suddenly
41 Sudden heart attack (abbr.)
42 Beats per minute
43 First heart sound
47 Type of fiber
48 Department of Transportation (abbr.)

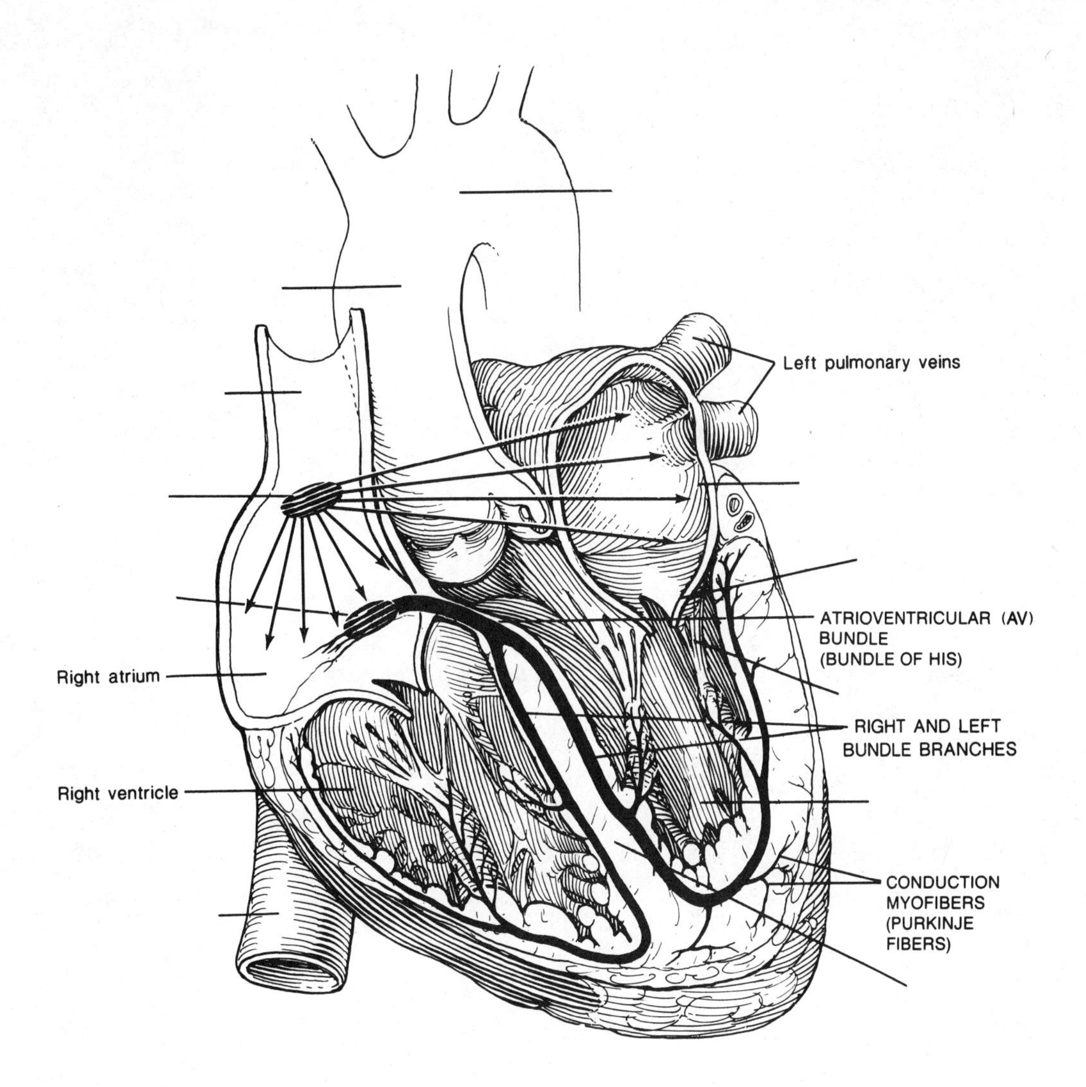

Anatomic Artwork

1. Label the area of the conducting system which self-excites most rapidly.
2. Label the area of the conducting system which slows down the impulse.
3. Label the chamber that receives oxygenated blood from the lungs.
4. Label the valve that has only two cusps.
5. Label the chamber that pumps into the aorta.
6. Label the vessel leading to the systems.
7. Label the vessel bringing blood back from the legs.
8. Label the vessel bringing blood from the head and neck.
9. Label the structures that prevent the valves from "flipping" back into the atrium.
10. Label the wall that prevents well oxygenated blood from pouring into a neighboring chamber with poorly oxygenated blood.

Chapter 16

The Cardiovascular System: Blood Vessels

Introduction

The vessels that carry blood in the body have only one ultimate function: to deliver nutrients to the cells and to pick up wastes from the cells. In essence, the goal is *exchange*. In order to effect this, a set of arteries, which are thick walled and elastic, carry the blood to smaller vessels called *arterioles*. These vessels are ringed with smooth muscle. The muscle serves to constrict and dilate the arterioles, and thereby they provide one of the major means of regulation of blood pressure. The arterioles are called *resistance* vessels. They lead into a vast network of tiny vessels called *capillaries*, across whose thin endothelial walls exchange occurs with the tissues of the body. The blood is then carried back to the heart through larger vessels called *venules*, which then empty into thin-walled, compliant vessels called *veins*. They can hold a great deal of blood and so are called blood *reservoirs*.

After reading the textbook concerning the vessels, and understanding how the body regulates both the direction and pressure of the blood flow, you should be able to answer the following questions and work on the activities.

Multiple Choice

___ 1. The last area of control over whether or not blood will enter a particular capillary bed is called:
a. arteriolar series of smooth muscles
b. precapillary sphincter
c. venule musculature
d. all of these

___ 2. Vasodilation works on the peripheral circulation:
a. at the precapillary sphincter
b. to increase blood flow
c. at the smooth muscles of the arterioles
d. all of these

___ 3. The vessels that hold the greatest amount of blood at rest are the:
a. veins
b. arteries
c. arterioles
d. capillaries

___ 4. The vessels that provide the greatest control over the resistance to blood flow are the:
   a. veins
   b. arteries
   c. arterioles
   d. capillaries

___ 5. The area of the brain that houses the vasomotor center is the:
   a. pons
   b. medulla
   c. cerebellum
   d. hypothalamus

___ 6. The main purpose of blood flow is to:
   a. meet the metabolic needs of the tissues
   b. create a pressure gradient between systolic and diastolic values
   c. allow exchange at the arterioles
   d. none of these

___ 7. Which of these affects blood pressure?
   a. resistance to flow
   b. blood viscosity
   c. amount of blood
   d. all of these

___ 8. Blood pressure control is carried out by regulating smooth muscle via:
   a. parasympathetic stimulation
   b. parasympathetic and sympathetic stimulation
   c. sympathetic stimulation
   d. none of these

___ 9. The areas of the body that sense blood pressure are known as:
   a. baroreceptors
   b. chemoreceptors
   c. viscoreceptors
   d. none of these

___ 10. Which of these aids in venous return?
   a. skeletal muscle massage
   b. valves
   c. the pumping action of the heart
   d. all of these

True/False

__________ 1. The vasomotor center is located in the *cerebellum*.

__________ 2. Blood pressure is controlled by *sympathetic and parasympathetic* impulses to the smooth muscles of the arterioles.

__________ 3. *Metabolic needs* are the greatest controllers of how much blood flows into a particular capillary bed.

__________ 4. Local *vasodilator* substances are CO_2, lactic acid, and histamine.

__________ 5. Norepinephrine and epinephrine bring about *vasoconstriction* of arterioles in the skin and viscera.

__________ 6. Characteristics of shock are *warm skin*, clammy skin, and sweating.

__________ 7. A weakening in the arterial wall is known as a *thrombosis*.

____________ 8. The velocity of blood flow is *greatest* in the capillary bed.

____________ 9. Flabby and dilated veins are known as *varicose veins*.

____________ 10. The amount of blood ejected by the heart per minute is known as *stroke volume*.

Completion

1. When the lumen of a blood vessel increases in size, it has ________________.
2. The circulatory route that carries blood from the intestines directly to the liver is the ________________.
3. High blood pressure is known as ________________.
4. A temporary loss of consciousness is called ________________.
5. The intermittent flow of blood into the capillary bed is known as ________________.
6. Opposition to blood flow in the peripheral circulation, due to friction, is known as ________________.
7. The area of the medulla that is responsible for control of the diameter of blood vessels is known as ________________.
8. Neurons located in blood vessels, which are sensitive to blood pressure, are known as ________________.
9. Blood pressure is measured manually by an instrument known as a(n) ________________.
10. Chemoreceptors that sense the level of CO_2 are located in ________________.

Matching

Set 1

___ 1. reservoir

___ 2. exchange

___ 3. atherosclerosis

___ 4. single layer of endothelium only

___ 5. delivers directly to capillaries

a. capillaries

b. veins

c. arteries

d. arteriole

Set 2

___ 1. pressure due to resistance in vessels

___ 2. difference between systolic and diastolic pressure

___ 3. ratio of 3:2:1; what order of a, b, and c?

___ 4. pressure when heart contracts

___ 5. pressure when heart relaxes

a. systolic pressure

b. diastolic

c. pulse pressure

Double Crosses

1. *Across*:
   property of large arteries

   Down:
   tubes that carry blood

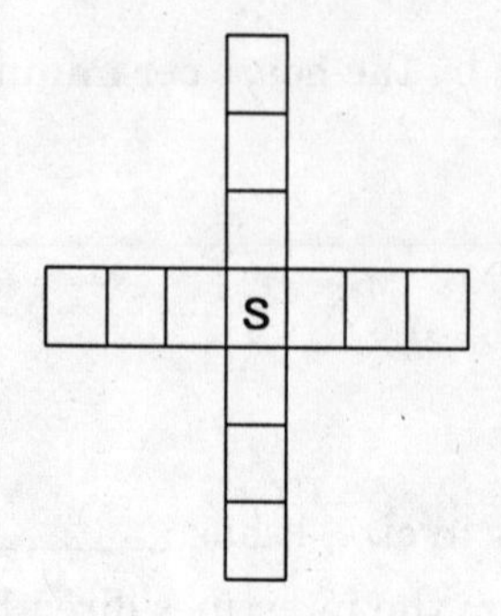

2. *Across*:
   area of brain that houses vasomotor center

   Down:
   substance dissolved in blood which creates osmotic pressure

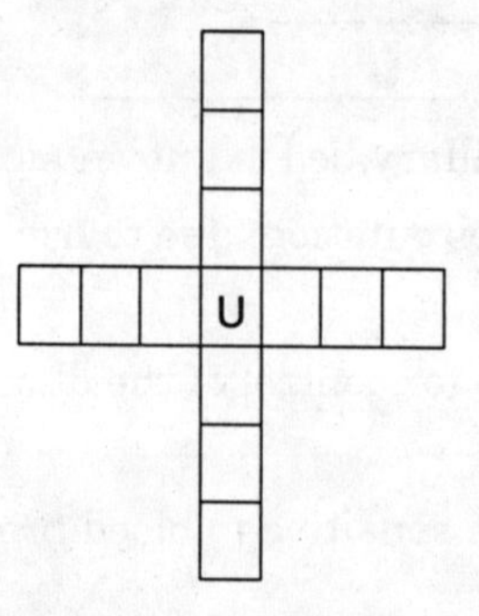

3. *Across*:
   property of "thickness" of blood

   Down:
   a. "the great returner" (2 words)
   b. pressure during contraction
   c. "the great passenger"
   d. lack of O_2 to tissues
   e. ________ manometer

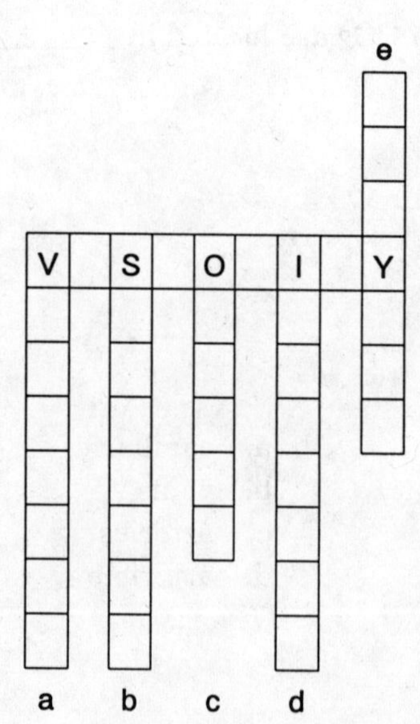

Word Scrambles

1. a. slow rate — ABARDCYDRIA — ☐ ☐ ☐ _ _ ☐ _ _ _ _ _
   b. what the venous beds are — SIRVREEORS — ☐ ☐ ☐ ☐ _ _ ☐ _ _ _
   c. abbreviation for exchange vessels — SCAP — _ _ ☐ _
   d. major artery of neck area — TARDOCI — _ _ ☐ ☐ ☐ _ _

Total: don't press me

☐☐☐☐☐☐☐☐☐☐☐☐☐

2. a. leads to the lungs — MARYPOULN — ☐ ☐ ☐ _ _ _ _ ☐ _
   b. carries the blood — LESSEV — _ ☐ ☐ ☐ ☐ _
   c. skin tone in shock — LAPE — ☐ _ _ ☐
   d. pulse place — SWIRT — _ ☐ _ ☐ _
   e. aneurysm — GULBE — _ ☐ _ _ _

Total: oh what's the difference?

☐☐☐☐☐ ☐☐☐☐☐☐☐☐

Crossword

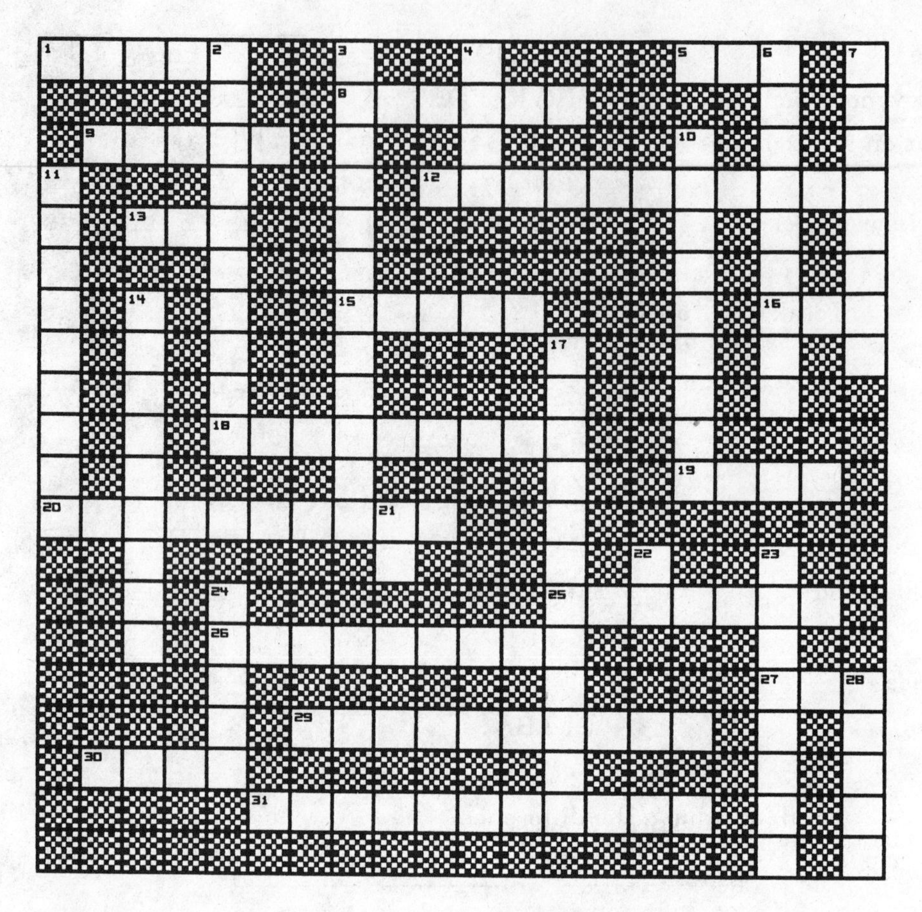

Across

1 ________ tension
5 Center that slows heart (abbr.)
8 Vessel carrying blood away from heart
9 One area where pulse is taken
12 Inner epithelium of vessels
13 Manual pumping
15 Inner hollow space in vessel
16 One factor affecting blood pressure
18 Muscle at entrance to capillary bed
19 How skin feels in shock
20 Friction of blood against vessel walls
25 Neck vessel
26 Vasodilator
27 Pressure during contraction (abbr.)
29 Intermittent flow in capillary bed
30 Pulse is ________ in shock
31 Excitement rate

Down

2 What veins act as
3 Increasing diameter of vessel
4 Return vessel
6 Exchange vessel
7 Change in this results from dilation or constriction of vessel
10 These needs must be met
11 Center controlling size of vessels
14 Thickness of blood
17 Pressure sensor
21 Ion that affects contractile strength (abbr.)
22 Resistance on the outskirts (abbr.)
23 Relaxation
24 Greatly decreased venous return
28 SBP – DBP = ________ pressure

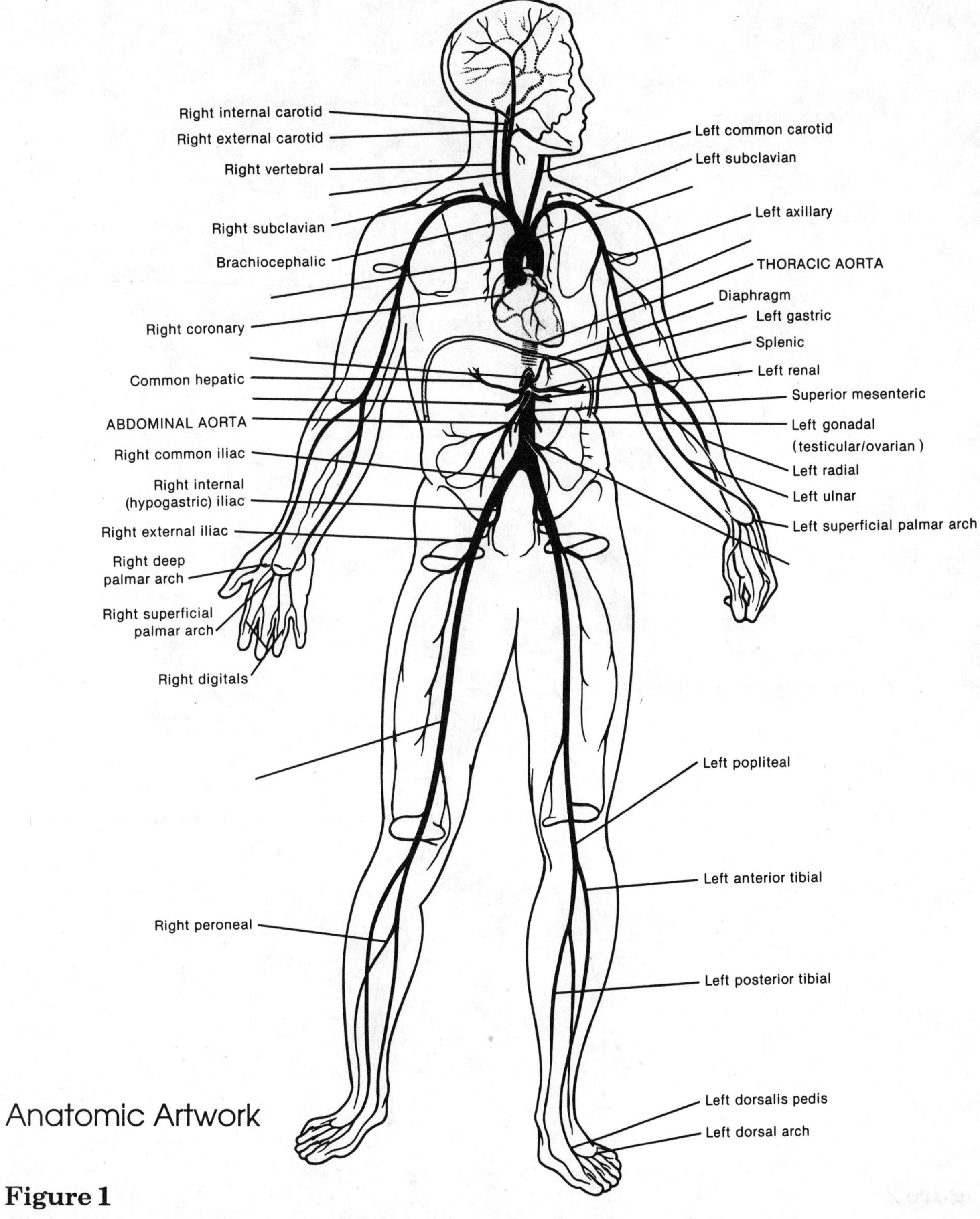

Anatomic Artwork

Figure 1

1. Label the vessel that is the first to carry oxygenated blood out of the heart.
2. Label the vessel that serves the quadriceps femoris muscle.
3. Label the vessel that serves the brachial area.
4. Label the vessel that serves the stomach, spleen, and liver.
5. Label the vessel where pulse is often monitored in CPR.

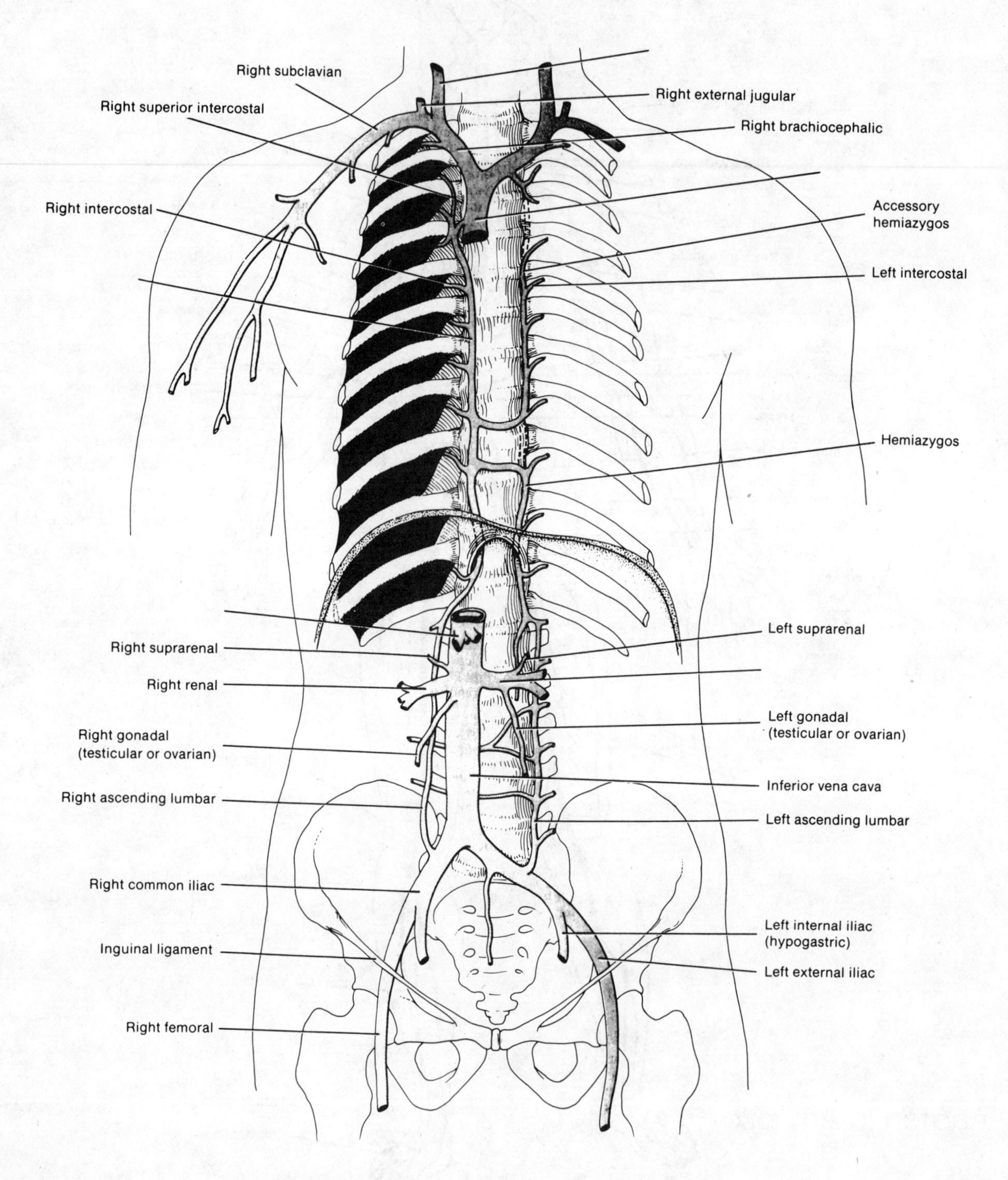

Figure 2

1. Label the vessel that drains the liver.
2. Label the vessels that drain into the right atrium.
3. Label the vessel that drains the neck area.
4. Label the vessel that drains the intercostal area.
5. Label the vessel that drains the kidney.

Chapter 17

The Lymphatic System and Immunity

Introduction

As blood flows through the capillary beds, the hydrostatic pressure forces some of the fluid out into the tissue spaces. Some of this fluid is returned to blood by the osmotic "pull." The remaining fluid is *recaptured* by the *lymphatic vessels,* which start as blind-ended capillaries in the tissues of the body. Ultimately, this fluid, referred to as *lymph,* is returned to the venous portion of the cardiovascular system. Along the route of return, lymph is filtered at *nodes,* which are interspersed on the lymph vessels.

If a foreign antigen is present in lymph, it can be phagocytized at the node by resident *macrophages.* In addition, the antigen can trigger the production of specific *antibodies.* They are produced by B lymphocytes. Or the antigen can be attacked by another population of lymphocytes, the T cells. Both antibodies and lymphocytes can enter the general circulation in blood. They represent the *specific immune response.*

The body also has several *nonspecific* mechanisms by which it *resists* disease and tissue destruction. Many mechanical and chemical means exist, in addition to the capacity for *phagocytosis* of foreign and dead materials.

All these mechanisms should become clear to you after reading the chapter in the textbook. Then you should be able to tackle the following activities (without resistance!)

Multiple Choice

___ 1. Which of these is *not* part of the nonspecific response?
   a. intact skin
   b. antibodies
   c. macrophages
   d. perspiration and oil

___ 2. Interferon is part of the nonspecific chemical response.
   a. It is produced by body cells infected by a virus
   b. It binds to the surface of nearby uninfected cells
   c. It inhibits viral replication in nearby cells
   d. All of these

___ 3. A group of antimicrobial proteins that normally exist in blood is called complement. Which of these statements is *not* true about complement?
a. It attracts phagocytes to the site of inflammation
b. It can cause opsonization
c. It initiates cytolytic reaction
d. It engages in phagocytosis

___ 4. Which is/are true about the process of phagocytosis?
a. It is engaged in by macrophages and granulocytes
b. Phagocytic cells are attracted by chemicals released at the site of damage
c. It involves adherence, ingestion, and digestion
d. All of these

___ 5. Which of these is *not* a symptom of inflammation?
a. cool, clammy skin
b. pain
c. swelling
d. loss of function

___ 6. Which of these is/are a stage of inflammation?
a. phagocytosis (by neutrophils and monocytes)
b. vasodilation and an increase in permeability
c. fibrin clot and pus formation
d. all of these

___ 7. Which of these is/are associated with histamine?
a. produced by mast cells
b. causes heat and swelling
c. causes arteriolar dilation and increased capillary permeability
d. all are associated

___ 8. Which of these is/are phagocytic?
a. agranulocytes
b. fibroblasts
c. macrophages and granulocytes
d. all are phagocytic

___ 9. Opsonization involves:
a. increased permeability
b. phagocytosis
c. adherence to an antigenic cell
d. all of these

___ 10. What occurs in the process of diapedesis?
a. increase in permeability
b. cells are chemically drawn to a site
c. white cells squeeze out of capillaries
d. none of these

___ 11. Which of these is *not* considered antigenic?
a. proteins and nucleoproteins
b. cholesterol
c. large polysaccharides
d. a microbe or its toxin

___ 12. Which is a function of a macrophage?
a. It phagocytizes the antigen
b. It presents the antigen to T cells and B cells
c. It secretes lymphokines
d. All of these

___ 13. Which of these is/are true about the cytotoxic T cell?
a. It makes holes in plasma membranes
b. It can increase phagocytosis by macrophages
c. It stimulates division of other nonsensitized lymphocytes
d. All are true

___ 14. Natural killer (NK) cells:
a. secrete antibodies
b. are highly specific
c. kill pathogens nonspecifically
d. all are true

___ 15. The cell that interacts with B cells, macrophages, and killer T cells is:
a. memory T cell
b. suppressor T cell
c. helper T cell
d. amplifier T cell

___ 16. The cells that leave lymphoid tissue and migrate to the area of the invading antigen are:
a. killer T cells
b. active B cells
c. memory B cells
d. none of these

___ 17. The lymphocyte that acts right in the lymph node is the:
a. T cell
b. B cell
c. wandering macrophage
d. none of these

___ 18. The origin of B and T cells in the embryo is:
a. bone marrow
b. thymus
c. gut associated lymphoid tissue
d. none of these

___ 19. How do antigens activate B cells?
a. bind to surface receptors
b. the T cells "present" them
c. the antigen differentiates
d. none of these

___ 20. How does a B cell response reach the invasion site?
a. Antibodies produced in lymphoid tissue travel to the site via blood and lymph
b. Macrophages carry B cells
c. Macrophages carry the antibodies
d. None of these

True/False

__________ 1. Plasma and interstitial fluid are *essentially the same.*

__________ 2. Excess tissue fluid is recaptured directly by *the thoracic duct.*

__________ 3. The process of neutrophils and monocytes squeezing through the capillary wall is called *margination.*

__________ 4. Nodes, which are a collection of lymphoid tissue, filter *blood.*

__________ 5. The spread of cancer may occur by way of the lymphatic system; the spread is called *metastasis.*

__________ 6. The largest mass of lymph tissue is the *spleen.*

__________ 7. The spleen *produces B cells.*

__________ 8. The spleen stores *blood and lymph.*

__________ 9. The thymus *increases* in size with age.

__________ 10. B cells begin life in bone marrow and mature in *the spleen.*

__________ 11. When antibodies attack antigens directly, the inactivated result is then *phagocytized* by macrophages.

__________ 12. When an antibody or substance adheres to a cell to make the cell more readily phagocytized, it is called *opsonization.*

__________ 13. *Antibody–antigen complexes* can activate blood proteins called "complement."

__________ 14. The lymphocyte that responds more forcefully and readily upon the second exposure to an antigen is the *"natural killer"* cell.

__________ 15. The skin acts as a physical and *immune barrier* to infection.

__________ 16. The AIDS virus *has not yet* been isolated.

__________ 17. A cell that appears to act as a reservoir for AIDS virus is the *B cell.*

__________ 18. The fetus receives *naturally acquired* immunity against a disease like rubella from its mother.

__________ 19. The natural killer cell is an example of a *specific response* to infection.

__________ 20. *Inflammation* is a nonspecific response, which confines and destroys the microbe and ends in pus formation and repair.

Completion

1. How do interstitial fluid (ISF) and lymph compare? __________
2. Plasma contains more __________ than lymph.
3. Lymphatic vessels drain any excess fluid that has escaped into the __________ .
4. From where does the fluid in the tissue spaces come? __________
5. In the reticular tissue in the lymph node, lymph fluid is filtered as it passes through the __________ of the node.
6. Foreign substances are __________ by macrophages in the lymph node.
7. The __________ are a group of lymph nodules that are arranged in a ring at the oropharyngeal junction.
8. The largest mass of lymph tissue in the body is the __________ .

9. The main collecting duct of the lymphatic system, draining the entire body below the ribs, is the ________________.
10. An excess accumulation of fluid in the tissue spaces is called ________________.
11. Ingestion and digestion of foreign matter by macrophages are called ________________.
12. A specific resistance to a disease is called ________________.
13. A foreign substance causing a specific response in the body is called a(n) ________________.
14. The body's own molecules are generally recognized as ________________.
15. The B cell's response to an antigen is to produce a(n) ________________.
16. The designation "Ig" in terms such as "IgG" and "IgA" refers to the word ________________.
17. Cell-mediated immunity is carried out by ________________.
18. Bacteria that infect us elicit an antibody response from ________________ cells.
19. The purpose of lymphoid tissue is to intercept a foreign agent before it enters the ________________.
20. ________________ process and present antigenic substances to the T and B cells.

Lost Sheep

1. redness, coolness, swelling, histamine
2. capillary, thoracic duct, lymph node, aorta
3. spleen, liver, thymus, tonsils
4. pus, fibrin clot, margination, immunity
5. red cell, margination, diapedesis, neutrophil
6. lymphocyte, neutrophil, B cell, T cell
7. plasma cell, T cell, B cell, antibody
8. skin, antibody, mucus, tears
9. properdin, histamine, interferon, complement
10. diapedesis, adherence, ingestion, digestion

Matching

Set 1

___ 1. produces antibodies
___ 2. leaves lymph node to destroy antigen
___ 3. presents antigen on its surface
___ 4. produces memory cells
___ 5. does not leave lymphoid tissue to confront antigen
___ 6. phagocytizes the antigen
___ 7. produces plasma cells
___ 8. part of the specific response
___ 9. produces "perforin" that lyses foreign cells
___ 10. produces IgG and IgA, for example

a. T cells
b. B cells
c. macrophage
d. a and b

Set 2

Use any items from the right-hand column that apply to the left-hand column

___	1. histamine causes this	a.	margination
___	2. traps the antigen	b.	diapedesis
___	3. WBCs line up in capillary	c.	chemotaxis
___	4. WBCs are attracted by chemicals to site of antigen	d.	vasodilation
		e.	fibrin clot in tissue
___	5. WBCs squeeze out of capillary	f.	increase in permeability
___	6. causes area to become warm		
___	7. causes redness of area		
___	8. causes swelling of area		
___	9. allows antibodies to flow out of blood		
___	10. allows phagocytes and defensive substances to flow out of blood		

Double Crosses

1. *Across:*
   ingester, digester

   Down:
   a. foreign organism
   b. covering on lymph node
   c. flow from node to bloodstream
   d. foreign molecule
   e. outward path

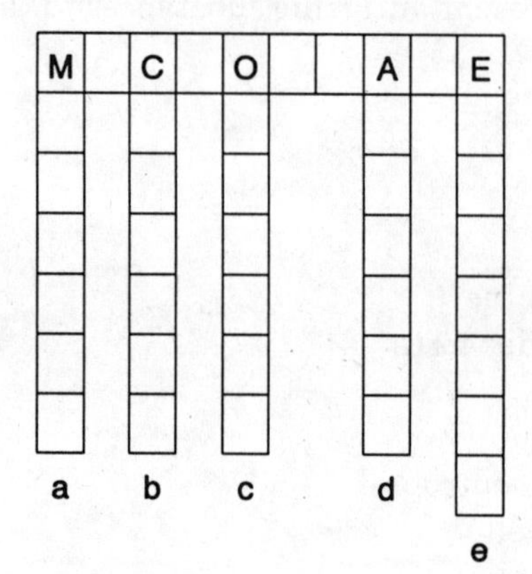

2. *Across:*
   immunity carried by the blood and body fluids

   Down:
   description of immune system in AIDS

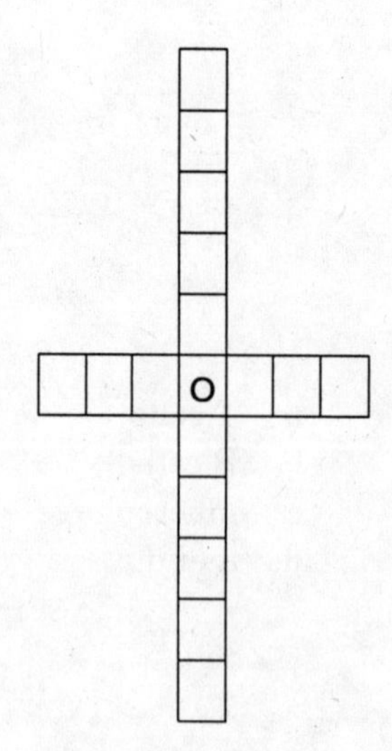

Sleuthing

1. Mrs. B had a radical mastectomy of her left breast last year. Since then she has had a problem with her left arm swelling.
   a. What is the cause of this swelling?

   b. Why was the procedure in "a" necessary?

Word Scrambles

1. a. lymph can "rescue" protein from tissue ________ PASCES ☐ ☐ _ _ ☐ _
   b. when a salt dissociates into electrolytes it ________ ZIONESI ☐ ☐ ☐ _ ☐ _ _
   c. "corpus" that houses resistance DYOB _ ☐ _ _

   Total: to make "tasty" for a phagocyte

   ☐☐☐☐☐☐☐☐

2. a. the study of cells TYGYCOLO ☐ ☐ ☐ ☐ _ ☐ _ _
   b. big eater CHAMPOERAG _ _ _ _ _ ☐ ☐ ☐ ☐ _
   c. area in lymph node NIUSS ☐ ☐ _ _ ☐

   Total: big mac's business

   ☐☐☐☐☐☐☐☐☐☐☐☐

Addagrams

1. a. what the spleen does to a quantity of blood 4, 17, 2, 12, 13, 14,
   b. proteins that ________ from capillaries are recaptured by lymphatic vessels 6, 16, 11, 18, 5, 21
   c. abbreviation for smallest vessels 20, 18, 5
   d. how many antibodies does a T cell make? 1, 2, 3, 21
   e. type of immunity induced by a vaccine 18, 12, 17, 8, 9, 10, 7, 15, 18, 0
   f. malignant growth 11, 18, 19, 20, 21, 12

   Total: a general fight

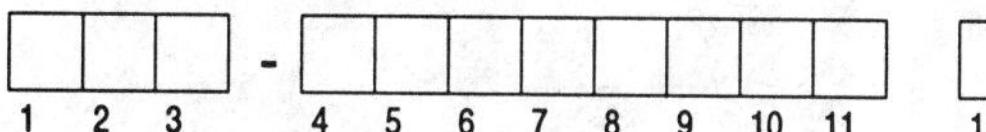

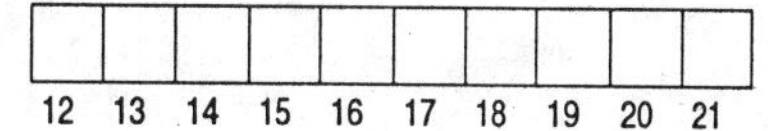

Crossword

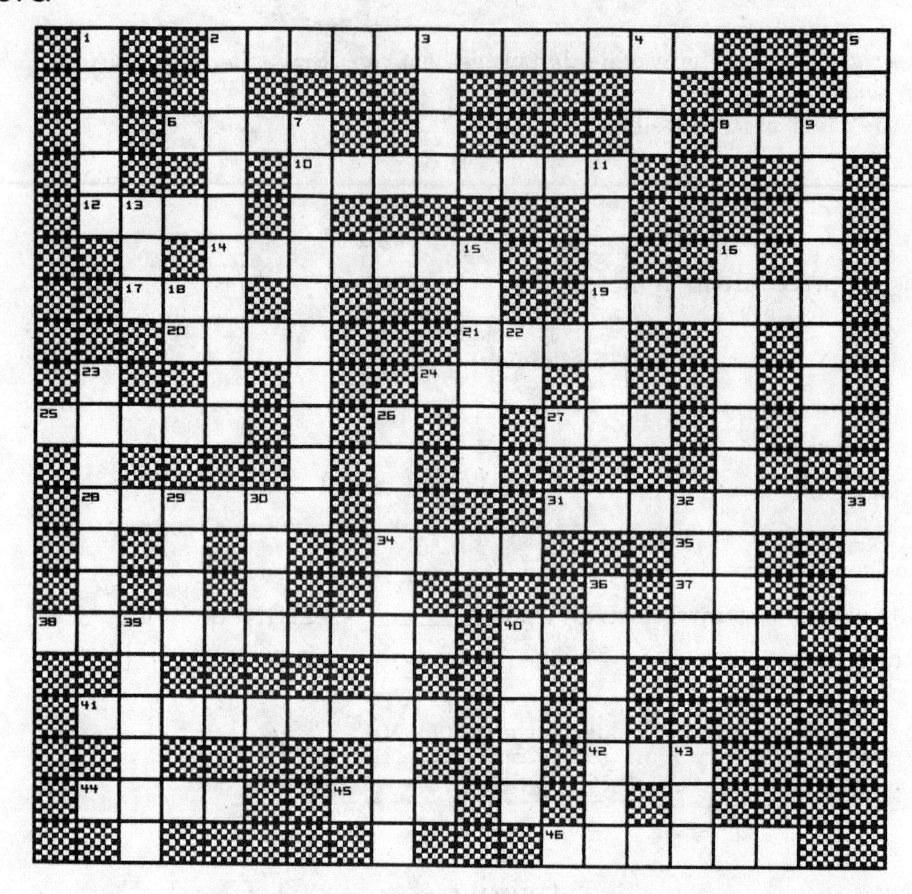

Across

2 Nonspecific response to tissue destruction
6 Type of cell that gives rise to mature blood cells
8 First barrier
10 Vessel entering lymph nodule
12 One of last steps in inflammation
14 What type of molecule is attacked in immune response
17 Even
19 Largest lymph organ
20 What lymph nodules do to antigens
21 ________ tasse
24 Accumulation of dead cells in fluid in body
25 What lymph system can do to tissue spaces
27 Fluid between cells (abbr.)
28 Lymph vessels drain excess fluid and protein from ________ spaces
31 ________ duct—drain
34 Barrier with Langerhans' cells
35 ________ and out
37 Southern East Coast state (abbr.)
38 Proteins that float and fight
40 Mucus ________ in respiratory tract
41 Lethal to cells
42 Immunoglobulin category
44 Thoracic ________—drain pipe
45 Initials of virus causing aids
46 B cells ready for another fight

Down

1 Type of microorganism
2 Antiviral chemical
3 Relative amount of protein in blood compared to lymph
4 Secreted onto skin as barrier
5 Inflammation is a ________ specific response
7 Cell that engages in phagocytosis
9 Results from lymphocyte action
11 Lymph ring in oral cavity
13 Anatomy animal
15 Diffuse lymph tissue
16 The fight against disease
18 Article
22 "It" to Jose
23 Organic compound that is collected from tissue spaces by lymph capillaries
26 The type of response that inflammation is
29 Half
30 "Do ________ others"
32 Skating ________
33 Type of activity for T cells (abbr.)
36 Type of immunity you get when you receive a gamma globulin shot
39 Gland through which T cells pass
40 Fluid that can carry humoral immunity
43 Time past

Chapter 18

The Respiratory System

Introduction

The respiratory system provides the surface for *gas exchange* between the lungs and circulatory system (*external respiration*) and also between the capillaries and cells (*internal respiration*). Through the mechanism of *breathing*, air is moved through the tracheobronchial tree to alveoli, where O_2 enters the body and CO_2 leaves by *diffusion*. The oxygen, transported by the blood, is dropped off at the cells for use in cellular respiration, generally oxidation of carbohydrates and fats. This utilization of oxygen produces CO_2, which in turn is carried by the blood to the lungs and is released. This "simple" transport of CO_2 serves an additional purpose: by releasing the volatile acid H_2CO_3 as CO_2 and H_2O, the respiratory system helps maintain *acid–base balance* in the body. Additionally, metabolic functions, such as *surfactant* production and activation of angiotensin I into angiotensin II, are also performed by the lungs.

After you have read the chapter on respiration in the textbook, you should be able to breathe easy and whip through these exercises dealing with respiration.

Multiple Choice

___ 1. Most of the carbon dioxide is carried as ________ in the blood.
- a. carboxyhemoglobin
- b. carbaminohemoglobin
- c. bicarbonate
- d. dissolved carbon dioxide

___ 2. Which factor(s) encourages hemoglobin to release its oxygen?
- a. a low partial pressure of oxygen
- b. an increase in temperature
- c. a decrease in pH
- d. all of these

___ 3. Which arterial blood gas level is correct?
- a. O2 - 105 mm Hg; CO2 - 46 mm Hg
- b. O2 - 105 mm Hg; CO2 - 40 mm Hg
- c. O2 - 40 mm Hg; CO2 - 46 mm Hg
- d. none of these

___ 4. When the carbon dioxide produced in the tissues combines with the water in the plasma and RBC what occurs next?
a. first carbonic acid is formed
b. H^+ and HCO_3^- are formed
c. the pH drops
d. all of these

___ 5. Which is/are descriptive of the larynx?
a. It connects the pharynx and trachea
b. It consists of nine pieces of cartilage
c. It contains the true and false vocal cords
d. All of these

___ 6. Which is true regarding the conchae?
a. They are all "shelves" projecting off the ethmoid bone
b. They are lined with mucous membrane
c. The epithelium lining them is squamous
d. All of these

___ 7. Which disappears in the bronchioles?
a. smooth muscle
b. cartilage
c. elastin
d. all of these

___ 8. The epiglottis:
a. covers the esophagus during breathing
b. is an elastic cartilage over the top of the larynx
c. is controlled voluntarily
d. all of these

___ 9. Which chemical appears to be most effective in controlling respiration?
a. oxygen
b. carbonic acid
c. carbon dioxide
d. none of these

___ 10. Peripheral chemoreceptors are located in the:
a. aorta
b. carotid arteries
c. both a and b
d. neither a nor b

___ 11. Chemoreceptors can respond to changes in:
a. oxygen levels
b. carbon dioxide
c. both a and b
d. neither a nor b

___ 12. Which is/are true concerning the stretch reflex in the lung?
a. Receptors for stretch are located in the bronchi and bronchioles
b. Impulses are carried by the vagus nerve
c. The effect of stimulation of the receptors is to inhibit inspiration
d. All of these

___ 13. When blood CO_2 levels rise:
a. they can affect the medulla
b. they can affect chemoreceptors in the aortic arch
c. both a and b
d. neither a nor b

___ 14. Which can close off the trachea?
a. crushing of the C-rings
b. aspiration of a foreign object
c. inflammation of the mucous membrane
d. all of these

___ 15. The wall of the trachea is characterized by:
a. C-shaped cartilages
b. mucociliary escalator
c. smooth muscle
d. all of these

___ 16. Which of these is/are a protective mechanism against foreign particles entering the tracheobronchial tree and alveoli?
a. macrophages
b. sneezing and coughing
c. mucociliary escalator
d. all of these

___ 17. Which increase(s) surface area and whirls the air around in the nasal cavity?
a. adenoids
b. conchae
c. nasal septum
d. nasopharynx

___ 18. Which is *not* part of the upper respiratory tract?
a. nose
b. throat
c. trachea
d. adenoids

___ 19. Which structure is part of the lower respiratory tract?
a. trachea
b. bronchi
c. lungs
d. all of these

___ 20. Which muscle(s) accounts for moving about 75–80 percent of the tidal volume?
a. external intercostals
b. internal intercostals
c. diaphragm
d. all of these

___ 21. In inspiration there is:
a. an increase in chest pressure
b. an increase in intrapulmonic pressure (in the airways)
c. a descent of the diaphragm
d. a relaxation of the external intercostal muscles

___ 22. Which area of the brain controls the basic cycling of inspiration and expiration?
a. pons
b. medulla
c. apneustic
d. hypothalamus

___ 23. The air that is left in the lung at the end of a normal expiration is the:
a. expiratory reserve volume
b. residual volume
c. functional residual capacity
d. none of these

___ 24. Surfactant:
a. reduces surface tension in the alveoli
b. is produced by cells of the alveoli
c. increases the compliance ("give") of the lung
d. all of these

___ 25. Which is/are characteristic of the lung?
a. There is a cardiac notch in the left lung
b. The right lung is shorter than the left
c. They extend from the diaphragm below to above the clavicle
d. All of these

True/False

__________ 1. Normal breathing is called *eupnea.*

__________ 2. DPG is a substance in red blood cells which causes oxygen to bind to hemoglobin more *tightly.*

__________ 3. In *emphysema* the alveolar walls lose their elasticity, and the resistance of the lung increases as thick fibrous tissue replaces the alveoli.

__________ 4. *Pneumonia* refers to an inflammation of the lung where fluid fills up the alveoli.

__________ 5. About *75 percent* of the oxygen in arterial blood is carried on the hemoglobin molecule.

__________ 6. There are *two* groups of neurons in the basic rhythm control centers in the medulla of the brain.

__________ 7. The most important factor determining how much oxygen is carried on the hemoglobin molecule is the *partial pressure* of oxygen.

__________ 8. Carbon monoxide is *1/200* as tenacious as oxygen in its binding to hemoglobin.

__________ 9. The diffusion of oxygen from *arterial blood to the tissues* involves a movement from a partial pressure of 105 mm Hg to 40 mm Hg.

__________ 10. Alveolar Po2 must be *higher* than that in the blood in order for oxygen to diffuse into the blood.

__________ 11. The vocal cords are attached to the *arytenoid* cartilages in the larynx.

__________ 12. The receptors for the sense of smell are located *on the inferior* nasal conchae.

__________ 13. The auditory (eustachian) tubes open into the *nasopharynx.*

__________ 14. Sound is made in the larynx *only* by the true vocal cords.

__________ 15. The *parietal* pleura lines the inner chest wall.

__________ 16. The trachea has cartilage *rings* surrounding the tubular structure.

___________ 17. The lungs have certain metabolic functions such as the *activation* of angiotensin I to angiotensin II.

___________ 18. The highest level of PCO_2 is found in the *atmospheric air.*

___________ 19. In RBCs the *lower* the level of DPG the more oxygen is released.

___________ 20. The pneumotaxic area in the pons *inhibits* inspiration.

___________ 21. The basic rhythm of respiration is controlled in the *medulla.*

___________ 22. Surfactant *decreases* the surface tension in the alveolus.

___________ 23. Carbon dioxide chemoreceptors are *more* sensitive than oxygen chemoreceptors.

___________ 24. Input from the *pneumotaxic* center on the medulla causes an inhibition of inspiration.

___________ 25. An increase in body temperature, such as during a fever, *slows up* inspiration.

Lost Sheep

1. alveolar sacs, macrophages, septal cells, bronchioles
2. hyoid, arytenoid, cricoid, thyroid
3. alveolar walls, alveolar duct, basement membrane, endothelial cells of capillary
4. tidal volume, vital capacity, expiratory reserve volume, inspiratory reserve volume
5. oxygen, carbon dioxide, carbonic acid, water vapor
6. carbaminohemoglobin, bicarbonate, dissolved CO_2, carboxyhemoglobin
7. lactic, carbonic, pyruvic, bicarbonate
8. apneustic center, baroreceptor, medulla, pneumotaxic center
9. conchae, apex, base, hilum
10. stretch receptors, chemoreceptors, cortical impulses, severe pain
11. pharynx, nares, septum, mucous membrane
12. parathyroid, cricoid, thyroid, epiglottis
13. vena cava, heart, trachea, diaphragm
14. hilus, root, trachea, apex
15. 10–20 breaths per minute, 500 ml tidal air, 150 ml anatomic dead space, expiration: lasts 1 second
16. surfactant, pleurisy, lowered surface tension, open alveoli
17. expiration, higher pressure in air compared to lungs, relaxation of intercostals, movement upward of diaphragm
18. $CO_2 + H_2O \leftrightarrow H_2CO_3 \leftrightarrow HCO_3^- + H^+$, formation of carboxyhemoglobin, CO_2 dissolves in plasma, carbaminohemoglobin formation
19. medulla, apneustic center, cerebellum, pneumotaxic center
20. subatmospheric intrapleural pressure, inhalation, pressure in lungs lower than 1 atmosphere, volume increases

Matching

Set 1

___ 1. drops to 758 mm Hg during inspiration
___ 2. is 760 mm Hg and a reference point
___ 3. is normally less than atmospheric pressure
___ 4. rises back up to 756 mm Hg by end expiration
___ 5. rises above 760 mm Hg during a cough

a. intrapleural pressure
b. intrapulmonic (airway) pressure
c. atmospheric pressure
d. both a and b

___ 6. is affected by diaphragmatic movement
___ 7. pressure decreases as volume increases at start of inspiration
___ 8. at start of inspiration intrapulmonic pressure is less than this
___ 9. is normally positive (greater than atmospheric) during expiration
___ 10. varies from 756 to 754 mm Hg from beginning inspiration to end of normal inspiration

Set 2

___ 1. air needed in lungs to allow them to float
___ 2. total volume of air that can be exhaled in a maximal ventilatory effort
___ 3. volume moved in eupnea
___ 4. volume that can be breathed in after normal inspiration
___ 5. volume left in lungs at end of normal expiration

a. tidal volume
b. inspiratory capacity
c. expiratory reserve
d. inspiratory reserve
e. vital capacity
f. minimal air
g. functional residual capacity

Set 3

___ 1. simple squamous epithelium
___ 2. level at which cartilage disappears
___ 3. arytenoid cartilage
___ 4. thinnest wall
___ 5. C-shaped cartilage

a. nasal cavity
b. larynx
c. trachea
d. bronchioles
e. alveoli

Double Crosses

1. *Across*:
center that stimulates inspiration when pneumotaxic center is inactive

Down:
membrane that lines lung

U

2. *Across*:
lowest cartilage in larynx

Down:
sense of smell

C

3. *Across:*
substance that lowers surface tension

Down:
a. forceful exhalation
b. sensor
c. microscopic ending
d. referring to C-ringed tube
e. disease characterized by fluid accumulation

S R A T N

a b c d e

Sleuthing

1. Mrs. Farrell has delivered a baby girl prematurely, at the end of 23 weeks of gestation. The baby weighed 2.2 pounds and exhibited extreme difficulty in inflating her lungs. She was immediately put on a respirator and was diagnosed as having RDS.
   a. What is another name for this condition?

   b. What is the deficiency?

   c. What cells are responsible for the disorder?

   d. What is the function of the deficient substance?

   e. About how long will the baby have to be on the respirator?

Word Scrambles

1. a. breathe in — HENAIL — ☐ ☐ _ ☐ ☐ ☐
   b. amount of air — MULEOV — ☐ _ _ _ _ ☐
   c. air hunger — NYPDSEA — _ ☐ _ ☐ ☐ _ ☐
   d. three-lobed lung — TIRGH — ☐ ☐ _ _ ☐
   e. lymphatic tissue at back of mouth — STILNO — ☐ ☐ _ _ _ ☐

Total: heavy breathing

☐☐☐☐☐☐☐☐☐☐☐☐☐☐☐☐

2. a. oxygen enters cells from here PRILLACAY _ ☐ ☐ ☐ _ ☐ ☐ ☐ _
   b. emergency in the nursery DISS ☐ _ _ ☐
   c. consisting of only oxygen ERUP ☐ ☐ ☐ _
   d. must be done to breathe deeply TERXE ☐ _ ☐ ☐ ☐

Total: what oxygen exerts all by itself in air

☐☐☐☐☐☐☐ ☐☐☐☐☐☐☐☐

Addagrams

1. a. hyaline membrane disease 1, 5, 3
   b. kind of control describing chemical and neural 5, 6, 7, 8
   c. second route of entry into respiratory tract, besides nasal 10,1, 7, 8
   d. compare how much oxygen there is in arterial blood vs. venous blood 13,10,1, 2
   e. breathing movements indicate you are ________ 7, 11, 4, 9, 14
   f. letter for uracil in genetic code 6
   g. letter opposite "A" in making RNA 12

Total: all the air at the end of breathing out normally

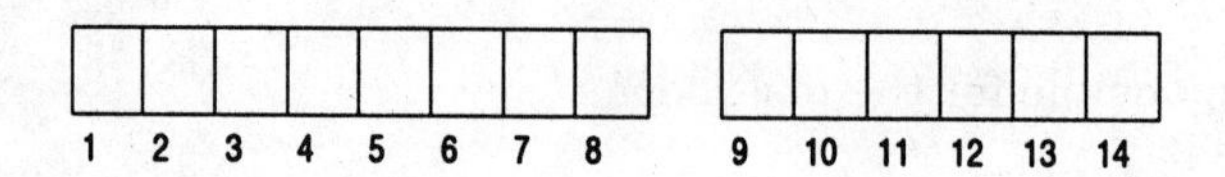

2. a. symbols for water's elements 1, 11
   b. what you breathe 4, 10, 7
   c. insertion of a tube in airway 10, 12, 9, 6, 2, 8, 5, 10, 11, 12
   d. suffix meaning "inflammation" 10, 5, 10, 3

Total: assessing the load (two terms)

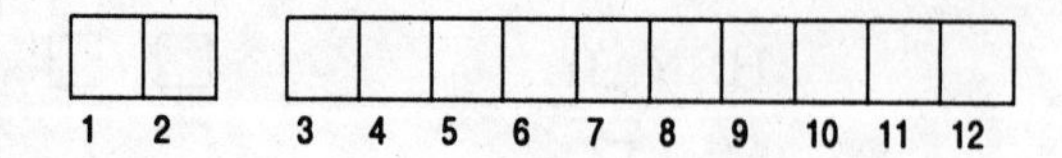

Crossword

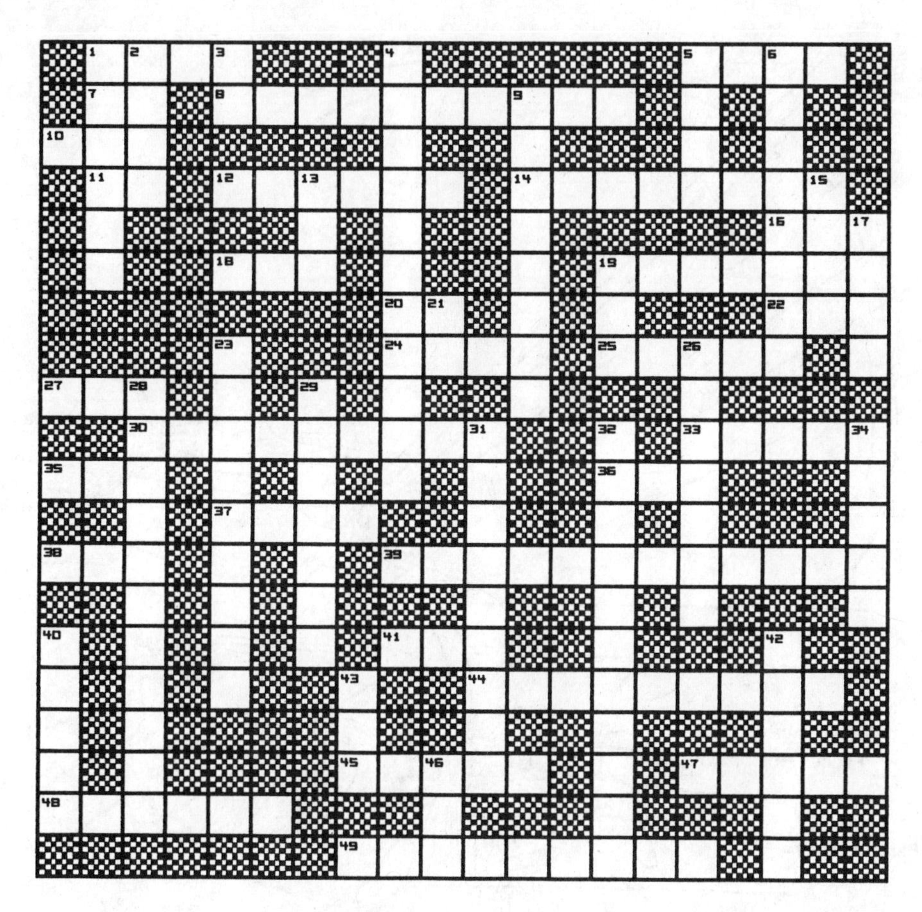

Across

1 End of tracheobronchial tree—alveolar ________
5 Entrance to respiratory tract
7 North Pole (abbr.)
8 Reduces surface tension
10 Envision
11 Prefix meaning "past"
12 Membranous lining around lung
14 Refers to "organ"
16 "Me" to Pierre
18 Air you can breathe in after a tidal inspiration (abbr.)
19 Large cartilage in front of larynx
20 Abbreviation for oxygen carrier
22 Utilize
24 Civil liberties group (abbr.)
25 3 on right, 2 on left
27 What you breathe
30 Disease where elastin is destroyed in lung
33 Absence of breathing
35 Respiratory Distress Syndrome (abbr.)
36 Prefix meaning "not present"
37 White flower
38 Prefix for "on top" or "outer"
39 Within chest cavity
41 Normal breathing rate—12 breaths per ________ (abbr.)
44 Sense of smell
45 True and false vocal ________
47 House for bronchioles
48 Whole reason for respiratory system
49 Disease characterized by fluid accumulation in alveoli

Down

1 Sudden, forceful expulsion of air through nose
2 Tip of lung
3 Social Security (abbr.)
4 Housekeepers down in alveoli
5 Where external nares are
6 Shape of cells in alveolar wall
9 Microscopic sac at end of "tree"
13 All the air you can breathe out after a tidal breath (abbr.)
15 Woman's name
17 Thought
19 T-cell (abbr.)
21 Time before A.D.
23 What picks up oxygen from alveolus
26 How the mucus layer in the trachea is described
28 System to expel carbon dioxide
29 ________ membrane disease
31 Cartilage that holds vocal cords
32 "Breathe in"
34 Adam's ________
40 Type of receptor
42 Cartilaginous structures around the trachea
43 All the air in the lung after a tidal exhalation (abbr.)
46 Dark bread

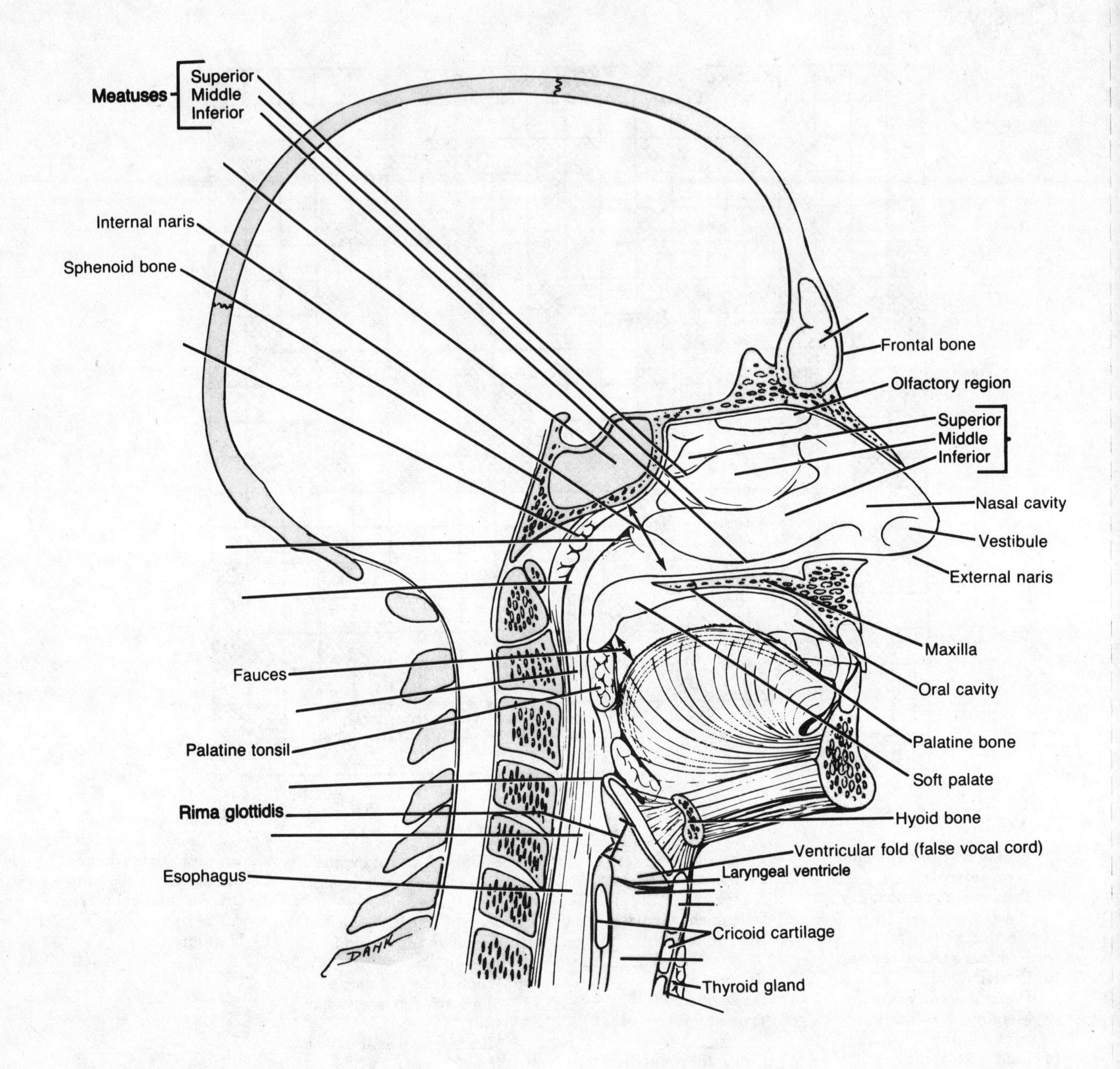

Anatomic Artwork

1. Label the structures that increase the surface area of the nasal cavity.
2. Label the structure that closes off the larynx.
3. Label the structure involved in initially producing sound.
4. Label the structure that divides into the tubes that enter the lung.
5. Label the structures that maintain a patent airway between the larynx and bronchi.
6. Label the area that, when swollen, causes a child to be a mouth breather.
7. Shade in and label that portion of the respiratory tract which is common to the digestive tract.
8. Label the area where the Eustachian tubes open and the specific opening.
9. Label the paranasal sinuses visible in this section.
10. Label the cartilage known as the Adam's apple.

Chapter 19

The Digestive System

Introduction

As good as it tastes, food has yet another function! In *ingesting, digesting,* and *absorbing* the breakdown products of food, the body is given the wherewithal to build and repair tissues. It is also given the energy to drive the myriad reactions that constantly occur. Every reaction of the various metabolic pathways is catalyzed by a protein *enzyme* with a specific pH range.

Enzyme release, specific to the digestive process, is carefully controlled by *neural, hormonal,* and *direct* means. An advantage of the neural control is that the stomach, for example, can prepare for the arrival of food by stimulating the secretion of digestive juices in advance of the food's arrival. Hormonal and direct control help to ensure that secretion of digestive juices continues, particularly while food is present in the organ. In addition, impulses traveling along neural pathways stimulate the movement of food through the lumen of the digestive tube.

In this unit, the overall *mechanical* and *chemical digestion* of food as it travels through the digestive system is covered. The subsequent absorption of the end products of this digestion is also included, along with the anatomy of the system. After completing the reading in your text, you should be able to answer the following questions and complete the activities included.

Multiple Choice

___ 1. Which of these two glands secretes enzymes with the same activity?
a. parotid gland and gallbladder
b. submandibular gland and liver
c. sublingual gland and pancreas
d. all of these

___ 2. Haustra are:
a. projections in the small intestine
b. folds in the stomach wall
c. sphincters
d. sections in the colon

___ 3. Deciduous teeth:
a. generally begin to appear by six months of age
b. are temporary teeth
c. both a and b
d. neither a nor b

___ 4. Mechanical digestion includes:
a. deglutition
b. mastication
c. peristalsis
d. all of these

___ 5. Rugae are characteristic of the:
a. pancreas
b. pharynx
c. stomach
d. colon

___ 6. The rounded mass of food that enters the stomach is called a:
a. chyme
b. bolus
c. cementum
d. none of these

___ 7. Which of these statements is true concerning bile?
a. It is produced in the gallbladder
b. It digests fats into glycerol and fatty acids
c. It helps solubilize fats and increase surface area for enzyme attack
d. All of these

___ 8. The ileocecal sphincter guards the junction of the:
a. esophagus and stomach
b. stomach and duodenum
c. small intestine and large intestine
d. large intestine and rectum

___ 9. Secretions below the stomach are usually:
a. highly acidic
b. strongly basic
c. weakly basic
d. mildly acidic

___ 10. Which is correctly paired for the stomach?
a. enteroendocrine cells–gastrin
b. parietal cells–B intrinsic factor
c. both a and b
d. neither a nor b

___ 11. Enzymes that break down lactose, sucrose, and maltose are secreted by the:
a. pancreas
b. liver
c. small intestine
d. stomach

___ 12. Pepsinogen is activated by:
a. mucus
b. gastrin
c. secretin
d. HCl

___ 13. Proteolytic enzymes digest proteins into:
a. fatty acids
b. disaccharides
c. amino acids
d. all of these

___ 14. When the mucus barrier is penetrated by acid in the stomach, which of the following results?
a. a hernia
b. an ulcer
c. diverticulitis
d. none of these

___ 15. The tissue layer that contains the columnar epithelium which absorbs nutrients is the:
a. mucosa
b. submucosa
c. muscularis
d. serosa

___ 16. The peritoneum extends onto the viscera as the:
a. mucosa
b. serosa
c. falciform ligament
d. none of these

___ 17. The principal pigment in bile is:
a. hemoglobin
b. bilirubin
c. jaundice
d. melanin

___ 18. Segmentation occurs in the:
a. colon
b. jejunum
c. stomach
d. esophagus

___ 19. Projections of the small intestine which increase the surface area for absorption include:
a. microvilli
b. villi
c. circular folds
d. all of these

___ 20. Contraction and relaxation of the smooth muscle of the muscularis produce:
a. peristalsis
b. churning
c. mixing
d. all of these

___ 21. Which of the following statements is true?
a. Most water absorption takes place in the rectum
b. Bacteria in the large intestine aid in the synthesis of several vitamins
c. Movements of the colon mix chyme with digestive juices and aid the absorption of food
d. All statements are true

___ 22. Which of these is a function of the liver?
a. phagocytosis of bacteria and old red blood cells
b. metabolism of nutrients
c. production of bile
d. all of these

___ 23. Which substances are produced in the exocrine portions of the pancreas?
a. bicarbonate and lipase
b. secretin and gastrin
c. insulin and glucagon
d. all of these

___ 24. Secretion of the stomach's juices is controlled by:
a. hormonal regulation
b. neural regulation
c. both a and b
d. neither a nor b

___ 25. What does the small intestine release to control digestive activity in the stomach?
a. gastrone
b. enterocrinin
c. gastrone
d. prostaglandin

True/False

__________ 1. *Accessory* structures of the digestive system include teeth, tongue, salivary glands, liver, gallbladder, and pancreas.

__________ 2. After digestion, triglycerides travel in *chylomicrons* in blood.

__________ 3. The openings from esophagus to stomach, stomach to small intestine, and small intestine to large intestine are regulated by *sphincters.*

__________ 4. Mechanical digestion *facilitates* chemical digestion.

__________ 5. *Mucous* cells of the stomach secrete intrinsic factor.

__________ 6. The liver is divided into two main lobes, each of which in turn is divided into functional units called *lobules.*

__________ 7. Fingerlike projections on the surface of the cells that line the small intestine are called *fauces.*

__________ 8. The appendix is attached to the *cecum.*

__________ 9. The function of bile is to *hydrolyze* fats.

__________ 10. Taenia coli are found in the *stomach.*

__________ 11. Enlargement of the veins of the rectal columns is characteristic of *hemorrhoids.*

__________ 12. Contraction of the *muscularis* is responsible for the movement of food through the digestive tube.

__________ 13. *Cementum* is a bonelike substance that covers the dentin of the teeth.

__________ 14. The tunica serosa and *visceral peritoneum* are synonomous.

__________ 15. Salivary amylase begins the digestion of *carbohydrate* in the mouth.

__________ 16. *Fungiform* papillae are conical projections on the tongue which contain no taste buds.

__________ 17. Mumps is an inflammation of the *parotid* gland.

_____________ 18. *Cholecystokinin* is released by the duodenum and stimulates the contraction of the gallbladder to expel bile.

_____________ 19. Stimulation of the vagus nerve innervating the stomach *inhibits* secretion of gastric juices.

_____________ 20. Bile is produced in the *gallbladder.*

_____________ 21. Generally, secretions superior to the *pyloric* sphincter are acidic and those inferior are alkaline.

_____________ 22. *Secretin,* produced by the small intestine, stimulates release of enzyme-rich pancreatic juice.

_____________ 23. A combined activity of the esophagus, pharynx, and mouth is *deglutition.*

_____________ 24. *Rennin* and calcium act on milk to produce a curd in the baby's stomach.

_____________ 25. The bicarbonate released by the *pancreas* aids in neutralizing the acidity of the food as it enters the small intestine.

Completion

1. In the human, the appendix is attached to the _______________.
2. The entrance to the stomach from the esophagus is regulated by the _______________.
3. The type of epithelium lining the small intestine is _______________.
4. The _______________ are the exocrine glands of the pancreas which secrete digestive enzymes.
5. The secretions from the pancreas and liver enter the digestive tract in the _______________ of the small intestine.
6. The salivary secretions of the _______________, _______________, and _______________ glands are released into the buccal cavity.
7. Saclike outpouchings of the wall of the colon are called _______________.
8. In an infant's stomach, which is less acidic than the adult's, both _______________ and _______________ are active in digestion.
9. In general, after the initial neural phase of digestive control, the hormone _______________, secreted by the stomach, stimulates secretion of gastric glands.
10. Mixing of the contents of the small intestine is effected by _______________ contractions.
11. The gases produced from undigested food in the colon are the result of _______________.
12. The macerated food mixed with digestive juices, which leaves the stomach, is called _______________.
13. Carbohydrate digestion in the adult begins in the _______________.
14. Pancreatic juice contains enzymes that digest which food types? _______________
15. When triglycerides are reformed in the intestinal mucosa, they travel (in blood) coated with protein, cholesterol, and phospholipid as _______________.
16. The major absorption of digested foods occurs in the _______________.
17. Foods that enter the circulatory system immediately upon digestion and absorption travel to the _______________ by way of the _______________ vein.
18. The union of the hepatic and cystic ducts is called the _______________.
19. Low density lipoproteins transport _______________ to tissues for use in hormone synthesis and manufacture of cell membranes.

20. Intrinsic factor is produced by the ________________.
21. Ulcers formed by exposure to acidic gastric juice are called ________________ ulcers.
22. A gastric enzyme that curdles milk in the infant's stomach is called ________________.
23. Parasympathetic innervation to the stomach and gallbladder is via the ________________ nerve.
24. Hepatitis type ________________ can be spread through contaminated food.
25. The wave of muscular contraction that passes along the gastrointestinal tract is called ________________.

Lost Sheep

1. glucose, maltose, sucrose, lactose
2. amino acid, fatty acid, amylase, glucose
3. gallbladder, hepatic portal vein, bile duct, appendix
4. micelles, dipeptide, chylomicrons, fats
5. duodenum, jejunum, ileum, rugae
6. amylase, lipase, gastrin, trypsin
7. parietal, parotid, sublingual, submaxillary
8. gastric, haustra, pylorus, parietal cells
9. HCl, pepsin, amylase, B intrinsic factor
10. ileocecal, mesenteric, pyloric, cardiac
11. sucrose, fructose, disaccharides, proteins
12. liver, cecum, parotid, submandibular
13. ileum, villi, antrum, jejunum
14. amylase, mucus, lysozyme, pepsin
15. vagal stimulation, secretin, gastrin, food in stomach
16. mastication, salivation, taste, acid secretion
17. cystic duct, islets of Langerhans, beta cells, trypsin
18. rugae, Kupffer's cells, lobules, central vein
19. production of bile, metabolism of nutrients, phagocytosis, churning
20. haustra, taenia coli, circular folds, colon

MATCHING

Set 1

___	1. ingestion	a. large intestine
___	2. deglutition	b. liver activity
___	3. emulsification	c. food intake
___	4. segmentation	d. occurs in small intestine
___	5. mass peristalsis	e. swallowing

Set 2

Indicate all the answers that apply

___	1. stimulates secretion of gastric juice	a.	cholecystokinin
___	2. digests carbohydrates	b.	chymotrypsin
___	3. enzyme secreted by intestinal glands	c.	lactase
___	4. secreted by pancreas	d.	amylase
___	5. stimulates ejection of bile from gallbladder	e.	gastrin

Set 3

___	1. characterized by villi	a.	large intestine
___	2. produces bile	b.	mouth
___	3. parietal cells secrete HCl	c.	liver
___	4. secrete sucrase	d.	stomach
___	5. first digestion of starch	e.	gallbladder
___	6. characterized by haustra	f.	pancreas
___	7. has rugae	g.	esophagus
___	8. stores bile	h.	small intestine
___	9. has bacteria that synthesize certain vitamins		
___	10. absorbs fat-soluble vitamins		
___	11. most absorption occurs here		
___	12. secretes bicarbonate ions		
___	13. gluconeogenesis occurs here		
___	14. secretes pepsinogen		
___	15. completion of most digestion		

Double Crosses

1. a. wave of muscular contraction
   b. a ruptured appendix can cause this
   c. chain of amino acids

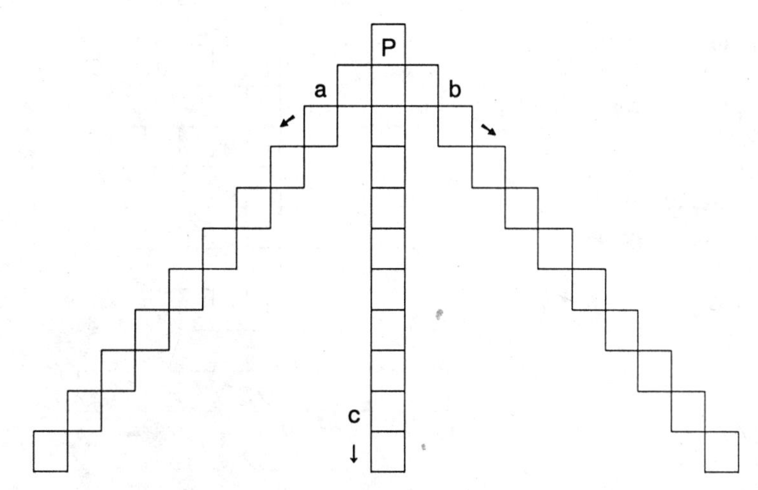

2. a. to chew
   b. type of papilla
   c. ________ appendix
   d. proteolytic enzyme

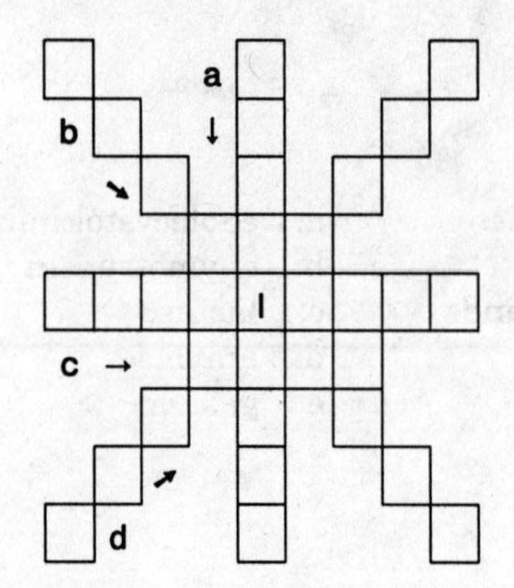

3. a. ileocecal and pyloric
   b. general term for carbohydrate unit
   c. gland beneath the tongue

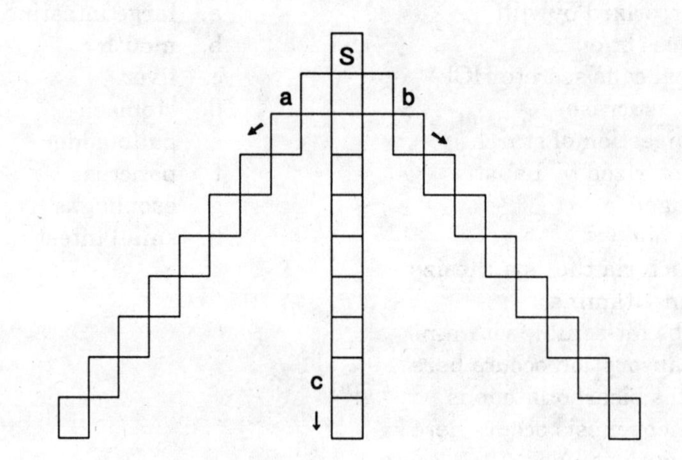

4. *Across*:
   a. "bulk" of food
   b. orifice regulator
   c. middle of small intestine
   d. what a vein has that an artery doesn't
   e. bile pigment
   f. control of chewing
   g. location of taenia coli
   h. fat transporter

   Down:
   fold of mucosa under the tongue (2 words)

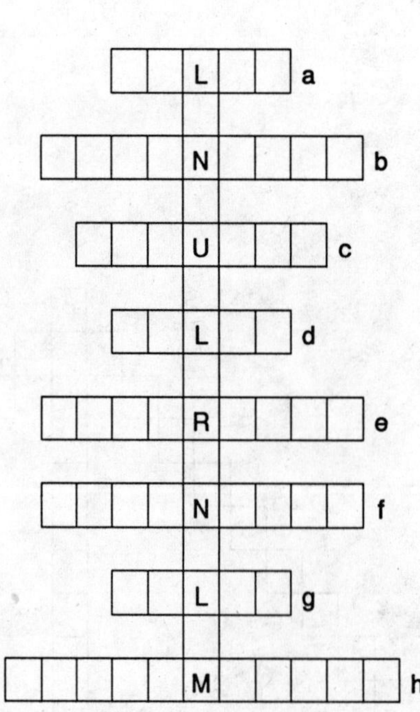

Sleuthing

1. Ming Toy is a 28-year-old Chinese woman who has recently arrived in the United States from Hong Kong. Since her arrival here, her diet has changed and one of the main differences is that she has been drinking large quantities of milk. She repeatedly experienced a problem of cramping pains, flatulence, and diarrhea. The doctor diagnosed it as a lactose intolerance.
   a. What does the term "intolerance" mean in this case? Explain.

   b. What is lactose and what would normally happen if there were no intolerance?

   c. What is the route of absorption for the end products of lactose digestion?

   d. What is the cause of the flatulence?

Word Scrambles

1. a. superior to stomach — SAPHEGOSU — ☐ ☐ _ _ _ ☐ _ _ ☐
   b. double-duty passage — PRNAYXH — ☐ _ _ ☐ _ _ _
   c. colon's final section — IMIOSGD — ☐ ☐ _ _ _ ☐ _
   d. nutrient laden vessel — RLAOTP — _ _ _ ☐ _ ☐

Total: function of muscularis

☐☐☐☐☐☐☐☐☐☐☐

2. a. food form entering stomach — LOBSU — ☐ _ ☐ _ _
   b. folds in small intestine — ILVLI — _ ☐ _ _ ☐
   c. neutralizes HCl in — DOSMIU — _ _ _ ☐ ☐ _
   d. small intestine (2 words) — IABNBROTCAE — ☐ ☐ _ _ ☐ _ _ ☐ _ _ _

Total: pigment in bile

☐☐☐☐☐☐☐☐☐

3. a. large intestine portion — OLCNO — □ □ _ □ _
   b. stores a liver product (2 words) — LGAL RDLDABE — _ _ □ _ / □ _ _ _ □ _ _
   c. digestion catalyst — ZMEYNE — □ □ _ _ □ _
   d. protects stomach lining — USMCU — □ □ □ _ _
   e. lipid category — TAFS — _ _ □ _
   f. secretes pancreatic juice — CIANI — _ _ □ _ _

Total: hepatic and cystic union

Addagrams

1. a. first phase in control of gastric juice secretion — 6, 21, 20, 1, 13, 8, 11, 12
   b. area of stomach near junction with small intestine (2 words) — 22, 2, 8, 9, 10, 15, 14 4, 28, 27, 24, 26, 25
   c. saccharides — 12, 17, 10, 0, 5, 7, 2, 3, 4,13, 0, 28, 23
   d. anterior palate — 1, 17, 10, 16
   e. secreting structure — 27, 8, 17, 18, 19
   f. bonelike substance beneath tooth enamel — 19, 21, 29, 0, 11, 18

Total: stomach's partners

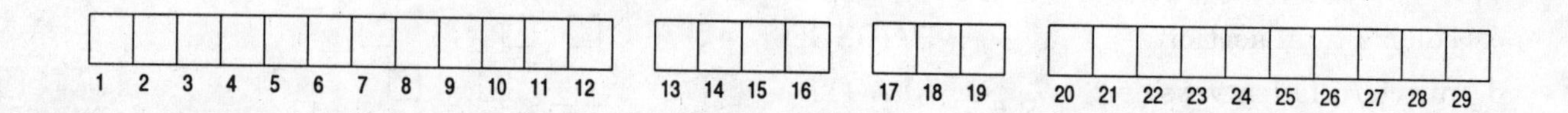

1 2 3 4 5 6 7 8 9 10 11 12 | 13 14 15 16 | 17 18 19 | 20 21 22 23 24 25 26 27 28 29

2. a. binge and purge — 8, 7, 9, 16, 15, 16, 14
   b. side of the body where the descending colon is located — 6, 10, 1, 3
   c. secretion of parotid gland — 4, 2, 9, 16, 11, 14
   d. small and large ________ — 12, 17, 13, 10, 18, 3, 16, 17, 10, 4
   e. acids that link to form protein — 2, 15, 16, 17, 5

Total: A, D, C, K

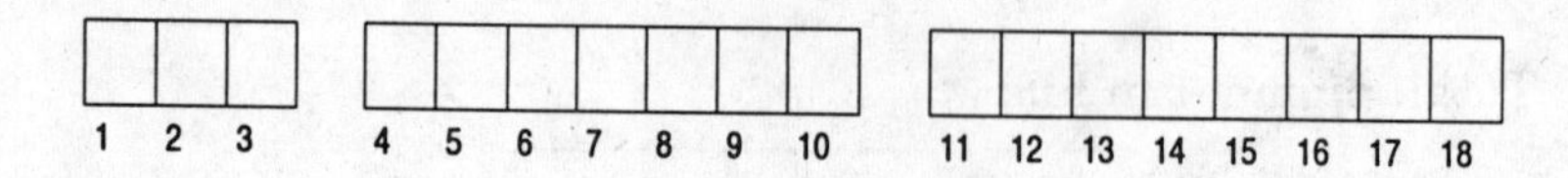

1 2 3 | 4 5 6 7 8 9 10 | 11 12 13 14 15 16 17 18

3. a. nerve plexus in submucosa — 4, 2, 10, 1, 20, 6, 5, 18, 26
   b. refers to cavity lined by peritoneum — 16, 17, 18, 19, 21, 11, 12, 5, 8, 23
   c. hormone which promotes secretion of gastric juice — 3, 13, 24, 9, 18, 25, 14
   d. "active" ingredients in bile: bile ________ — 24, 22, 23, 7, 1
   e. mechanical and chemical processes that break down food — 15, 10, 3, 2, 20, 21, 10, 11, 12

Total: mechanical actions of small intestine

1	2	3	4	5	6	7	8	9	10	11	12	13	14	15	16	17	18	19	20	21	22	23	24	25	26

Crossword

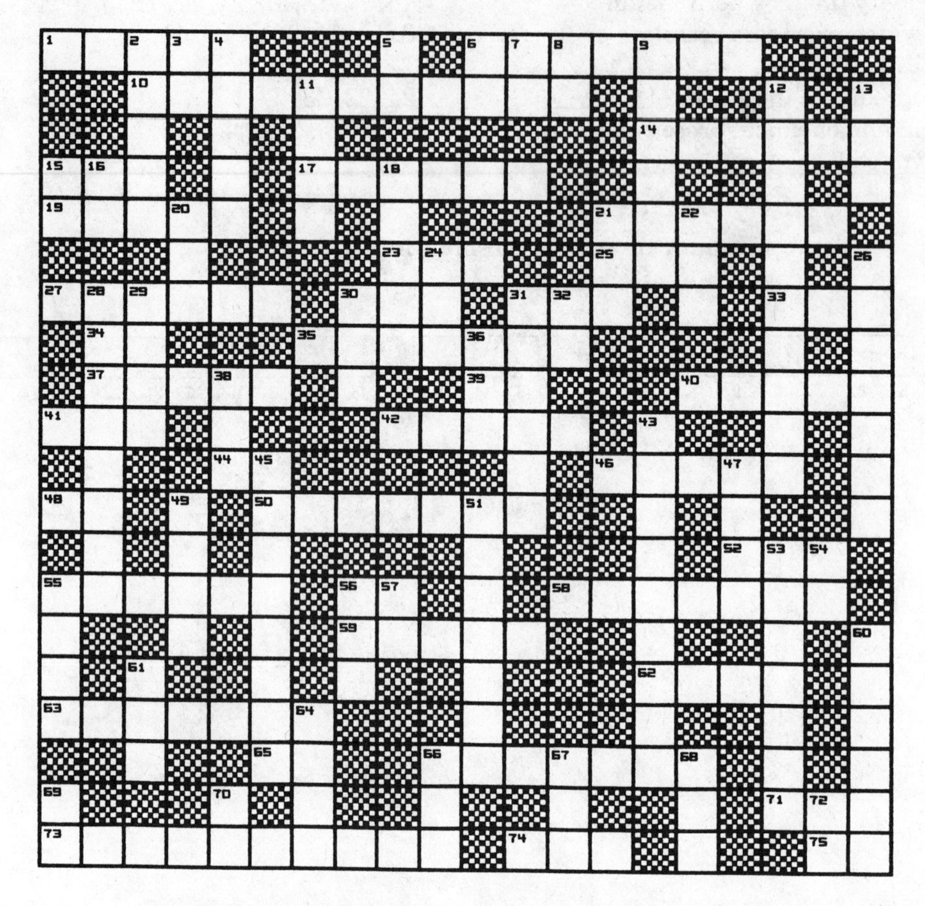

Across

1 Pancreatic secreting units
6 Substrate for pepsin
10 Stores bile
14 Contains amylase
15 Very
17 Functional liver unit
19 Sweeping apparatus
21 Lymph nodules or patches
23 Transports cholesterol
25 Time period
27 Where large intestine terminates
30 Baby flower
31 What blood levels of cholesterol should be
33 By way of
34 Short for "mother"
35 Hormone that stimulates gastric secretions
37 Results from too much acid in stomach
39 Not down
40 Has ascending, transverse, and descending sections
41 Not well
42 Glucose or lactose, for example
44 Refractory period (abbr.)
46 Folds of intestinal mucosa
48 Prefix meaning two
50 An enzyme that digests carbohydrate
52 What the *Concorde* is
55 Duct that drains gallbladder
56 Advanced placement (abbr.)
58 One of the salivary glands
59 What leaves the stomach
62 Term for the stream of light
63 Lymphatic capillary
65 Preposition
66 Common passage for food and air
71 Daughter of the American Revolution (abbr.)
73 A process carried out in the small intestine
74 High-energy compound
75 Title for a woman

Down

2 Eskimo home
3 Sodium (abbr.)
4 Terminal portion of small intestine
5 What works with rennin to curdle milk (abbr.)
6 Police Department (abbr.)
7 In regard to (abbr.)
8 Either/________
9 Irving Berlin's song: "________ Parade"
11 An emulsifier
12 Cellular projections to increase surface area
13 Substrate for lipase
15 Lung disease (abbr.)
16 Surgical room (abbr.)
18 What enters the stomach
20 Not in
21 Church seat
22 Sweet potato
24 Banned insecticide
26 Straps of colon
28 What bile will do to fat
29 Telephone a friend
30 Where one can have a drink
31 Found in pancreatic juice
32 Not off
36 Floor covering
38 Hearing organ
43 Bile pigment
45 This organ is endocrine and exocrine
47 What the ancient city of Atlantis is
49 Singing voice
51 Where parietal cells secrete HCl
53 "S" shaped portion of colon
54 What often happens on Monday night football (abbr.)
55 Basic unit of living matter
56 Highest card in a deck
57 This is very low in the stomach
60 Lactose, sucrose, glucose
61 Secreted by parietal cells
64 The phosphate is ______ when ATP is broken down to ADP
66 Often used with a needle
67 Lab animal
68 Thirty to a Roman
69 What you get from the digestion of protein (abbr.)
70 Medical man (abbr.)
72 Morning (abbr.)

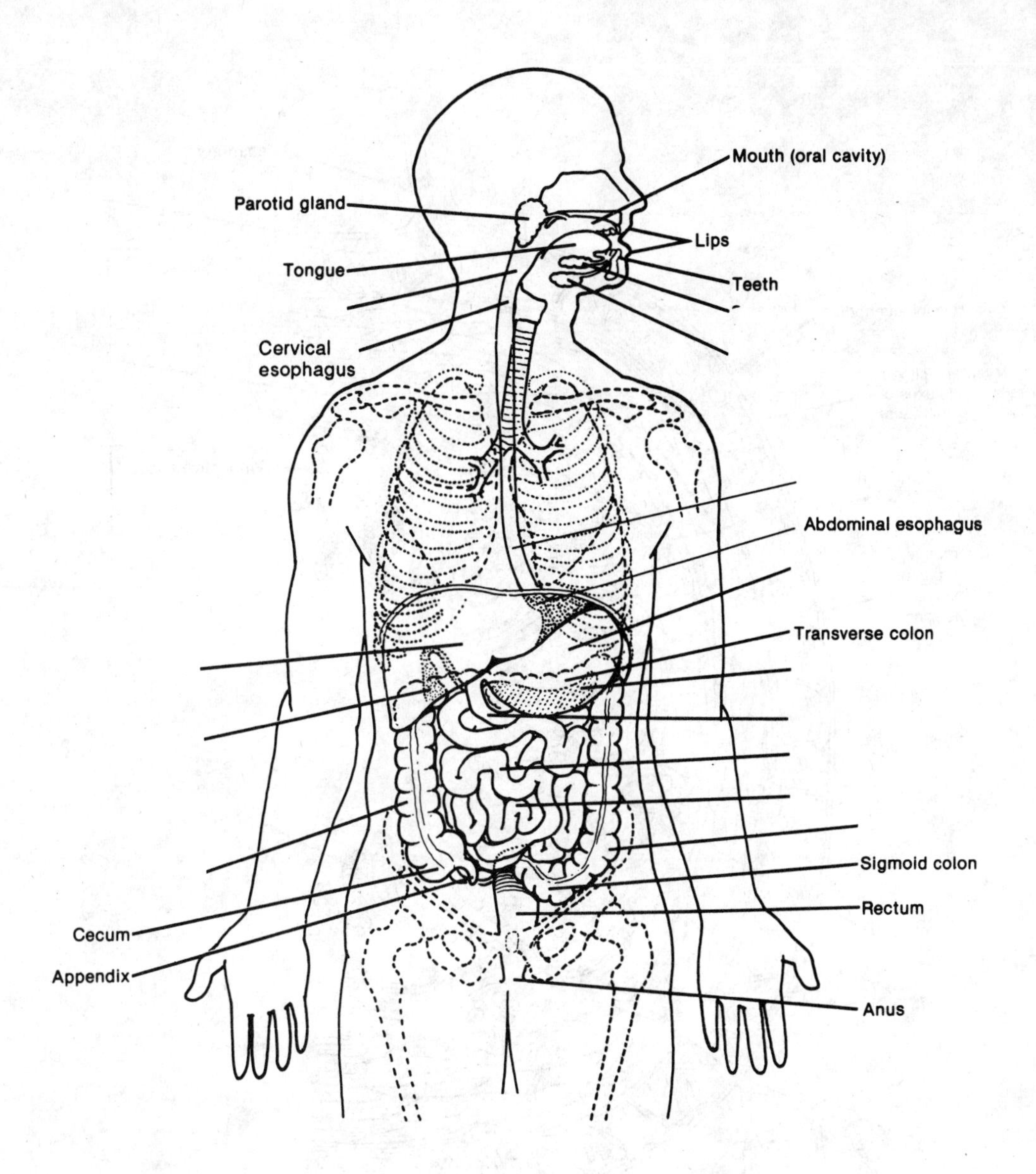

Anatomic Artwork

Figure 1

1. Label the organ that serves as a common passage for food and air.
2. Shade in and label the organ that is primarily responsible for the interconversion of foodstuffs.
3. Label the organ that has acini.
4. Label the first area with haustra.
5. Label the structure that stores and concentrates bile.
6. Label the organ that is responsible for the greatest amount of absorption of nutrients.
7. Label the structure that contains rugae.
8. Label the accessory glands that secrete enzymes into the mouth.

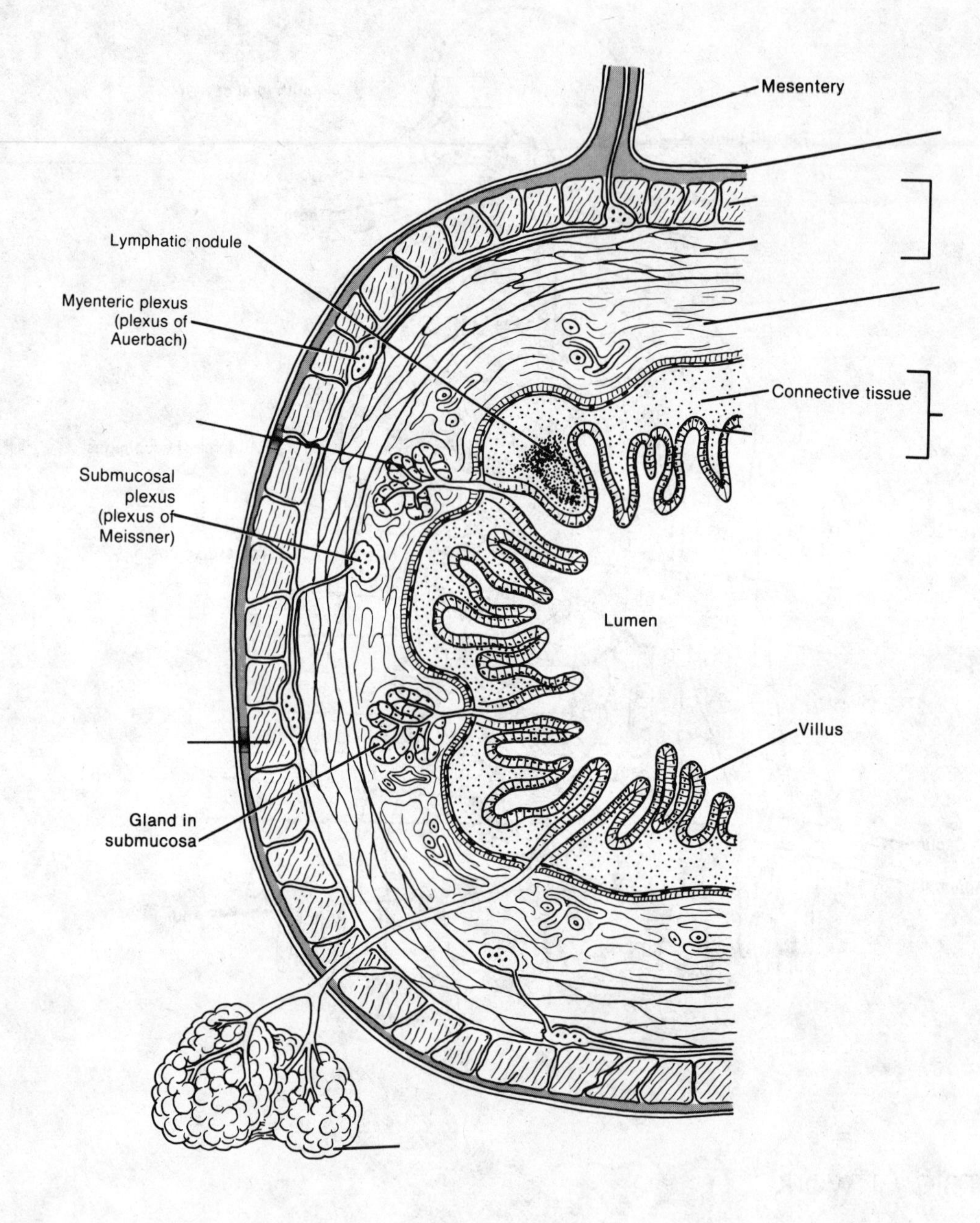

Figure 2

1. Label the tissue which first absorbs nutrients.
2. Label the specific layer of muscle that, when it contracts, causes a shortening of the tube.
3. Label the layer that is also called the visceral peritoneum.
4. Label the accessory glands that secrete digestive enzymes.
5. Label the layer that contains numerous blood vessels and lymphatic vessels.
6. Label the coat that has an additional layer in the stomach.

Chapter 20

Metabolism

Introduction

Once food has been digested and absorbed across the wall of the small intestine, it is eventually carried to the cells of the body for *metabolic use*. The basic nutrients can be synthesized into larger macromolecules. This is called *anabolism*. Or, nutrients, such as glucose, can be broken down and the energy released can be used to form ATP. This breakdown is known as *catabolism*. Ultimately, the three foodstuffs—proteins, carbohydrates and lipids—can be interconverted, depending on the body's needs.

The liver carries out many essential metabolic reactions, among which are deamination of amino acids, release of glucose into the bloodstream from stored glycogen, and interconversion of simple sugars. The ultimate "use" of the substances in the body is to derive energy from the catabolic reactions. The energy released is measured in *calories* and is an estimate of what is known as the *metabolic rate* of an individual.

In addition to the large complex *protein, carbohydrate,* and *lipid* nutrients, the body also requires *minerals* and *vitamins* for survival and the maintenance of homeostasis.

After completing the reading in your text, you should be familiar with these concepts and be able to answer the questions that follow and carry out the activities.

Multiple Choice

___ 1. These inorganic substances comprise about 4 percent of the body weight and are needed for homeostasis:
   a. proteins
   b. vitamins
   c. minerals
   d. carbohydrates

___ 2. Which of the following is carried out in the liver?
   a. emulsification
   b. insulin production
   c. deamination
   d. all of these

___ 3. Which of these is associated with fat metabolism?
   a. beta oxidation
   b. ketone body formation
   c. lipogenesis
   d. all of these

___ 4. An enzyme:
a. is a protein catalyst
b. has an active site
c. forms a temporary complex with the substrate
d. all of these

___ 5. Essential amino acids:
a. cannot be synthesized by the body
b. require a coenzyme for catabolism
c. have substrates that are essential
d. none of these

___ 6. The Kreb's cycle is:
a. a series of oxidation–reduction reactions
b. carried out in the mitochondria
c. part of the aerobic metabolism of glucose
d. all of these

___ 7. When might glycogenolysis occur?
a. as a result of high blood glucose levels
b. right after a meal
c. in starvation
d. all of these

___ 8. The kilocaloric content of food is:
a. a measure of the heat it releases upon oxidation
b. a measure of its anabolic rate
c. an estimate of the nutritional value
d. an estimate of the heat lost due to radiation

___ 9. Which of these factors would affect metabolic rate?
a. food intake
b. vigorous exercise
c. hormone levels
d. all of these

___ 10. Which is a water-soluble vitamin?
a. A
b. D
c. B
d. K

___ 11. A reduction reaction is one in which:
a. electrons are added to a molecule
b. electrons are removed from a molecule
c. water is used to hydrolyze molecules
d. none of these

___ 12. Ketogenesis:
a. occurs in the liver
b. is the formation of ketone bodies
c. is part of normal fatty acid catabolism
d. all of these

True/False

_______ 1. *Fats* stored in adipose tissue constitute the largest reserve of energy for the body.
_______ 2. Abnormally high levels of ketone bodies in blood is referred to as *ketogenesis.*
_______ 3. *Glycogenolysis* usually occurs between meals.
_______ 4. The Kreb's cycle and electron transport chain are *anaerobic* processes.
_______ 5. The conversion of glucose into glycogen is an example of a *catabolic* process.
_______ 6. *Triglycerides* are ultimately digested into short and long chain fatty acids and monoglycerides and glycerol.
_______ 7. *Ketosis,* if untreated, can lead to acidosis and possibly death.
_______ 8. The conversion of glycogen into glucose is called *glucogenesis.*
_______ 9. Transference of heat to a substance or object such as clothing is called *convection.*
_______ 10. *Increased* secretion of thyroxine can result in an increase in body temperature.
_______ 11. In the *first* stage of starvation, the carbohydrate stores of the body are depleted.
_______ 12. *Calcium* is one of the macrominerals.

Completion

1. Vitamin C and B complex vitamins are soluble in _______.
2. The satiety center is located in the _______, and when it is stimulated feeding _______.
3. All the chemical reactions of the body are collectively referred to as _______.
4. A(n) _______ is a chemical substance that alters the rate of a reaction.
5. The oxidation of glucose is also known as cellular _______.
6. The formation of the most commonly used monosaccharide from noncarbohydrate sources is called _______.
7. Each gram of fat produces about 9.0 _______ when oxidation occurs.
8. Normal fat catabolism involves the conversion of some _______ molecules into ketone bodies.
9. The removal of nitrogen groups from amino acids is called _______.
10. To balance the anabolism and catabolism of protein, the recommended guideline for protein consumption is in the range of _______ grams per day.
11. The BMR can be used as a measure of production of the hormone _______ since it regulates the rate of food breakdown.
12. The greatest percentage of body heat is lost through the process of _______.

Lost Sheep

1. vitamin A, vitamin B, vitamin D, vitamin E
2. ketone bodies, lipogenesis, beta oxidation, splits off acetic acid
3. glycolysis, aerobic, lactic and pyruvic acids, 2 ATPs
4. protein synthesis, deamination, urea, catabolism
5. anabolism, glucogenesis, beta oxidation, lipogenesis

6. secretes bicarbonate, glycogen storage, interconversion of food, liver
7. PKU, phenylalanine, vitamin E, Kreb's cycle
8. iron, magnesium, manganese, cobalt
9. carbohydrate, deamination, amino acid, liver
10. satiety center active, hypothalamus, results from elevated blood glucose, hunger center inhibition

MATCHING

Set 1

___ 1. glycolysis
___ 2. gluconeogenesis
___ 3. glycogenolysis
___ 4. glycogenesis
___ 5. lipolysis
___ 6. ketogenesis
___ 7. lipogenesis
___ 8. deamination

a. anabolism
b. catabolism

Set 2

___ 1. riboflavin
___ 2. ascorbic acid
___ 3. micromineral
___ 4. macromineral
___ 5. fat-soluble
___ 6. provitamin
___ 7. coenzymes

a. vitamin K
b. potassium
c. iron
d. vitamin B
e. carotene
f. vitamin C
g. derived from vitamins

Double Crosses

1. *Across*:
excess ketone bodies

Down:
with oxygen

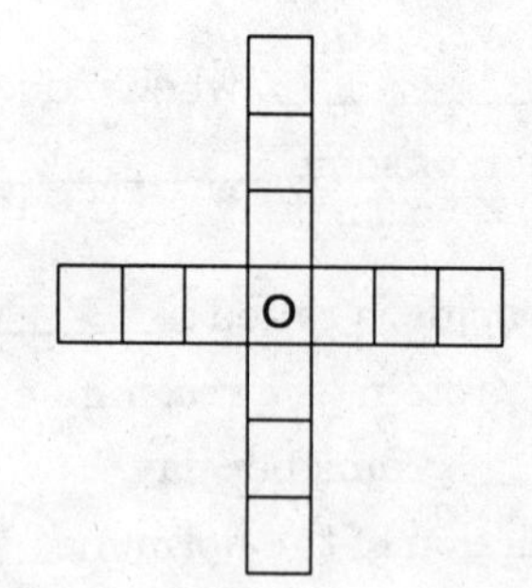

2. *Across:*
   a. metabolic cycle
   b. brain center for eating control
   c. what an enzyme interacts with
   d. amino acids we can't make

   Down:
   occurs in the liver

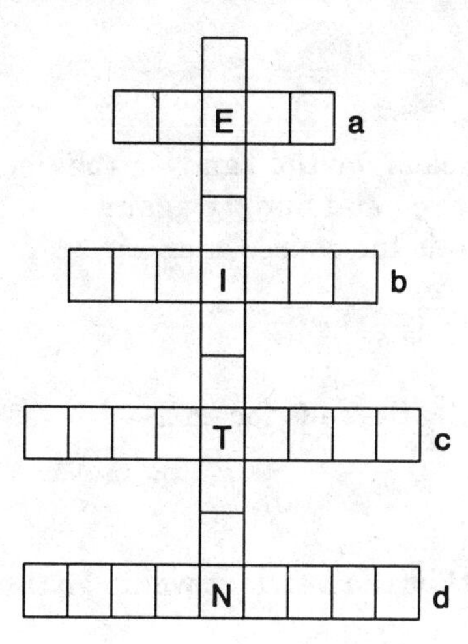

3. *Across*:
   without oxygen

   Down:
   a. synthesis
   b. goes with reduction
   c. cellular breakdown of glucose
   d. degradation reactions are ________
   e. what some vitamins are

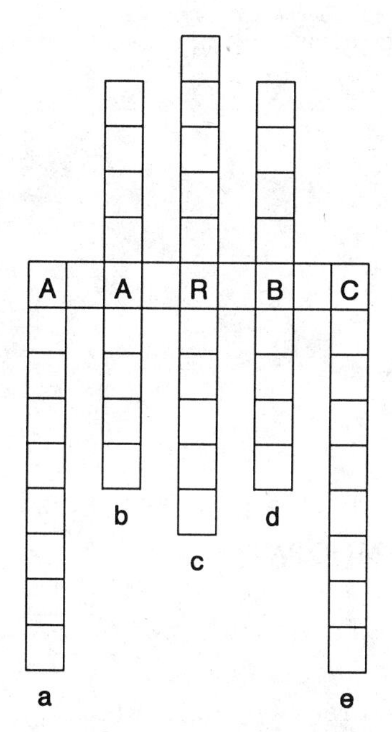

Sleuthing

1. When Jim gets tired at work, he heads for the candy machine. The chocolate bar gives him an immediate lift, but shortly he feels tired and hungry again.
   a. What type of chemical compound is the source of energy?

   b. What specific compound is used by the cells for energy?

   c. Outline the metabolic pathway that the nutrient would go through to provide the energy source.

   d. What hormones would be involved in controlling these reactions and what is their action?

   e. What would Jim be better off eating and why?

Word Scrambles

1. a. deficiency causes pellagra	NCINAI	☐ ☐ ☐ _ ☐ ☐
b. good source of vitamin C	MOOTATES	☐ _ ☐ ☐ _ ☐ ☐ _
c. vitamin that is abundant in milk	D	☐

Total: first action in breakdown of amino acid

☐☐☐☐☐☐☐☐☐☐☐

2. a. endocrine gland which affects metabolic rate	HIDOTRY	☐ _ ☐ ☐ _ ☐ _
b. where hunger center is located	UYPTHHLMAOSA	☐ _ _ ☐ _ ☐ _ _ ☐ ☐ _ _
c. transfer of heat by movement of gas between areas of different temperature	CNCEOTONVI	☐ _ ☐ _ _ ☐ _ _ ☐ _
d. how some vitamins function	ZONECYME	☐ _ ☐ _ _ _ _ _

Total: terminal respiratory pathway with oxygen as the final acceptor

☐☐☐☐☐☐☐☐☐☐ ☐☐☐☐☐

Addagrams

1. a. one of the B vitamins — 5, 7, 8, 16, 12, 13
   b. can result from lack of B — 10, 5, 1, 11, 17, 15
   c. programs to lose weight — 18, 12, 4, 6, 3
   d. can result if ketosis occurs — 10, 16, 12, 18, 14, 19, 7, 2
   e. results from deficiency of vitamin D — 14, 2, 6, 4, 14, 11, 15, 9, 8, 16, 17, 10

Total: tryptophan, lysine and valine for example

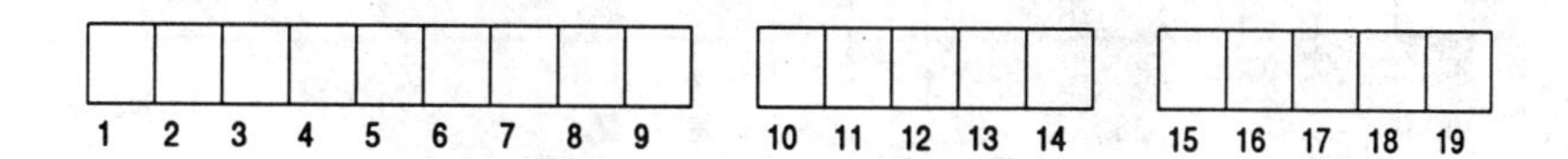

2. a. process of degradation — 14, 2, 8, 9, 1, 11, 5, 13, 3, 6
   b. vitamin C acid — 4, 3, 14, 11, 15, 10, 13, 14
   c. chemical class of cortisol — 3, 17, 7, 15, 11, 12, 0
   d. "with air" — 16, 18, 15, 11, 1, 13, 14
   e. one diets to ________ weight — 12, 11, 3, 18

Total: the body's minimal energy expenditure

1 2 3 4 5   6 7 8 9 10 11 12 13 14   15 16 17 18

Crossword

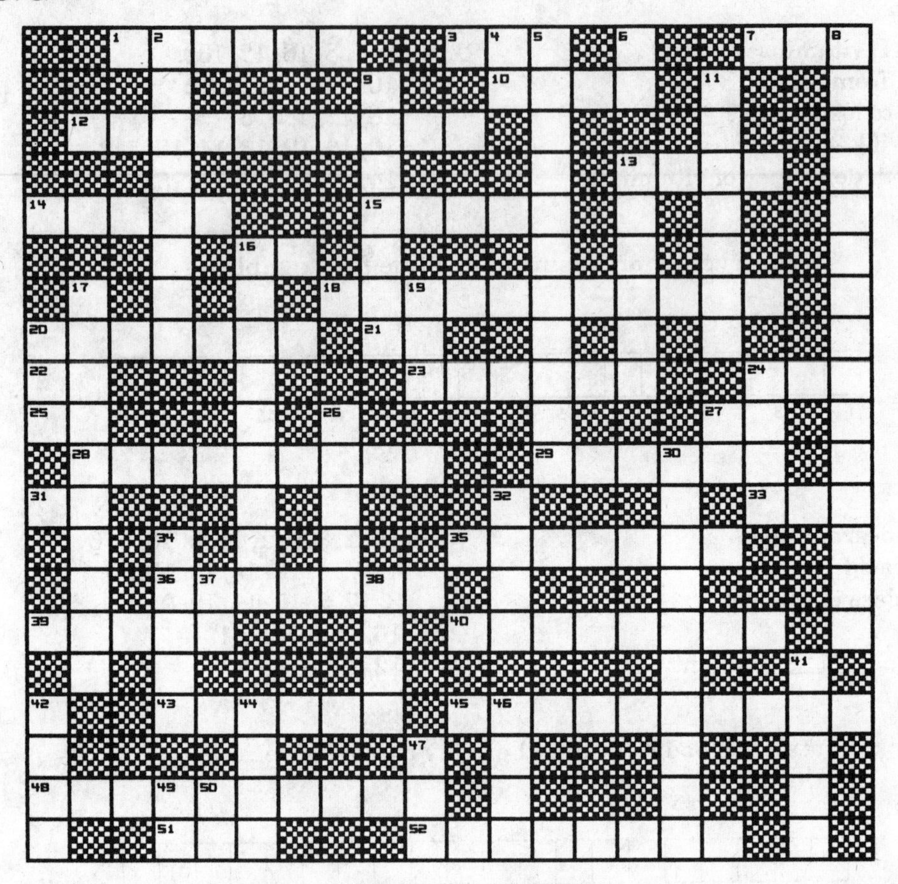

Across

1 Biological catalyst
3 Means of transportation
7 Cup
10 Cells that secrete insulin
12 Total of the body's chemical reactions
13 Boy's name
14 Steroid's chemical class
15 Pancreatic secretors
16 Animal of burden
18 Unit of measurement associated with food's value
20 Made of amino acids
21 Negative reply
22 In regard to (abbr.)
23 Three times
24 Boat that pulls others
25 Medical specialty dealing in deliveries (abbr.)
27 Physician (abbr.)
28 What an enzyme acts on
29 A B vitamin
31 Alternating current (abbr.)
33 Enzyme ending
35 Hypothalamic center that affects feeding
36 Nutritional requirement
39 Gold rush marker
40 Vitamin C acid
43 Needed for aerobic respiration
45 Anaerobic phase of glucose catabolism
48 Term for sugar
51 Drinks are served here
52 Some vitamins have this as a function

Down

2 Glucose, protein, or fat
4 College degree (abbr.)
5 Complete oxidation of glucose is cellular ________
6 Calcium (abbr.)
8 Formation of carbohydrate from other sources
9 A, D, E, or K is a ________
11 With oxygen
13 Formed during beta oxidation
16 Beta ________
17 Part of carbohydrate catabolic path (2 words)
19 Acre
20 Paid athlete (abbr.)
24 Brings amino acids to site of protein synthesis
26 Most common chemical compound in body
27 Princess of Wales
30 Degradation reaction
32 "Without" in France
34 Acids in protein
37 I am (contraction)
38 Prayer ending
41 Cholesterol is one
42 Seafood
44 January through December
46 Crippled
47 Room where games are played (abbr.)
49 Citizen Band (abbr.)
50 Expression of laughter

Chapter 21

The Urinary System

Introduction

Normally about one-fourth of the cardiac output is pumped to the human *kidneys* at rest. The filtering of the blood that is carried out by the kidneys is tremendous, for if the blood flowing to the kidneys were reduced tenfold, many blood constituents would still be at a nontoxic level. A person can lose one and three fourths of her/his kidneys and still maintain acceptable solute levels in blood.

The three main processes that occur in the kidney are *filtration* of blood (from glomerulus to Bowman's capsule) and *reabsorption* and *secretion* (along the kidney tubules). Virtually anything in true solution in the blood will flow through into the filtrate, but large molecules (like proteins bigger than about 70,000 molecular weight) and blood cells normally remain totally behind in the blood. Since filtration is a nonspecific process, adjustments are made as the filtrate travels through the nephron toward collection in the calyx. Thus, substances like glucose and sodium ions are actively transported back into the body, and water follows due to the change in osmotic pressure across the tubule wall. Control over sodium retention occurs by *aldosterone,* a hormone released in response to blood pressure changes and feedback information from the filtrate. Water reabsorption is monitored by the hypothalamus, which secretes *ADH* in response to osmotic pressure and blood pressure information.

Secretion by kidney tubule cells rids the body of certain *wastes* (like drugs) and is the mechanism for *acid–base* control. Specific excretion or retention of H^+ ions helps maintain body fluid pH at a homeostatic level of 7.4. The regulating of pH works in concert with the respiratory system and blood buffers.

After you have read the chapter in the textbook which deals with the kidneys and their function, you should be able to work through these questions and activities.

Multiple Choice

___ 1. Which of these correctly describes the plasma in the glomerulus and the filtrate in Bowman's capsule?
   a. The filtrate has about 200 times more protein than plasma does
   b. The plasma contains certain inorganic ions that the filtrate does not
   c. The filtrate contains less than 1 percent of the protein that the plasma does
   d. Plasma and filtrate are about the same

___ 2. Which of these passes through the filtration barrier from glomerulus to Bowman's capsule?
   a. proteins
   b. RBCs
   c. electrolytes
   d. all of these

___ 3. Which of these is the proper sequence in the kidney?
a. glomerulus, Bowman's capsule, PCT, loop of Henle
b. Bowman's capsule, glomerulus, PCT, loop of Henle
c. glomerulus, PCT, Bowman's capsule, loop of Henle
d. none is correct

___ 4. Which of these represents the proper sequence for excretory system function?
a. pelvis, calyx, urethra, bladder, urethra
b. calyx, pelvis, ureter, bladder, urethra
c. calyx, pelvis, urethra, bladder, ureter
d. calyx, pelvis, bladder, urethra, ureter

___ 5. The normal capacity for an adult bladder is about:
a. 200 ml
b. 500 ml
c. 800 ml
d. 1200 ml

___ 6. Normal urine output averages about:
a. 200 ml/day
b. 500 ml/day
c. 1000–1500 ml/day
d. 2000–2500 ml/day

___ 7. Voluntary control over micturition is effected by:
a. internal sphincter muscle
b. trigone
c. bladder wall muscle
d. external sphincter muscle

___ 8. Which of these does *not* appear in the urine of a normal fasting adult?
a. Na^+
b. urea
c. creatinine
d. glucose

___ 9. Which of these is a normal constituent of urine?
a. albumin
b. white blood cells
c. uric acid
d. hemoglobin

___ 10. Sodium reabsorption into the body by the kidney is controlled by:
a. ADH
b. aldosterone
c. epinephrine
d. all of these

___ 11. In the collecting duct, water reabsorption is regulated by:
a. renin
b. aldosterone
c. angiotensin
d. ADH

___ 12. Renin is released by the:
a. juxtaglomerular cells
b. hypothalamus
c. adrenal cortex
d. none of these

___ 13. Renal autoregulation of blood flow for moderate changes in systemic arterial pressure includes:
a. maintenance of a constant GFR
b. maintenance of a constant glomerular blood pressure
c. control over the smooth muscle in the afferent arteriole
d. all of these

___ 14. If blood pressure is low:
a. GFR is low
b. Renin is released by the JGA
c. The angiotensin–aldosterone system will operate
d. all of these

___ 15. In the proximal convoluted tubule:
a. sodium is actively transported back into the body
b. ions move out of the tubule into peritubular blood
c. chloride follows sodium and the osmotic pressure of blood increases
d. all of these

___ 16. Which structure(s) is(are) located in the medulla of the kidney?
a. renal capsule
b. pyramids
c. ureter
d. all of these

___ 17. Which is *not* a structure of the nephron?
a. glomerulus
b. loop of Henle
c. the DCT
d. the pyramid

___ 18. Which is *not* part of the macroscopic structure of the kidney?
a. cortex and medulla
b. glomerulus
c. papilla
d. major calyx

___ 19. The net filtration pressure at the glomerulus is due to:
a. high capsular hydrostatic pressure, lower glomerular hydrostatic pressure
b. high glomerular hydrostatic pressure, lower capsular hydrostatic pressure
c. both a and b
d. neither a nor b

___ 20. The hydrostatic pressure in the glomerulus is about 50–70 mm Hg versus about 30 mm Hg in a normal tissue capillary. The anatomical factor keeping this pressure high is:
a. an efferent arteriole that is more constricted than the afferent arteriole
b. an afferent arteriole that is more constricted than the efferent arteriole
c. a glomerular capillary that is more constricted than the efferent arteriole
d. none of these

___ 21. Which of these is secreted by the kidney tubule?
a. H^+ ions
b. Na^+ ions
c. glucose
d. none of these

___ 22. In hemodialysis:
a. blood is pumped from the radial artery
b. the loss of 500 ml, which is in the machine, is compensated by vasoconstriction and an increased cardiac output
c. substances equilibrate across the dialyzing membrane
d. all of these

___ 23. Normally the pH of the urine of a fasting person is about:
a. 1.5
b. 4
c. 6.5
d. 8.5

___ 24. Which is present in the filtrate but not in urine?
a. sodium
b. glucose
c. chloride
d. all of these

___ 25. When blood pressure and blood volume drop:
a. the kidney secretes renin
b. angiotensin is activated
c. aldosterone is released
d. all of these

True/False

___________ 1. If body fluids are acidic, *H^+ ions* are secreted into the filtrate as it passes through the kidney tubule.

___________ 2. When *insufficient* amounts of aldosterone are secreted, low amounts of sodium will be retained and high amounts of potassium will be excreted.

___________ 3. Micturition occurs when *parasympathetic* impulses from the lower spinal cord cause a relaxation of the internal sphincter muscle.

___________ 4. As filtrate passes through the kidney tubules about *25 percent* is reabsorbed back into the body.

___________ 5. Angiotensin II is a powerful vasoconstrictor of the *efferent arteriole.*

___________ 6. ADH is secreted in response to *low* osmotic pressure of the extracellular body fluids.

___________ 7. Substances like Na^+, Cl^-, and glucose *would* pass through the filtration barrier between the glomerulus and Bowman's capsule.

___________ 8. The *kidneys and adrenals* are located retroperitoneally.

___________ 9. Blood pressure in the glomerulus is maintained by constriction of the *afferent* arteriole.

___________ 10. The NFP (net filtration pressure) is about *60* mm Hg.

___________ 11. The filtrate in Bowman's capsule is similar to plasma, except that it contains only about *1/200* of the protein that plasma does.

__________ 12. Dilation of *afferent* arterioles causes an increase in the GFR (glomerular filtration rate).

__________ 13. Most of the water reabsorption by the kidney takes place in the *proximal convoluted tubule.*

__________ 14. Glycosuria occurs in untreated diabetes *insipidus.*

__________ 15. The kidney is assisted by the *skin and lungs and GI tract* in maintaining a homeostasis of volume, pH, and the chemical composition of body fluids.

__________ 16. For the most part, urine is moved by the peristaltic movements of the *urethra.*

__________ 17. *Glucose* does not appear in the urine of a normal fasting person because all of it is reabsorbed by the kidney tubules.

__________ 18. In the human male, excretory and reproductive tracts are separate throughout the *entire* systems.

__________ 19. When blood volume is *low,* angiotensin II stimulates the secretion of aldosterone, which in turn increases sodium and water retention.

__________ 20. The opening from the bladder into the *urethra* is guarded by the internal sphincter muscle.

__________ 21. About *180–200 mg/ml* is the normal level of glucose in the blood.

__________ 22. The internal sphincter muscle at the base of the bladder is normally *contracted.*

__________ 23. Stones may form in urine because of low water intake or an overactive *parathyroid* gland.

__________ 24. Renal failure is often characterized by *oliguria.*

__________ 25. Blood is filtered in the nephron from the *glomerulus* to Bowman's capsule.

Completion

1. For micturition to occur, __________ impulses must stimulate the detrusor muscle to contract.
2. In diabetes mellitus, excessive amounts of __________ appear in the urine due to a deficiency of insulin.
3. __________ is the process of cleansing the blood by the use of an artificial kidney.
4. The most abundant normal constituents of urine are __________ , __________ , and __________ .
5. Hydrogen ion is __________ as the kidney adjusts the pH of body fluids.
6. The average capacity of the bladder is __________ ml of urine.
7. The barrier between the glomerulus and Bowman's capsule is the __________ membrane.
8. When sodium and chloride move out of the tubule, water follows by the process of __________ .
9. The movement of water in #8 above is considered a(n) __________ flow.
10. The movement out of the kidney tubule in response to the presence of ADH is called a(n) __________ flow.
11. Normally when the bladder is about __________ ml full, the desire to urinate is experienced.
12. The tubule of the excretory system which leads outside the body is the __________ .
13. About __________ percent of the cardiac output flows through the kidneys at rest.

14. Sodium is removed from the proximal convoluted tubule by means of ______________.
15. Penicillin, potassium, and H^+ are removed from the body by being ______________.
16. Proteins that do not filter through the glomerulus into Bowman's capsule are held back by ______________.
17. The hydrostatic pressure in the glomerulus is normally ______________ (greater, less) than systemic capillary hydrostatic pressure.
18. The functional unit of the kidney is the ______________.
19. Failure of the kidneys to produce urine is called anuria and may lead to a toxic state called ______________.
20. The pH range of urine is about ______________.
21. If body fluids are becoming acidic, ______________ ions may be eliminated in urine.
22. Water accounts for about ______________ percent of the total volume of urine.
23. Two hormones that act in response to angiotensin II are ______________ and ______________.
24. With considerable sympathetic stimulation the ______________ are constricted in the kidney.
25. The tissue type lining the bladder, which is adapted for stretching, is ______________.

Lost Sheep

1. renal tubule, Bowman's capsule, glomerulus, calyx
2. PCT, afferent arteriole, peritubular capillaries, glomerulus
3. urea, creatinine, glucose, uric acid
4. pyramid, calyx, pelvis, capillaries
5. blood colloidal pressure, net filtration pressure, glomerular blood pressure, hydrostatic pressure
6. sebaceous glands, sweat glands, GI tract, lungs
7. vasodilation, angiotensin, blood pressure rise, strong sympathetic stimulation
8. filtration, glomerulus, secretion, hydrostatic pressure
9. capillaries, Bowman's capsule, glomerulus, pelvis
10. major calyx, proximal convoluted tubule, loop of Henle, distal convoluted tubule
11. glomerulus, afferent arteriole, efferent arteriole, renal artery
12. glucose, Tm, glycosuria, sodium
13. juxtaglomerular apparatus, renin, ADH, angiotensin
14. constriction of efferent arterioles, constriction of afferent arterioles, increase of glomerular pressure, increase net filtration pressure
15. red blood cells, electrolytes, amino acids, urea
16. H^+, drugs, glucose, K^+
17. PCT, collecting duct, DCT, capsule
18. internal sphincter muscle, trigone, rugae, calyx
19. ADH, FSH, aldosterone, angiotensin
20. hypothalamus, JGA, adrenal glands, thyroid gland

Matching

Set 1

Use as many letters from the right-hand column as apply

___ 1. glucose
___ 2. sodium
___ 3. H^+ ion
___ 4. ammonia
___ 5. penicillin

a. is secreted by tubule
b. is reabsorbed by tubule
c. is filtered at glomerulus

Set 2

___ 1. has rugae
___ 2. has cortex and medulla
___ 3. exhibits involuntary wave of contraction
___ 4. contains trigone
___ 5. duct common to reproductive tract in male
___ 6. microscopic unit of urine formation
___ 7. connects kidney to bladder
___ 8. exit from here is guarded by two sphincters
___ 9. where calculi often lodge: in pelvis of ________
___ 10. contains calyces

a. bladder
b. ureter
c. nephron
d. urethra
e. kidney

Set 3

___ 1. cells that secrete enzyme that activates angiotensin
___ 2. keeps large proteins out of filtrate through here
___ 3. blood from glomerulus is forced through here
___ 4. contains first filtrate
___ 5. stimulates afferent arteriole to dilate in low blood pressure
___ 6. is more constricted than afferent arteriole
___ 7. contains blood
___ 8. has more protein than filtrate
___ 9. constricts upon angiotensin II stimulation
___ 10. is stimulated to secrete renin if blood pressure is low

a. glomerulus
b. juxtaglomerular cells
c. endothelial–capillary membrane
d. efferent arteriole
e. Bowman's capsule
f. macula densa

Double Crosses

1. *Across*:
   how excess H^+ ion is eliminated

   Down:
   stoppage of urine flow

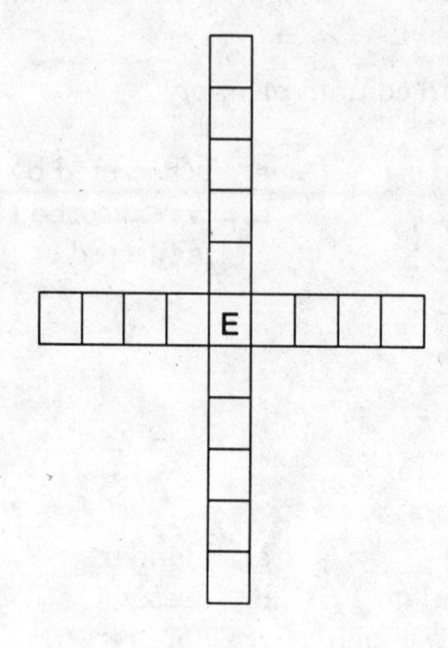

2. *Across*:
   a. region of kidney where interlobar vessels are found
   b. urination
   c. hormone ultimately activated by renin

   Down:
   process from glomerulus to Bowman's capsule

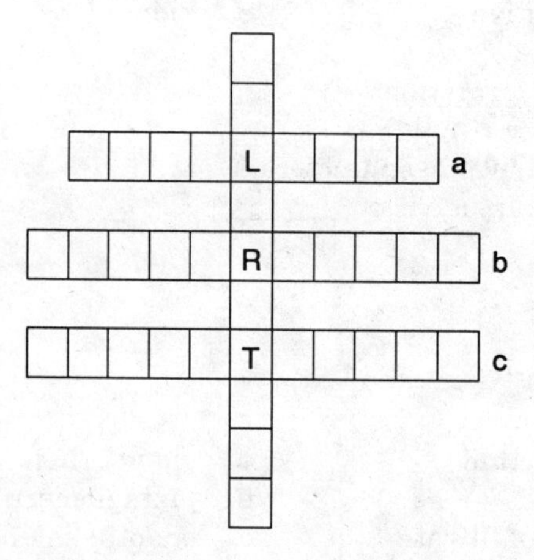

3. *Across*:
   concentration of body fluids which inhibits ADH release

   Down:
   condition of blood in urine

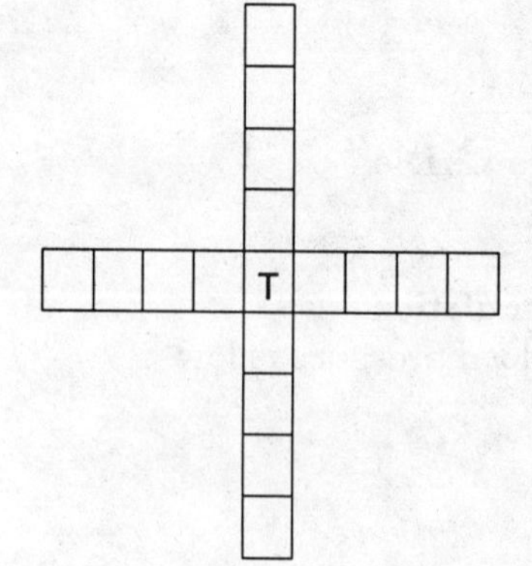

4. *Across:*
   a. inorganic breakdown product of urea
   b. kidney location: behind membrane

   Down:
   cells secreting renin

O a

R b

Sleuthing

1. Bobby is 7 years old. He has come into the hospital emergency room with his breath smelling sweet and a complaint of frequent urination. His mother confirms the frequent urination and adds that he is constantly thirsty. The doctor tentatively diagnoses his condition as Type I diabetes mellitus..
   a. Why is the sugar appearing in his urine?

   b. What is the sweet smell on his breath?

   c. What is one explanation for the excessive urination?

Word Scrambles

1. a. what is filtered in the kidney	DOLOB	☐ _ _ _ _
b. what is reabsorbed in the DCT	AWTRE	☐ ☐ _ _ _
c. endothelial–capsular ________ : filtration barrier	BAMENERM	☐ _ _ _ _ _ ☐ _
d. curly capillary	MUROLLEGUS	_ _ ☐ _ _ _ _ _ _ ☐

Total: capsule

2. a. houses male excretory organ — SENIP — ☐ _ _ ☐ _
   b. blood vessel — EVIN — ☐ ☐ _ _
   c. ADH is responsive to the concentration of ________ in body fluids — SLATS — _ _ ☐ _ ☐

Total: fed by the calyces

☐☐☐☐☐☐

3. a. structure joining bladder to outside — TARUHER — ☐ _ _ ☐ _ _ _
   b. outer "shell" of kidney — TOXERC — _ _ ☐ _ ☐ _
   c. term referring to kidney — LANER — ☐ ☐ _ _ _

Total: tube that undergoes peristaltic contractions

☐☐☐☐☐☐

4. a. bones protecting pelvic cavity: os ________ — AXECO — _ ☐ ☐ ☐ _
   b. appears in urine in diabetes mellitus — SOCULGE — ☐ ☐ ☐ _ _ ☐ _
   c. inner segment of kidney — MALLEDU — ☐ ☐ _ ☐ ☐ ☐ _
   d. excess bilirubin leads to this — ANICJDUE — ☐ _ _ _ _ _ ☐ ☐
   e. vessel on either side of glomerulus — TOILERERA — ☐ ☐ ☐ _ ☐ _ _ ☐ _

Total: renin secretor

☐☐☐☐☐☐☐☐☐☐☐☐☐☐☐ ☐☐☐☐☐

Addagrams

"Marathoner"

1. a. tubular process that concentrates ammonia — 15, 1, 12, 19, 21, 5, 9, 4, 2
   b. part of nephron where "adjustments" are made — 5, 16, 23, 16, 8, 7
   c. branch off segmental artery: inter ________ — 11, 4, 23, 13, 24
   d. inner segment of kidney — 20, 27, 3, 16, 17, 8, 18
   e. loop of nephron — 6, 27, 26, 17, 7
   f. cells that stimulate JG cell: ________ densa — 22, 10, 12, 16, 8, 25
   g. fluid that gets filtered — 14, 17, 18, 15, 22, 25

Total: "crossing guard"

1	2	3	4	5	6	7	8	9	10	11	-	12	13	14	15	16	17	18	19

20	21	22	23	24	25	26	27

Crossword

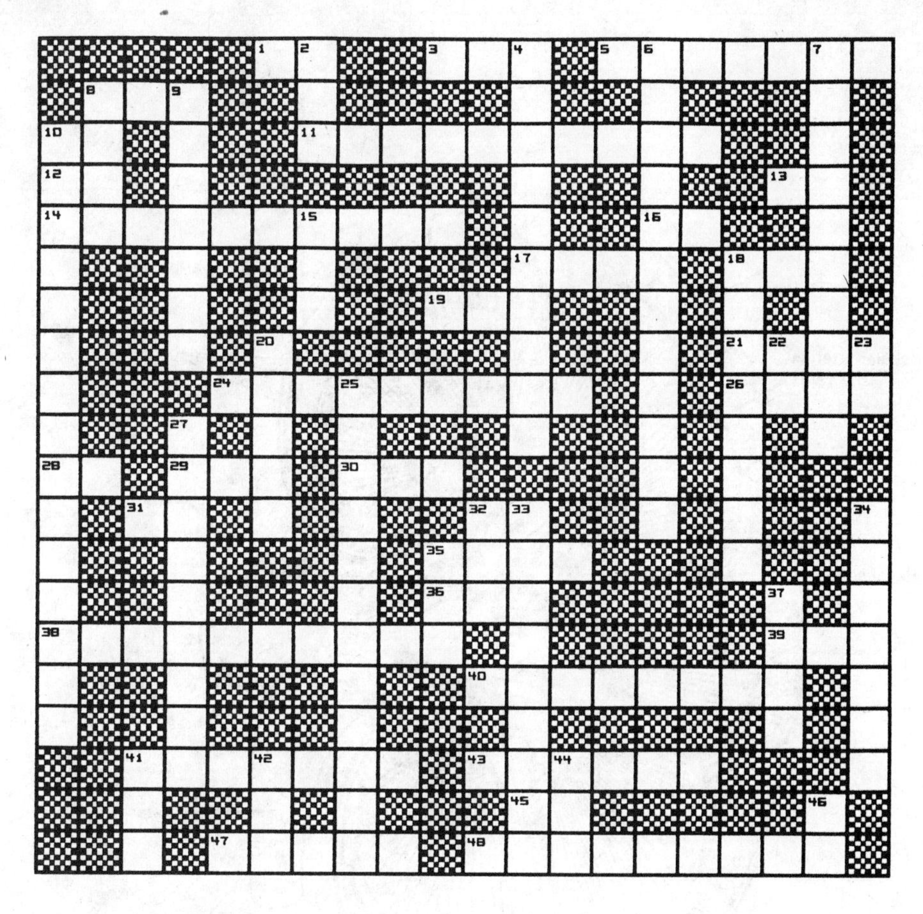

Across

1 Master of Science (abbr.)
3 What does "afferent" mean?
5 Kidney unit
8 Weep
10 American optical (abbr.)
11 Where most of the reabsorption occurs: proximal ________ tubule
12 Greek letter
13 ________ tract—intestine
14 Filtration unit
16 Transcendental Meditation (abbr.)
17 Gel medium for bacterial growth
18 Federal intelligence organization (abbr.)
19 Tubule close to glomerulus (abbr.)
21 Halt
24 Means of expelling hydrogen ions in kidney
26 Car's rubber roller
28 Measure of total solids (abbr.)
29 ________ favor (to Juanita)
30 Hormone involved in water retention (abbr.)
31 Scale to measure acidity
32 United Artists (abbr.)
35 12 deciders
36 Intestine
38 Condition of glucose in the urine
39 Cheer
40 Force
41 What does not filter through glomerulus
43 Outer shell of kidney
45 "The" to Francoise
47 Shape of kidneys
48 Blood in the urine

Down

2 1/60 of a minute (abbr.)
4 Process at the glomerulus
6 Ions
7 Water flow that must occur
8 Inner essence/conscience
9 Capsule into which filtrate pours
10 Molecule that gets activated by renin
15 Infection of excretory system (abbr.)
18 Bladder infection
20 Prefix meaning "behind"
22 Note on musical scale
23 Gym class (abbr.)
25 Process to regain glucose
27 Muscle at exit of bladder
32 Code on RNA, for example
33 Tiny blood vessel leading into glomerulus
34 Renal processing unit
35 Area that secretes renin
37 Chemical carrying nitrogenous waste in blood
41 Mendel's vegetable
42 Golfer's pin
44 Dream sleep
46 North America (abbr.)

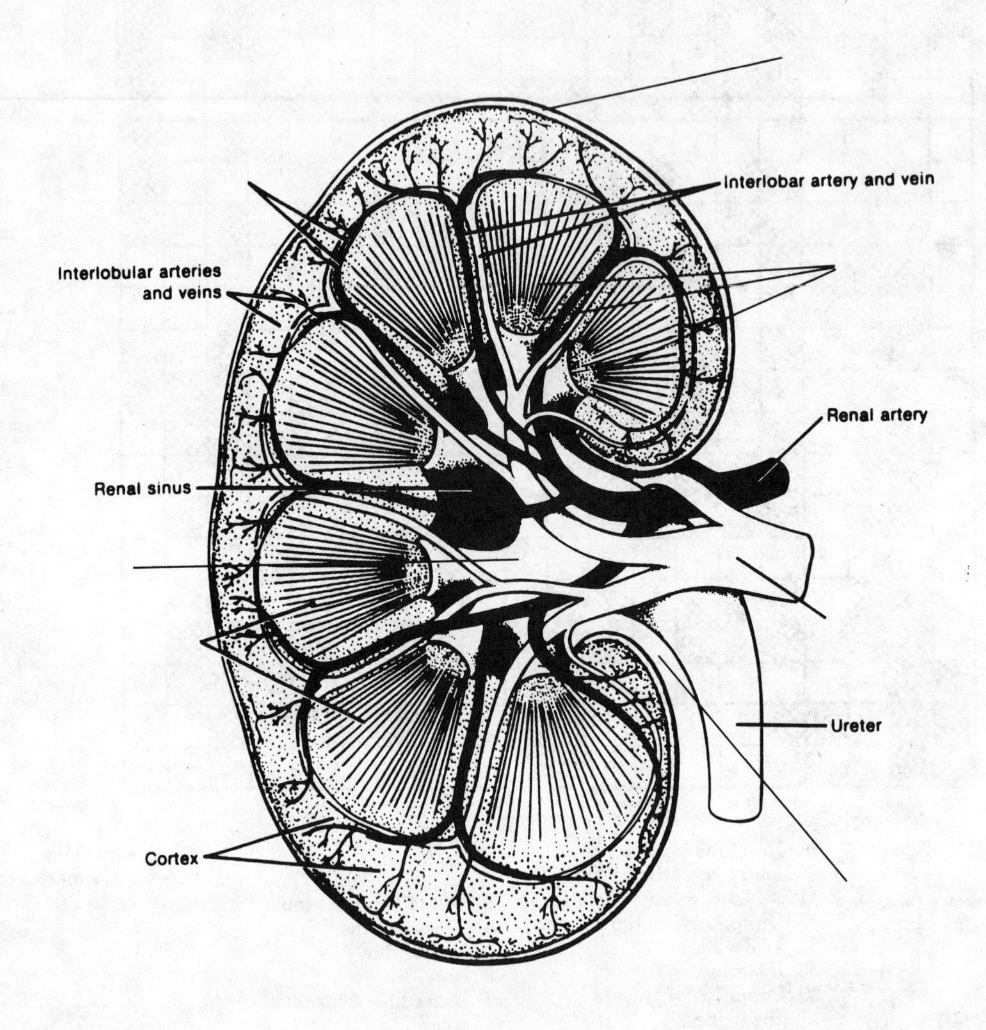

Anatomic Artwork

Figure 1

1. Label the renal saclike structure, the edge of which contains cuplike extensions that drain the pyramids.
2. Shade in the striated structures composed of straight collecting ducts.
3. Label the fibrous outer covering of the kidney.
4. Label the area of the kidney where blood vessels enter and leave.
5. Label the arc-shaped blood vessel that ultimately branches into the afferent arteriole.
6. Label the blood vessel that ultimately drains the peritubular network.
7. Shade in the area of the medulla.

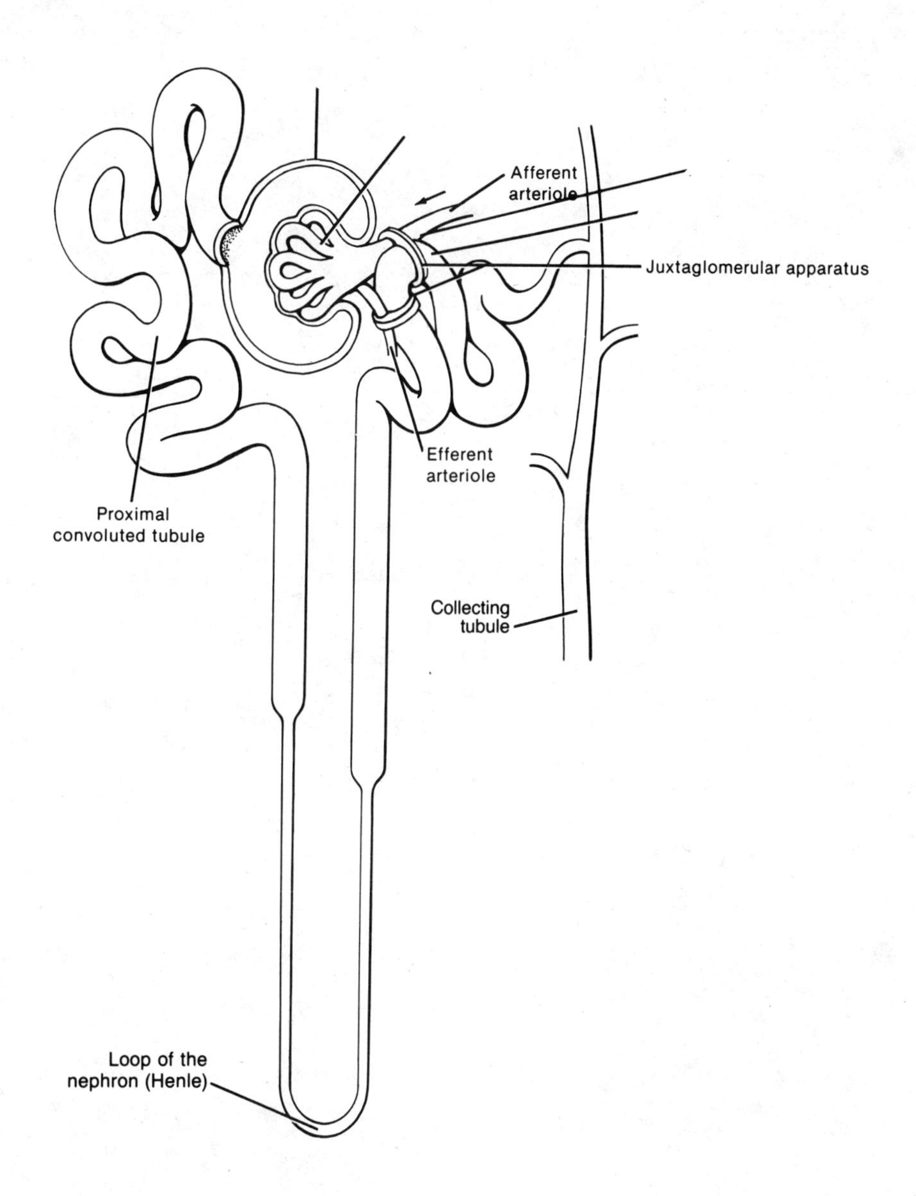

Figure 2

1. Label the structure whose cells become modified as the macula densa.
2. Label the cells that become modified to secrete renin.
3. Label the structure that contains blood and is allowing it to be filtered.
4. Label the structure into which the filtrate first flows.

Chapter 22

Fluid, Electrolyte, and Acid–Base Balance

Introduction

The blood serves the important purpose of keeping the body in health and in homeostatic balance. In doing so, it must be in ready contact with the cells and the various other fluid compartments of the body. The *fluid balance* refers to the proportion of water in each of these compartments, and the main determinant of how and where this fluid moves is the *concentration of solutes* in blood (plasma), bathing the cells (interstitial fluid), and within each cell (intracellular compartment).

After you have become familiar in the textbook with these fluids, and their constituents and movements, you should test your understanding with the following activities.

Multiple Choice

___ 1. Which of these is/are involved in controlling fluid output?
  a. ADH
  b. aldosterone
  c. sodium levels
  d. all of these

___ 2. Severe overhydration is also called:
  a. dehydration
  b. water intoxication
  c. interstitial pressure
  d. none of these

___ 3. Hypokalemia refers to:
  a. high potassium in blood
  b. low potassium in blood
  c. high calcium in blood
  d. low calcium in blood

___ 4. Osmotic pressure of blood is mainly created by:
  a. protein and sodium ion
  b. chloride ion
  c. aldosterone
  d. ADH

___ 5. When carbon dioxide combines with water it leads (immediately and eventually) to:
a. carbonic acid production
b. hydrogen ion production
c. bicarbonate ion production
d. all of these

___ 6. Which of these acts as a buffer?
a. carbonic acid – bicarbonate
b. phosphate
c. proteins
d. all of these

___ 7. A person who has become acidic due to metabolic reasons will:
a. hypoventilate
b. hyperventilate
c. both a and b
d. neither a nor b

___ 8. The body fluids of a person whose respirations are shallow will become:
a. acidic
b. alkaline
c. neither a nor b
d. there is no relationship

___ 9. Which would occur as acidosis sets in?
a. hypoventilation to compensate
b. kidney would excrete hydrogen ions
c. the pH would move up to near 7.5
d. none of these

___ 10. Proteins can act as buffers by:
a. combining with hydrogen ions
b. releasing hydrogen ions
c. both a and b
d. neither a nor b

True/False

__________ 1. Plasma contains *many* protein anions compared to interstitial fluid.

__________ 2. *Hyperkalemia* can cause death from fibrillation of the heart.

__________ 3. Aldosterone acts on the *adrenal medulla*.

__________ 4. Hypernatremia may be caused by *excess* water loss.

__________ 5. Calcium is largely an *extracellular* ion.

__________ 6. Calcium concentration in the blood is controlled by *ADH and aldosterone*.

__________ 7. Negatively charged chloride ions frequently follow *sodium* ions across membranes.

__________ 8. The number of solute particles in a glucose solution is *the same* as an equal concentration of a sodium chloride solution.

__________ 9. Magnesium levels are regulated by *aldosterone* secretion.

__________ 10. Blood hydrostatic pressure *opposes* blood osmotic pressure.

Completion

1. The primary source of body fluid comes from ________________.
2. Most of the fluid in the body is located in the ________________ compartment.
3. The movement of water from an area with fewer particles in solution to one with a greater concentration of particles is called ________________.
4. Chemicals that function to prevent drastic changes in pH are called ________________.
5. The body usually produces more ________________ compounds than basic ones.
6. The pressure of water against the capillary wall is known as ________________.
7. A positive electrolyte is known as a(n) ________________, while a negative one is a(n) ________________.
8. The amount of fluid is regulated in the body by the hormones ________________ and ________________.
9. The major intracellular cation is ________________.
10. The hormone that controls sodium, potassium, and magnesium levels is ________________.

Lost Sheep

1. acidosis, hyperventilation causes, excess H^+,increased ketone production
2. hypoventilation, buildup of CO_2, excess H^+, respiratory alkalosis
3. ketones, carbonic acid–bicarbonate, phosphate, protein
4. +8 mm Hg pressure, net arterial-end capillary pressure, fluid moves out, fluid moves from interstitial spaces to plasma
5. aldosterone, sodium regulation, potassium regulation, net filtration pressure
6. regulated by calcitonin, most abundant ion in body, regulated by PTH, anion
7. chloride, cation, potassium, intracellular
8. magnesium, extracellular, activates enzymes controlling carbohydrate metabolism, regulated by aldosterone
9. water release, ADH, decrease in osmotic pressure, adjusts high electrolyte content of body fluids
10. marrow, aldosterone, distal convoluted tubule, increases sodium reabsorption

Matching

Set 1

___	1. activates the sodium–potassium pump	a.	sodium
___	2. major extracellular anion	b.	potassium
___	3. most abundant ion in body	c.	calcium
___	4. most abundant cation in extracellular fluid	d.	magnesium
___	5. creates most of the osmotic pressure in extracellular fluid	e.	chloride

Set 2

Choose as many items from right-hand column as apply

___ 1. two-thirds of body fluid
___ 2. extracellular fluid
___ 3. in vessels
___ 4. contains significant quantities of phosphate
___ 5. highest protein content

a. synovial
b. plasma
c. ICF

Double Crosses

1. *Across*:
helps create osmotic pressure in blood

Down:
major solvent in body

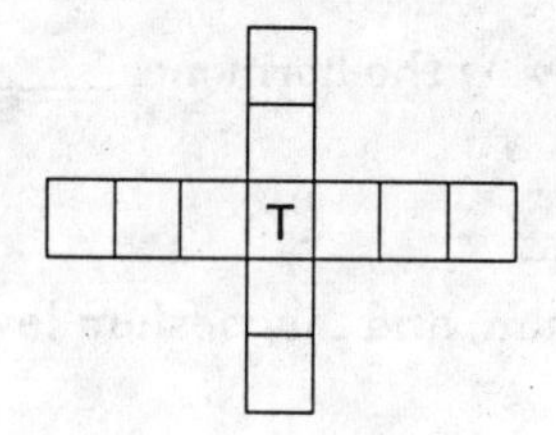

2. *Across*:
to lose water

Down:
hormone that controls water balance and total concentration of electrolytes

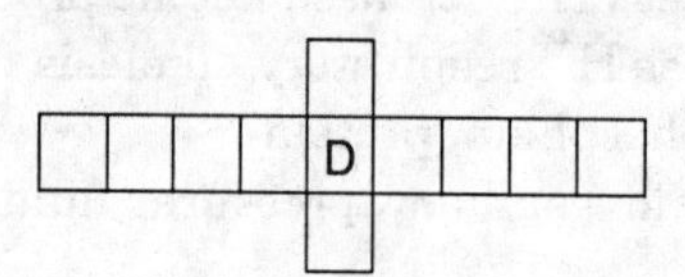

3. *Across*:
an ion

Down:
cellular ions

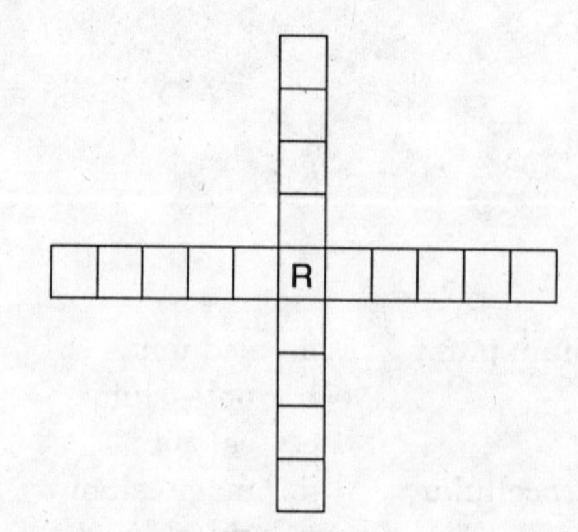

Sleuthing

1. A young child comes to the hospital suffering from malnutrition and presenting with a swollen belly.
   a. Why is her belly swollen?

   b. What pressure has dropped?

   c. Where is the fluid accumulating in her body?

Addagrams

1. a. high-energy molecule	17, 19, 16
b. pressure created in blood by proteins	14, 6, 15, 14, 23, 1, 13
c. referring to kidney	18, 21, 22, 17, 12
d. ________ lyte	4, 12, 21, 13, 7, 5, 14
e. physiological response to imbalance	13, 14, 20, 16, 21, 2, 11, 9, 3, 4
f. symbol for element needed to make thyroxin	8
g. symbol for element actively transported into thyroid	10

Total: nice warm bath

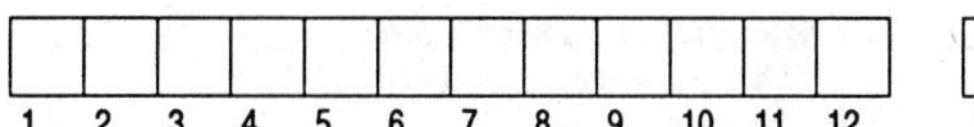

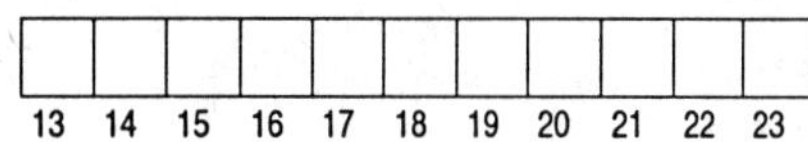

Crossword

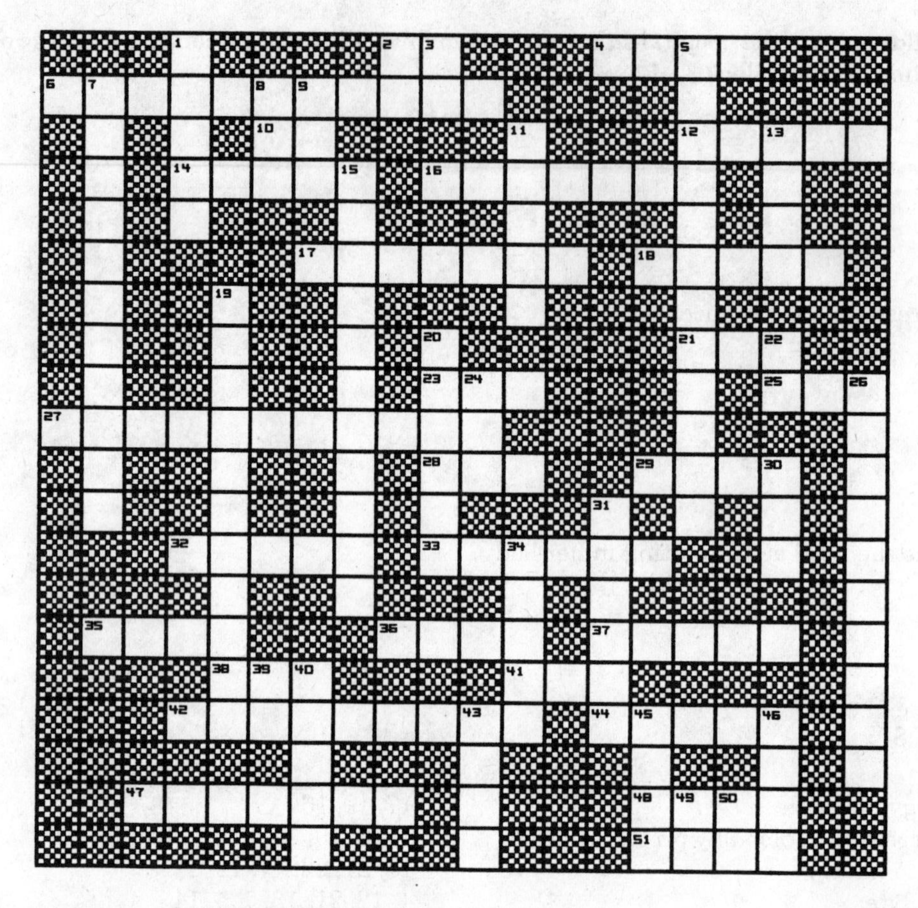

Across

2 Medicine man (abbr.)
4 Lowers pH
6 Loss of water
10 Preposition
12 ________iodothyronine: T4
14 Go in
16 What a buffer helps maintain, acid-base ________
17 Pressure created by higher concentration
18 ________cyte, type of neurolgia in CNS
21 Fluid inside cell
23 Gluco__________genesis
25 King topper in cards
27 Pressure of water against its container
28 ________ again, on again
29 Raises pH
32 Prevent
33 Major positive extracellular ion
35 Prefix meaning "blood"
36 Posterior
37 Liquid
38 Immediately
41 Solid state of water
42 A type of acidosis
44 Referring to the kidney
47 Creates osmotic pressure in plasma
48 What pH will do as solution becomes more acidic
51 Upon

Down

1 Prefix meaning "higher"
2 Heart attack (abbr.)
3 Enact
5 Space in-between cells
7 What sodium or chloride or potassium or calcium is
8 Lab animal
9 Yesterday's lunch activity
11 Aldosterone helps keep it in balance in body
13 Damage to muscle
15 A type of alkalosis
19 Hormone involved in fluid retention
20 Negative ions
22 Fourth note of musical scale
24 Fluid outside cell (abbr.)
26 One of four major tissue types
30 What compartment is plasma part of? (abbr.)
31 Helps maintain balance of pH
34 Type of column
39 Occupational Therapist (abbr.)
40 The solvent in the body
43 god
45 Prefix meaning meaning "within"
46 Prefix for molecules containing fats, steroids, etc.
49 Best friend in hospital (abbr.)
50 Friend to a P.T.

Chapter 23

The Reproductive Systems

Introduction

The systems you have studied thus far have dealt with the *homeostatic maintenance* of the human organism. Uniquely, the reproductive system ensures the *continuity* of the human species. We have an anatomical and physiological mechanism that can provide both the material (reduced DNA in the *egg* and *sperm*) and the mechanism (*copulation, fertilization,* and *gestation*) for this function of continuity.

The structures of the reproductive systems of the male and female can be grouped according to simple categories of function: *gonads* (preparation of genetic material), *tubules* (transport), *copulatory structures* (means of insemination), and *accessory structures* (facilitation). Females and males are alike in the production of *gametes* (egg and sperm), which have half the chromosome number, but the timing and control of these events are quite different.

After completing the reading in your text, you should have an understanding of the overall structure and physiology of the reproductive system. The events of copulation and gestation are included in the next chapter.

Multiple Choice

___ 1. Which of the following structures is/are homologous?
   a. clitoris–glans penis
   b. labia majora–scrotum
   c. Bartholin's glands–Cowper's glands
   d. all are homologous

___ 2. Which of the following cells is haploid?
   a. spermatogonium
   b. primary spermatocyte
   c. spermatid
   d. all of these

___ 3. Enzymes required for sperm penetration of the secondary oocyte are concentrated in the:
   a. midpiece
   b. acrosome
   c. spermatid
   d. glans

___ 4. Sugar required for sperm mobility is secreted by the:
a. spermatozoans
b. prostate gland
c. Sertoli cells
d. seminal vesicles

___ 5. Testosterone is secreted by:
a. interstitial cells
b. Sertoli cells
c. seminiferous tubules
d. sperm cells

___ 6. Sex hormones:
a. stimulate normal development and maintenance of the reproductive tract
b. trigger the development of secondary sex characteristics
c. both a and b
d. neither a nor b

___ 7. FSH secretion is restricted due to the action of which hormone?
a. inhibin
b. estrogen
c. both a and b
d. neither a nor b

___ 8. Which of the following statements is/are true?
a. The rete testis is a network of ducts in the testis
b. Sperm maturation is completed in the ejaculatory duct
c. Seminal fluid is produced by the vas deferens and urethra
d. Seminiferous tubules contribute over 50 percent of the testosterone released by the testes

___ 9. Which structure is *not* found in the spermatic cord?
a. vas deferens
b. testicular blood vessels
c. epididymus
d. testicular nerves

___ 10. Semen contains:
a. sugar
b. mucus
c. enzymes
d. all of these

___ 11. The urethra:
a. passes through the prostate gland
b. runs through the corpus cavernosum
c. contributes the bulk of seminal fluid
d. all of these

___ 12. Erection:
a. is a parasympathetic reflex
b. is due to blood filling the sinuses of the penis
c. both a and b
d. neither a nor b

___ 13. A Graafian follicle:
a. is a mature follicle
b. secretes estrogens
c. under the influence of LH releases a secondary oocyte
d. all of these

___ 14. The funnel-shaped opening of the Fallopian tube is called the:
a. perineum
b. infundibulum
c. pudendum
d. none of these

___ 15. The endometrium:
a. has two layers: stratum basalis and functionalis
b. responds to hormonal messages and increases its thickness and vascularity
c. can grow in places outside the uterus causing endometriosis
d. all of these

___ 16. Progesterone:
a. works with estrogen to prepare the endometrium for implantation
b. promotes the development of the corpus luteum
c. when elevated in blood inhibits GnRH
d. all of these

___ 17. Which of the following is correctly paired?
a. low progesterone–inhibits estrogen
b. elevated inhibin–decreases in FSH
c. low estrogen level–inhibits FSH
d. elevated FSH–low GnRH

___ 18. Which hormone is primarily responsible for ovulation?
a. FSH
b. LH
c. inhibin
d. relaxin

___ 19. The major source of progesterone in a non-pregnant female is:
a. corpus luteum
b. corpus albicans
c. germinal epithelium
d. none of these

___ 20. The terms "*Treponema*," "chancre," and "tertiary stage" are associated with which disease?
a. trichomoniasis
b. herpes
c. syphilis
d. gonorrhea

___ 21. Which disease, if untreated, can eventually result in sterility?
a. gonorrhea
b. chlamydia
c. both a and b
d. neither a nor b

___ 22. Genital herpes is:
a. transmitted sexually
b. caused by the herpes simplex virus, Type II
c. a lifelong infection
d. all of these

___ 23. Which of the following is correctly paired?
a. impotence–infertility
b. gonorrhea–simplex virus
c. amenorrhea–painful menstruation
d. none of these

___ 24. Abnormal changes in shape, growth, and number of certain uterine cells is called:
a. toxic shock syndrome
b. cervical dysplasia
c. pelvic inflammatory disease
d. none of these

___ 25. Gonads:
a. produce gametes
b. produce seminal fluid
c. both a and b
d. neither a nor b

True/False

__________ 1. Toxic shock syndrome is caused primarily by *Trichomonas vaginalis.*

__________ 2. The onset of menstruation is called *menarche.*

__________ 3. Mammary glands are modified *sudoriferous* glands.

__________ 4. Fertilization of the egg takes place in the *fundus of the uterus.*

__________ 5. The *myometrium,* under hormonal influences, cyclically undergoes changes in thickness, vascularity, and glandular activity.

__________ 6. The *corpus luteum* is a capsule of connective tissue below the germinal epithelium of the ovary.

__________ 7. Shedding of the endometrium is called *menstruation.*

__________ 8. Removal of the prepuce is called *circumcision.*

__________ 9. Seminal fluid is produced in part by the *seminal vesicles, prostate gland,* and *bulbourethral glands.*

__________ 10. A sperm count below 20 million/ml is associated with *sterility.*

__________ 11. The spermatic cord passes through the *prostate gland.*

__________ 12. The *ejaculatory duct* is the terminal duct in the male reproductive system.

__________ 13. *Inhibin* is secreted by Sertoli cells and corpus luteum.

__________ 14. *Spermatogonia* are diploid cells with 46 chromosomes.

__________ 15. Two chromosomes that belong to a pair of chromosomes are called *heterologous* chromosomes.

__________ 16. Testosterone is synthesized from cholesterol and acetyl coenzyme A by the *interstitial cells of Leydig.*

__________ 17. When the egg and spermatozoon unite, the resulting cell is called a *zygote.*

__________ 18. Failure of the testes to descend is called *impotence.*

__________ 19. *Meiosis II* is a reduction division.

__________ 20. Developing spermatids are nourished by *Leydig* cells.

__________ 21. The *ejaculatory duct* is formed by the union of the vas deferens and duct of the seminal vesicle.

__________ 22. The seminal vesicles, prostate, and Cowper's glands are *accessory* glands.

__________ 23. The covering of the glans penis is called the *prepuce.*

__________ 24. Semen is slightly alkaline due to secretion of fluid from the *prostate gland.*

__________ 25. Once ejaculated, semen coagulates due to a clotting enzyme produced by the *prostate gland.*

Completion

1. The maturation of spermatid into spermatozoa is called __________.
2. Mitochondria, which provide energy for locomotion, are found in the __________ of the sperm.
3. __________ acts in the seminiferous tubules to initiate spermatogenesis.
4. Secondary sex characteristics begin to develop at __________.
5. The __________ expels sperm into the urethra.
6. The average life span of ejaculated sperm is __________ hours.
7. The sugar secreted into semen for sperm mobility is __________ and it is released by the __________.
8. The enlarged distal end of the corpus spongiosum is called the __________.
9. The pH range of semen is __________.
10. The formation of a haploid ovum is called __________.
11. The primary oocyte divides to produce a(n) __________ and __________.
12. The fingerlike projections that help gather the egg into the Fallopian tube are called __________.
13. The __________ is the constricted neck of the uterus which opens into the vagina.
14. The layer of the uterus that changes cyclically is called the __________.
15. Transverse folds in the vaginal mucosa are called __________.
16. The circular pigmented area of skin surrounding the nipple of the breast is referred to as the __________.
17. The reduction in frequency of menstrual cycles is known as the __________ and is associated with a decrease in the levels of __________ in blood.
18. If fertilization does not occur, the corpus luteum degenerates into the __________.
19. In an average 28-day menstrual cycle, ovulation occurs on day __________.
20. Early in the preovulatory phase __________ is the the dominant pituitary hormone.
21. Sexually transmitted diseases are also known as __________ diseases.
22. If untreated, __________ is a progressive venereal disease that can span years and invade the brain, cardiovascular system, and skeletal system.

23. Absence of menstruation, known as _______________, can be due to continuous rigorous athletic training.
24. Inability to achieve or maintain an erection is called _______________.
25. Extensive bacterial infection of the uterus, Fallopian tubes, and/or ovaries is called _______________.

Lost Sheep

1. cryptorchidism, sperm production, prostate, scrotum
2. Sertoli cells, Leydig cells, spermatogonia, Graafian cells
3. meiosis I, spermatids, tetrad, crossing-over
4. puberty, LH secretion surges, initiation of spermatogenesis, age 8
5. spermatic cord, rete testis, testicular artery, vas deferens
6. fructose, seminal vesicle, alkaline, 30 percent volume of semen
7. corpus spongiosum, penis, corpus cavernosa, Cowper's glands
8. circumcision, glans, foreskin, bulbourethral
9. semen, bulbourethral, prostate, epididymis
10. inhibin, lower FSH levels, decreases spermatogenesis, Leydig cells
11. stroma, Fallopian tube, germinal epithelium, follicle
12. oogonia, primary oocyte, $2N$, haploid
13. fundus, fimbriae, infundibulum, Fallopian tube
14. cervix, fornix, endometrium, broad ligament
15. Bartholin's glands, Skene's glands, Cowper's glands, vagina
16. alveoli, Cooper's ligaments, areolar, perineum
17. elevated estrogen, menarche, development of secondary sex characteristics, menopause
18. degenerating functionalis, menstruation, elevated estrogen, endometrial breakdown
19. follicle rupture, LH surge, ovulation, menstruation
20. necrosis of endometrium, corpus albicans formation, preovulatory phase, LH level decreases

Matching

Set 1

___ 1. secretes testosterone
___ 2. fructose-rich secretion
___ 3. progesterone secretion
___ 4. secretes inhibin
___ 5. citric acid-rich secretion
___ 6. estrogen secretion
___ 7. secretes FSH
___ 8. secretes relaxin

a. sustentacular cells
b. pituitary gland
c. corpus luteum
d. interstitial cells
e. follicular cells
f. prostate gland
g. seminal vesicle

Set 2

___ 1. syphilis
___ 2. gonorrhea
___ 3. herpes
___ 4. impotence
___ 5. PMS
___ 6. TSS
___ 7. infertility

a. *Neisseria*
b. inability to maintain erection
c. simplex virus
d. *Staphylococcus*
e. chancre
f. low sperm count
g physical and emotional stress during postovulatory phase

Set 3

___ 1. sperm production	a.	vas deferens
___ 2. Fallopian tube	b.	ovary
___ 3. uterus	c.	epididymis
___ 4. spermatic cord	d.	cervix
___ 5. tunica albuginea	e.	corpus spongiosum
___ 6. sperm maturation	f.	seminiferous tubule
___ 7. Bartholin's glands	g.	fimbriae
___ 8. mammary glands	h.	seminal vesicles
___ 9. accessory gland	i.	vaginal orifice
___ 10. urethra	j.	Cooper's ligaments

Double Crosses

1. *Across*:
   spermatid formation

   Down:
   a. location for seminiferous tubules
   b. place of enzyme storage
   c. pouch
   d. tube with fimbriae
   e. hormone for pubic symphysis flexibility
   f. ductus ________
   g. primitive egg "houses"
   h. physiological "coming of age"
   i. cellular "crown"
   j. urethra in relationship to birth canal
   k. level of LH the day prior to ovulation
   l. tubal pregnancy
   m. inhibits FSH

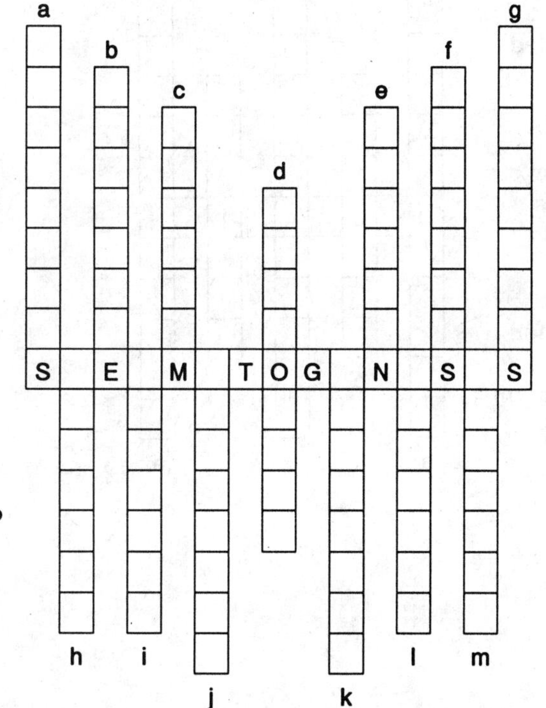

2. *Across*:
   lining

   Down:
   a. egg production
   b. female regulators
   c. "sex" cells
   d. spongy columns
   e. egg covering

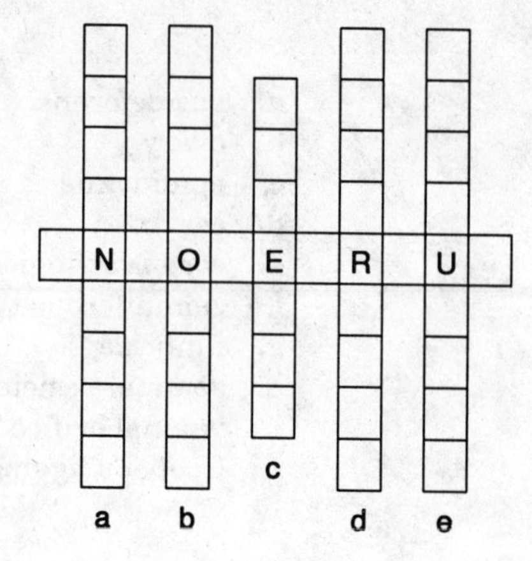

3. *Across*:
   female process controlled by hormones

   Down:
   a. information transporters
   b. testes compartments
   c. gametes formed continuously
   d. fluid that provides a transport medium for sperm
   e. what Pap smear or mammography is an example of: medical ________

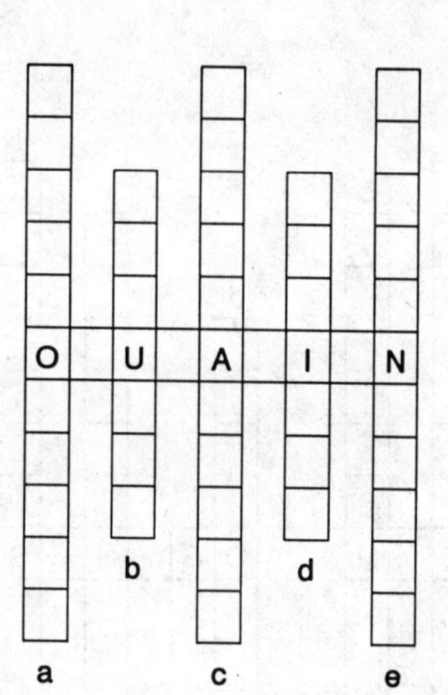

Sleuthing

1. Debbie has had unusual periodic abdominal pain associated with her menstrual cycle. When her gynecologist examined her, she palpated an enlarged ovary. Upon laparoscopic examination, she found a large endometrial cyst on Debbie's right ovary and removed the affected organ.
   a. Why is this called an "endometrial cyst"?

   b. What is happening to it each month?

c. Describe the cyclic changes that endometrial tissue normally undergoes in the uterus each month.

2. Mike had a vasectomy two years ago.
   a. What portion of the reproductive tract was ligated?

   b. Can he still have an erection? Explain.

   c. Can he have an ejaculation? Explain.

   d. Since males have a urogenital system, does this procedure affect the excretory system? Explain.

Word Scrambles

1. a. constricted neck	ERCXIV	□ _ □ _ □ □
b. a covering	OEKNIRFS	_ □ _ □ □ □ _ □
c. 2*N* or *N* depending on stage	EPMTCTEYOARS	□ _ _ _ □ _ _ □ □ □ _ _
d. produces gametes	GAODN	_ □ _ _ □
e. in males has a double purpose	RTRUAEH	_ _ _ □ □ _ _

Total: may be caused by virus-infected bacterium

□□□□□ □□□□□ □□□□□□□□

2. a. enzymes are found here	ASROCOEM	_ □ □ _ _ _ □ _
b. pea-sized glands	POECSWR	□ _ _ □ _ □ _
c. maturation site	DIYEPMIDSI	□ _ _ _ _ □ _ _ _ □
d. sometimes it disappears	TUCOLINAFNIS	_ _ _ _ □ _ □ _ □ _ □ _

Total: housing for vessels, duct, and nerves

□□□□□□□□□ □□□□

Addagrams

1. a. nurse cells — 4, 7, 4, 5, 2, 3, 5, 8, 10, 7, 9, 8, 6
   b. fluid-filled growth — 12, 11, 4, 5
   c. body that secretes progesterone: corpus ________ — 13, 7, 5, 14, 7, 1

Total: a timely variation

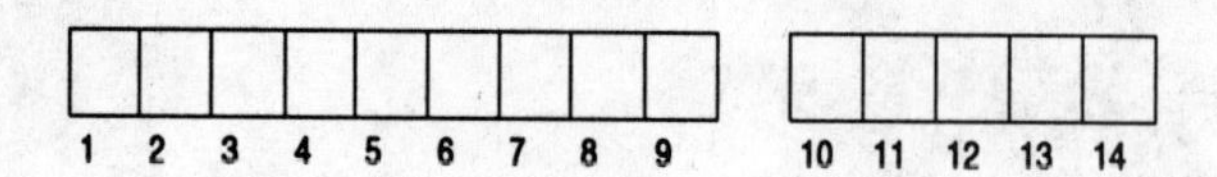

2. a. male or female problem — 1, 2, 19, 4, 5, 3, 8, 12, 10, 9, 22
   b. place where eggs "hatch" — 19, 18, 20, 16, 24, 13, 15, 21, 17
   c. a head — 25, 12, 11, 2, 6
   d. chromosome group that does the "twist" — 7, 14, 9, 5, 11, 23

Total: a man's best friends

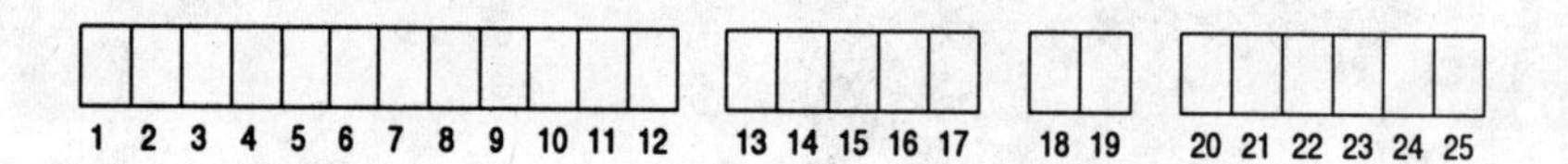

Crossword

Across

3 Holy book
6 Cry
9 Cells that secrete testosterone
12 Tubal pregnancy
14 Outer fat-filled flaps
16 World War II pilot with many "kills"
17 Mother (abbr.)
18 Promotes development of progesterone-secreting structure
20 What the chromosome number is in sperm and egg
21 What LH will do just prior to ovulation
22 Network of testis
23 Ballpoint pen
28 Promotes spermatogenesis (abbr.)
30 Near to
31 Constricted neck of uterus
32 Northeastern school (abbr.)
34 One of the endometrial builders
37 British Thermal Unit (abbr.)
41 Cloth
42 Movie cop
44 Birth canal
46 Where peas are found
48 Talking horse
49 Produces gametes
50 Provides for sperm motility
52 Cup
53 Contains enzyme for breakdown of egg covering
56 Where sperm cell mature
61 Tombstone initials
62 Toward
63 Classified unit
64 I am, you are, it ________
66 Copulatory male organ
67 What an archeologist does
69 In regard to
71 Ductus that travels in spermatic cord
73 Not the usual milk producers
75 Veterans' Administration (abbr.)
77 Modus Operandi (abbr.)
78 Writing instrument
79 Part of the nervous system which controls secretion
82 These vesicles produce the bulk of semen
85 Not out
86 First phase of female cycle
87 What muscles have
88 A term often found in a hypothesis

Down

1 Large animal
2 Womb
3 Blood pressure (abbr.)
4 Three to a Roman
5 Old Testament time frame (abbr.)
6 The chromosome count in a polar body and secondary oocyte is the ________
7 "Delivery" branch of medicine (abbr.)
8 Prefix meaning two
10 Affirmative answer
11 Visitor
13 Cardiac Care Unit (abbr.)
14 Parcel of land
15 Scar tissue
17 Periodic discharge
19 Not lo
24 Contains vas deferens and vessels
25 Has a slightly lower temperature
26 Source of estrogen
27 Tubal projections
28 Swirls air
29 Half the chromosome number state
33 Secreted by seminal vesicles for sperm mobility
35 Promotes the development of secondary sex characteristics in females
36 Toward
38 What the sun can cause in the skin
39 Area surrounding nipple
40 California (abbr.)
43 Cooking apparatus
45 Sperm or egg
47 Irish father
48 Contains a series of labyrinths
49 Something to chew
51 Interstitial cells of ________
52 Me, in Paris
54 Either/________
55 Produced in seminiferous tubules
57 Small horse
58 Run quickly
59 The one who contributes the Y chromosome (abbr.)
60 Nurse cells
64 Personality component
65 Contains alkaline solution, sugar, mucus, and sperm
68 Leave
70 Form of psychotherapy
72 Posted a letter yesterday
74 Doctors' organization (abbr.)
75 Ex-soldier (abbr.)
76 Latin to love
78 Math constant
79 "________ You Like It"
80 Mister in Mexico (abbr.)
81 Found in bone (abbr.)
83 Not out
84 Sympatomimetic hormone (abbr.)

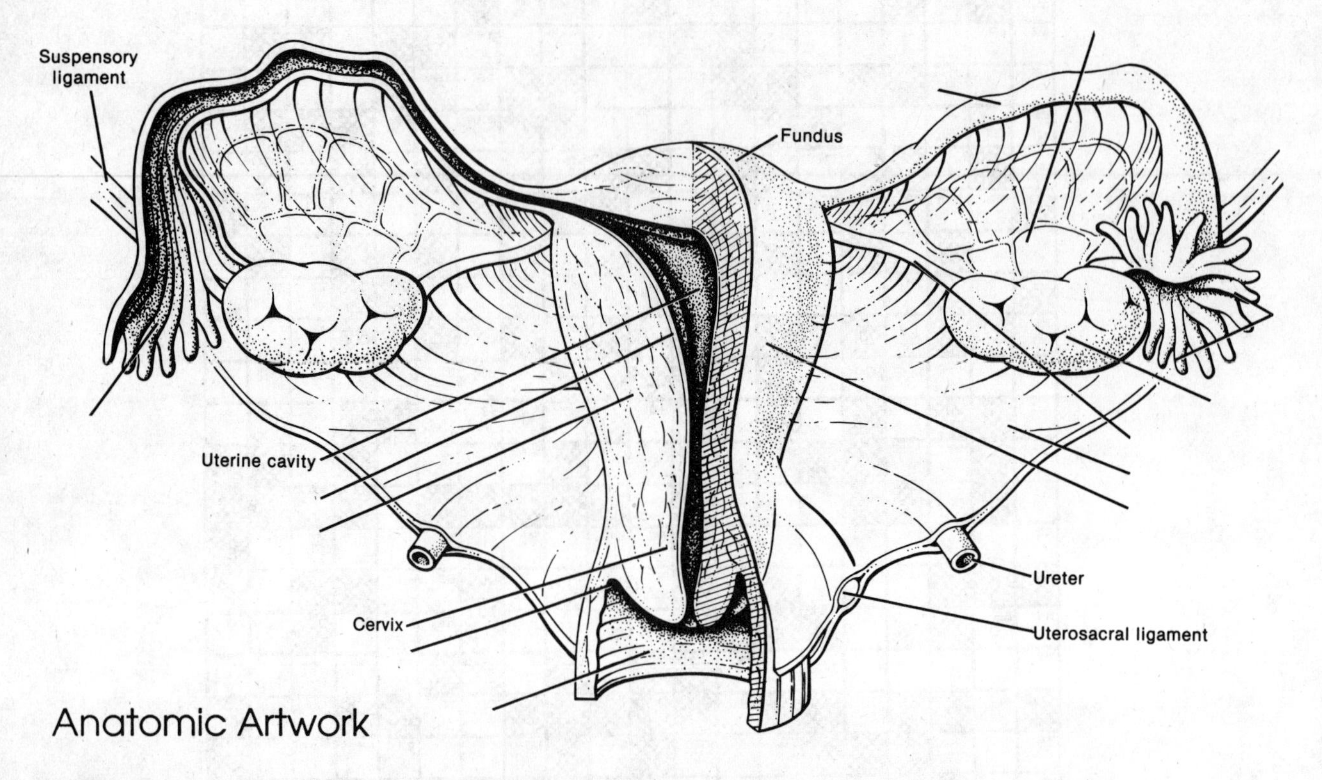

Anatomic Artwork

Figure 1

1. Label the structure where fertilization takes place.
2. Label the organ where the corpus albicans would be formed.
3. Label the structures that help to "catch" the egg upon ovulation.
4. Label all the ligaments that suspend the ovaries in place.
5. Label the structure where rugae are found.
6. Label the layer of the uterus that changes cyclically.
7. Label the organ where the fertilized egg normally implants.
8. Label the layer of the uterus that is responsible for the contractions of labor.
9. Label the enlarged opening of the Fallopian tube.
10. Label the structure that makes use of the contraceptive diaphragm possible.

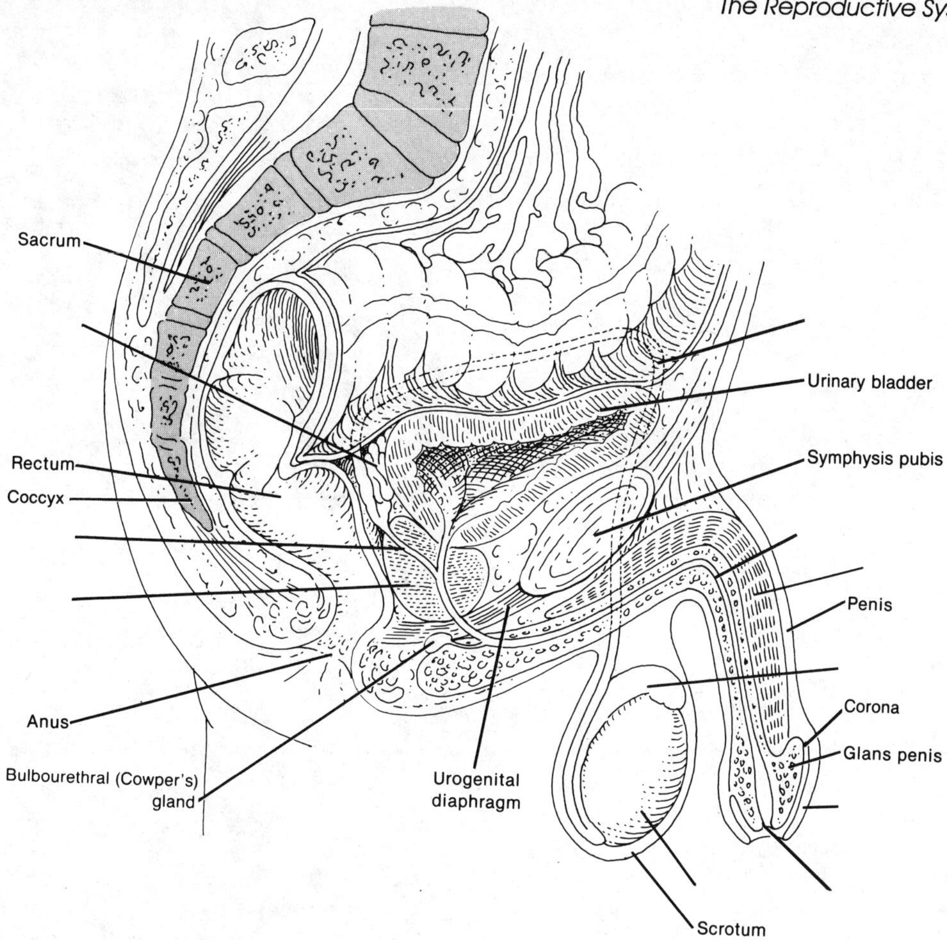

Figure 2

1. Label the structure where sperm "mature."
2. Label the structure that is ligated for contraceptive purposes.
3. Label the structures that produce the bulk of seminal fluid.
4. Label the passage that is common to the excretory and reproductive systems.
5. Label the structure that is removed during circumcision.
6. Label the structures that become engorged with blood during erection.
7. Label the organ where spermatogenesis takes place.
8. Label the structure that ejects sperm into the urethra.
9. Label the gland that produces a citric acid-rich secretion.
10. Label the final exit of sperm from the penis.

[illegible]

[illegible] the [illegible]

[illegible]

[illegible]

[illegible]

[illegible]

[illegible]

[illegible]

[illegible] the [illegible]

Chapter 24

Development and Inheritance

Introduction

The *zygote,* which is formed at the time of *fertilization,* has a full, unique complement of chromosomal material, having received 23 chromosomes from each parent. From the moment of fertilization onward, this single cell will grow to billions of cells by subsequent and repeated *mitoses.* Within just a week, the zygote grows from a single cell to a sphere of cells that encloses a fluid-filled cavity and implants into the uterus. Three basic *germ layers* develop, from which all body structures are derived. *Organogenesis* begins shortly and by three weeks the primitive heart is pumping an oxygen-carrying fluid through the *embryo.* The main organ systems are completed by three months and the lining of the uterus is maintained for the developing *fetus* by hormones secreted by the placenta, until *parturition* or birth.

After you have become familiar with the material on development and inheritance in your text, you should be able to answer these questions and work through the following activities.

Multiple Choice

___ 1. Which of these is not characteristic of the attachment of the fetus to the endometrium?
   a. Chorionic projections grow into the endometrium
   b. Chorionic villi are bathed in maternal blood sinuses
   c. Maternal and fetal blood mix
   d. None of these is true

___ 2. Which of these is/are correctly paired?
   a. mesoderm–backbone
   b. ectoderm–hair follicles
   c. endoderm–lining of urethra
   d. all

___ 3. hCG:
   a. is secreted by the placenta
   b. supports the corpus luteum
   c. both a and b
   d. neither a nor b

___ 4. The estrogens of pregnancy:
   a. are secreted by the corpus luteum throughout pregnancy
   b. are secreted by the placenta the first month of pregnancy
   c. influence the development of the mammary glands
   d. all of these

___ 5. The placenta:
   a. secretes estrogens
   b. secretes progesterone
   c. is released as "afterbirth"
   d. all of these

___ 6. Which of these is a normal development in the fetus?
   a. FAS
   b. anencephaly
   c. lanugo
   d. all of these

___ 7. Which is a significant factor for child development that is related to maternal smoking?
   a. cleft palate
   b. respiratory problems in year 1
   c. a tentative link to SIDS
   d. all of these

___ 8. Which of these is a natural method of birth control?
   a. barrier method
   b. depending on signs of ovulation
   c. both of these
   d. neither of these

___ 9. Which is true?
   a. The dominant gene is the trait that is never expressed
   b. The phenotype is the genetic makeup
   c. Someone who carries the same genes on homologous chromosomes is said to be homozygous for the trait
   d. All of these

___ 10. In pregnancy the:
   a. corpus luteum functions for 9–12 days
   b. mother's pituitary secretes LH
   c. corpus luteum releases FSH
   d. corpus luteum continues to secrete estrogens and progesterone for a short time

___ 11. Which is/are true?
   a. The genotype is the expression of genetic makeup
   b. Genes that are expressed differently when inherited from the mother or father are said to be examples of genomic imprinting
   c. The Punnett square is a device used to help the parents select the sex of the child
   d. All of these

___ 12. A person who has a recessive gene but does not express it, is said to be a(n):
   a. allelic
   b. imprinter
   c. carrier
   d. none of these

True/False

_____ 1. The layer under the corona radiata in the fertilized egg is known as the *corpus luteum.*

_____ 2. The attachment of the blastocyst to the *perimetrium* is called implantation.

_____ 3. The *"fetal period"* involves the time of life from the eighth week on.

_____ 4. The zygote undergoes successive divisions called *cleavage.*

_____ 5. The berrylike cluster of cells which forms in early cleavage is known as the *blastomere.*

_____ 6. Differentiation of the embryo occurs *after* implantation into the uterus.

_____ 7. The embryo's skeleton develops from the *mesoderm.*

_____ 8. In the placenta the baby's and mother's blood *intermingle.*

_____ 9. The ductus venosus bypasses the *liver* in the fetus.

_____ 10. The *foramen ovale* is an opening between the two atria in the fetal heart.

_____ 11. Estrogen and progesterone levels *suddenly drop* after fertilization.

_____ 12. The *umbilical artery* carries well-oxygenated blood.

_____ 13. The *placenta* secretes human chorionic gonadotropin, which stimulates the production of progesterone by the corpus luteum.

_____ 14. The process of birth is called *parturition.*

_____ 15. The *myometrium* of the uterus contracts to expel the fetus.

Completion

1. Fingerlike ingrowths from the developing embryo into the uterine wall are known as _____ .
2. The ectoderm of the embryo develops into _____ .
3. The endoderm of the embryo develops into _____ .
4. The mesoderm of the embryo develops into _____ .
5. During pregnancy, _____ and _____ , which are necessary for maintaining the uterus, are secreted by the placenta.
6. _____ , a hormone secreted by the hypothalamus, induces the rigorous contractions of labor.
7. The first fluid that emanates from the mother's breast is called _____ .
8. Taking a sample of epidermal cells in the fluid around a developing fetus is called _____ .
9. The sampling of chorionic villi by inserting a catheter into the uterus is called _____ .
10. The major hormone promoting lactation is _____ .
11. The contraceptive device that is inserted into the uterine cavity is called the _____ .
12. The contraceptive unit inserted into the upper vagina which blocks sperm entrance into the uterus is called the _____ .
13. The fertilized egg is surrounded by the _____ , which are several layers of follicular cells.

Lost Sheep

1. hollow, blastocyst, morula, fluid-filled
2. ectoderm, nerve tissue, epidermis, muscle
3. epithelium of glands, epithelium of respiratory tract, epithelium of skin, epithelium of vagina
4. villi, placenta, afterbirth, mesoderm
5. amnion, umbilical cord, chorion, yolk sac
6. estrogen, glucagon, relaxin, hCG
7. estrogen, hCG, FSH, progesterone
8. oxytocin, labor, uterine contractions, estrogens
9. relaxin, prolactin, lactation, estrogens
10. foramen ovale, ductus arteriosus, ductus venosus, fossa ovale
11. contraceptive sponge, pill, rhythm, cervical cap
12. ovulation, pregnancy, obstetrics, gestation

Matching

Set 1

___ 1. lining of respiratory tract
___ 2. epidermis
___ 3. nerve tissue
___ 4. heart muscle
___ 5. bones

a. ectoderm
b. mesoderm
c. endoderm

Set 2

___ 1. estrogen and progesterone
___ 2. covering over penis
___ 3. fits over cervix
___ 4. insets in uterus
___ 5. rhythm

a. IUD
b. condom
c. pill combination
d. diaphragm
e. natural means

Double Crosses

1. *Across*:
   a. factor or agent that causes birth defects
   b. cord that connects fetus to placenta
   c. secretion of milk

   Down:
   results in formation of a zygote

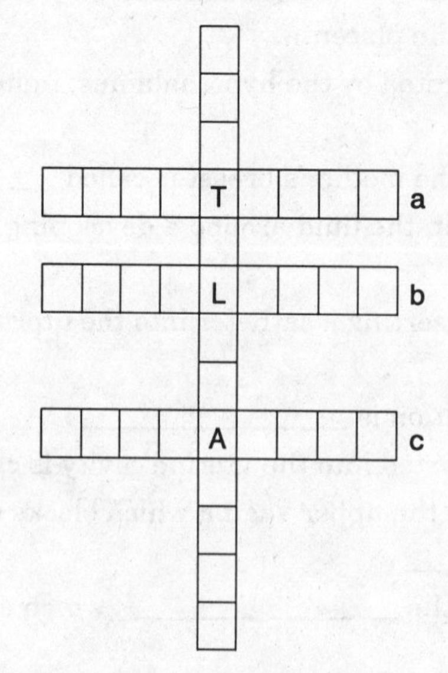

2. *Across*:
   a. first germ layer
   b. reduction division
   c. gestation
   d. first milk

   Down:
   process by which fetal defects can be detected

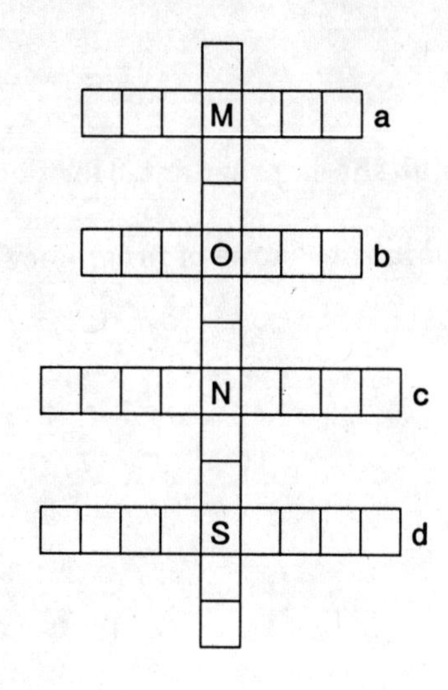

3. All words begin with the "E" at the center
   a. germ layer giving rise to lining of gut
   b. germ layer giving rise to epidermis
   c. period of first 8 weeks of life
   d. hormones aiding pregnancy

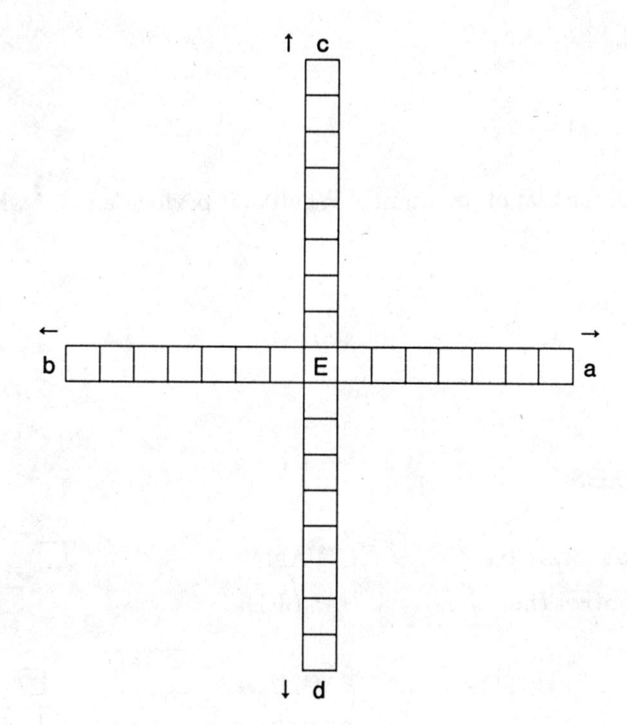

Sleuthing

1. Marie is 42 and has just found out that she is pregnant. Her doctor has insisted that she have amniocentesis.
   a. When is this test usually performed in the course of pregnancy?

   b. What does the procedure involve?

   c. What are some of the disorders that can be detected by amniocentesis?

   d. What is another test of pregnancy. When is it performed and what does it involve?

Word Scrambles

1. a. entrance into female tract	GIVANA	☐ _ _ ☐ ☐ _
b. genes that control the same traits	SALLEEL	_ _ ☐ _ ☐ _ _
c. tie and cut	TAGNIOLI	_ ☐ _ _ _ ☐ ☐ _
d. sometimes it works, sometimes it doesn't	THYRMH	☐ _ _ _ ☐ _
e. propulsion of semen	JONALTACIEU	_ _ _ ☐ _ _ _ _ ☐ _ _
f. male contraceptive sheath	DOMNOC	☐ ☐ _ _ _ _

Total: fingerlike projections into uterine wall

☐☐☐☐☐☐☐☐☐ ☐☐☐☐☐

2. a. "the connection" TANEACLP ☐ _ _ _ ☐ ☐ _ _
   b. polyurethane-containing spermicide GEPONS _ _ ☐ _ ☐ ☐
   c. refraining time THYMRH _ _ ☐ ☐ _ _

Total: the "perfect" combination

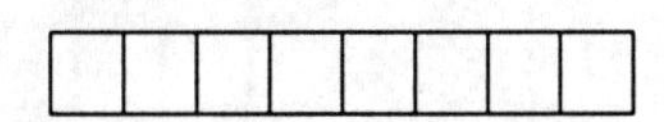

Addagrams

1. a. process that expels fetus from uterus 8, 9, 7, 3, 2
   b. hormone that stimulates milk secretion (abbr.) 4, 2, 8
   c. type of contraceptive 6, 2, 9, 8
   d. end of labors 7, 0, 2, 11, 5, 10
   e. type of cell in immunity 1

Total: differentiated outer layer of blastocyst

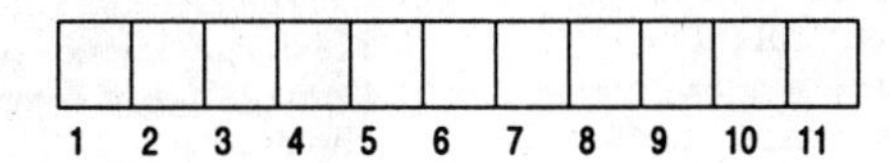

2. a. mass of cells formed by cleavage 2, 11, 12, 1, 9, 8
   b. contraceptive device 6, 1, 13
   c. inherited fluid "type" 3, 5, 11, 11, 13
   d. inherited characteristic of eyes 10, 11, 5, 11, 12
   e. sudden end: ________________ death 7, 12, 4, 3

Total: first apron string

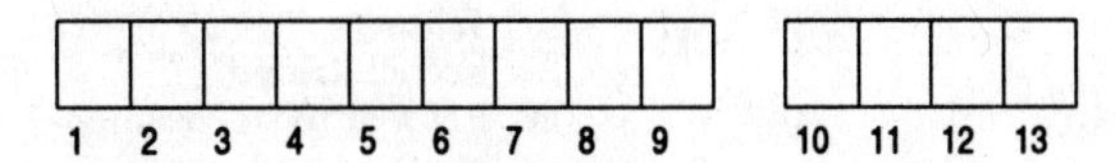

Crossword

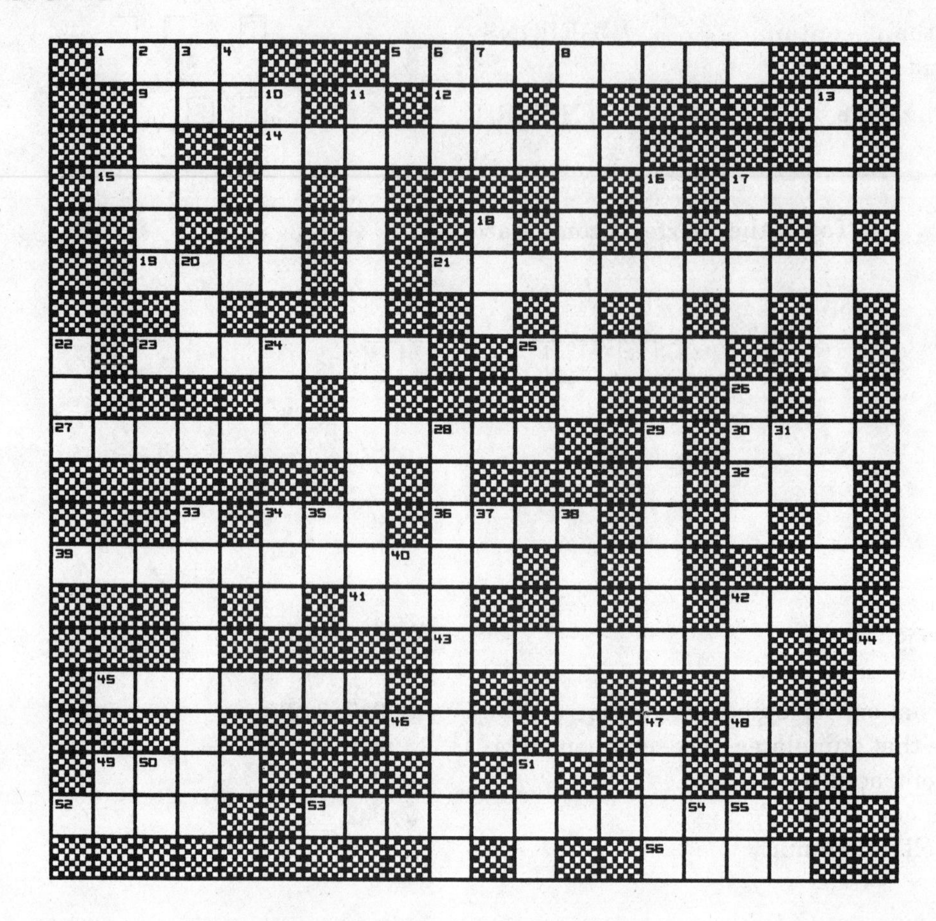

Across

1 Primitive layer of tissue
5 Cord carrying fetal blood
9 Owner of "XX"
12 Fit ________ a fiddle
14 One of women's contributions
15 Major hormone promoting lactation
17 ________ pellucida
19 Female contribution to the zygote
21 Process of giving birth
23 Hormone in female that eases the joint at pubic symphsysis
25 Ovum lives 24 ________ after ovulation (abbr.)
27 Process of sperm becoming able to fertilize
30 ________derm—layer that skin
32 Fish eggs
34 T-cell response (abbr.)
36 "Bull" to Senor Sanchez
39 Bodies that are made up of genetic material
41 Neither this ________ that
42 "XY's" ownership
43 Process of zygote undergoing mitosis
45 Reduction division of spermatogonium
46 Mechanical device inserted into uterus for birth control (abbr.)
48 Implantation occurs within ________ after fertilization
49 Lactation produces this
52 Who fertilizes the oocyte?
53 Fluid-filled cavity around inner cell mass
56 Inner cell ________ that develops into embryo

Down

2 First two months of human development, period of the ________
3 Northeast state (abbr.)
4 Master of Science (abbr.)
6 Contributor of sperm to the zygote
7 Boy Scouts of America (abbr.)
8 Process of producing milk
10 Male contribution
11 Process of sperm meeting "the right girl"
13 Taking a sample of fluid that surrounds fetus
16 Fuss
17 Greek letter
18 Abnormality produced when pregnant mother consumes alcohol (abbr.)
20 Compete
22 Blood returns from placenta to fetal heart by way of ___ ___ ___ (abbr.)
24 French friend
26 Ecto- or meso- or endo- ________
28 Coitus
29 "Raspberry" stage of embryonic development
31 Company (abbr.)
33 Cells that surround ovum
34 Small bed
35 Woman's title
37 ________ coxae—hip bone
38 Sexual climax
40 Way of doing things (abbr.)
42 Portion of sperm that breaks through egg in fertilization
44 What is performed to see if a woman is pregnant
47 Type of cell that develops into RBCs and WBCs
49 Mom's nickname
50 Midwest state (abbr.)
51 How big the egg is
54 West Coast city (abbr.)
55 "This" to Manuel

Answers

Chapter 1

Multiple Choice:

1.a, **2.** c, **3.** a, **4.** b, **5.** d, **6.** c, **7.** b, **8.** b, **9.** d, **10.** b, **11.** d, **12.** c, **13.** d, **14.** c, **15.** d

True/False

1. F, forward, **2.** T, **3.** F, distal, **4.** F, dorsal, **5.** T, **6.** T, **7.** T **8.** T, **9.** F, reverse, **10.** T, **11.** T, **12.** T

Completion

1. physiology, **2.** cellular, **3.** organs, **4.** metabolism, **5.** superior, **6.** deep, **7.** cross section, **8.** posterior, back, **9.** nine; four, **10.** internal, **11.** stimulus, **12.** increased

Lost Sheep

1. viscera (not found in dorsal cavity), **2.** appendix (not found in upper right abdomen), **3.** lungs (not in mediastinum) **4.** ICF (not found outside cells), **5.** output intensifies input (not true of negative feedback mechanism), **6.** transverse (not a lengthwise section), **7.** vertebral cavity (not associated with ventral cavity), **8.** lymphatic system (not a regulatory/control system) **9.** medial (does not refer to head region in humans), **10.** brachial (not a body cavity)

Matching

Set 1: **1.** e, **2.** a, **3.** c, **4.** j, **5.** g, **6.** g
Set 2: **1.** j, **2.** g, **3.** f, **4.** b, **5.** c, **6.** h

Double Crosses

1. across: frontal, down: cranial **2.** across: a) ventral, b) lateral; down: proximal **3.** across: mediastinum down: homeostasis **4.** across: contractile down: hypogastric

Word Cage

```
P A R A S A G I T T A L
C M M D R I S S E R T S
F I S F F O H D Y E I Y
L N T I G S Y S T E M R
A P R Y L V D G R O V E
R U E N D O C R I N E I
B T S P O C B Y D A R A
E I N P U T N A R C T D
T D O R S A L O T F E K
R E R N F S S E R A S L
E G R O W T H L A Y C Z
V M P E R I T O N E U M
```

Sleuthing

a. right, **b.** right upper quadrant, **c.** right hypochondriac, **d.** the appendix is located in the right iliac region or right lower quadrant. The location of the pain should be centralized in that area and therefore, would not be exactly in the same area as the gall bladder case.

Crossword

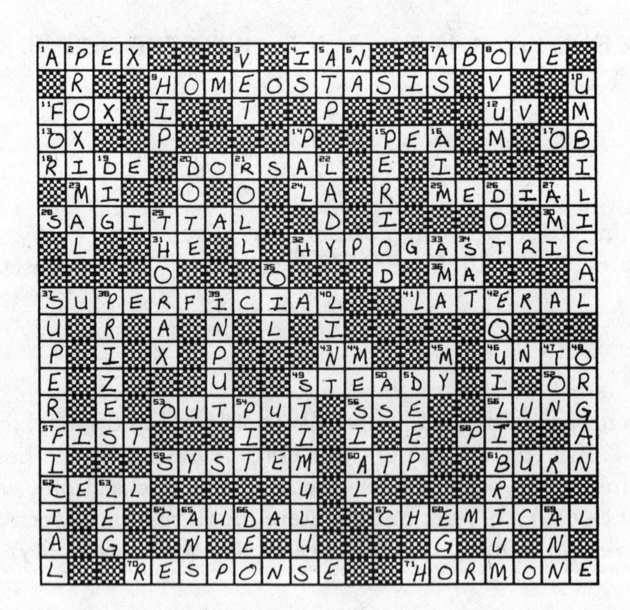

Anatomic Artwork

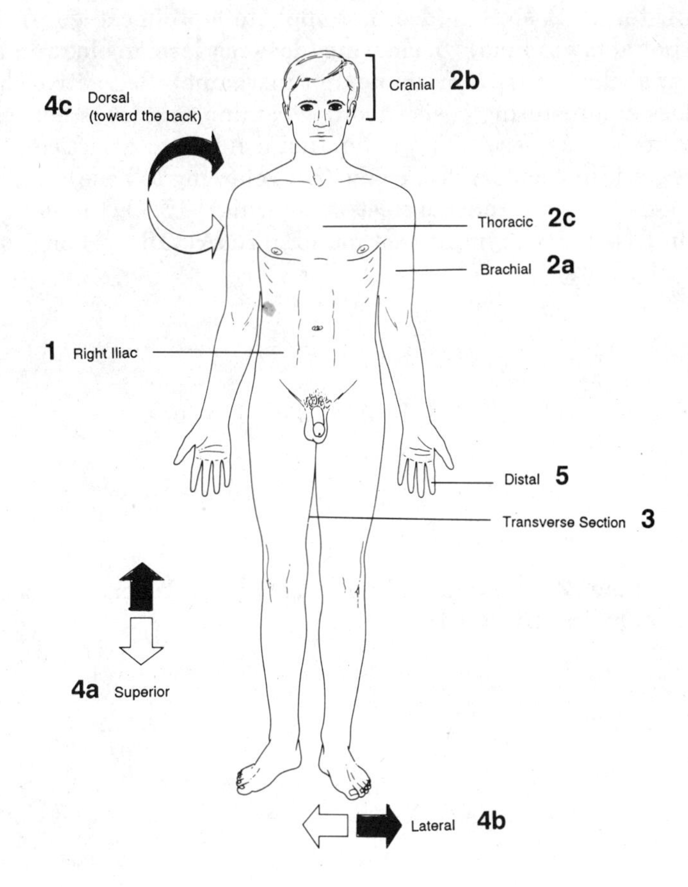

Chapter 2

Multiple Choice

1. c **2.** d **3.** d **4.** b **5.** a **6.** b **7.** c **8.** d **9.** d **10.** d **11.** a **12.** b **13.** c **14.** a **15.**d **16.** b **17.** b
18. c **19.** c **20.** c **21.** d **22.** c **23.** a **24.** c

True/False

1. F one of 3 **2.** T **3.** T **4.** F seven **5.** T **6.** F high **7.** T **8.** F C, H, O **9.** F not **10.** F proton
11. T **12.** T **13.** T **14.** F inorganic **15.** T **16.** F protein **17.** T **18.** T **19.** F ionic **20.** T **21.** T
22. T **23.** T **24.** F anabolic **25.** F negative

Completion

1. ionic **2.** a good solvent **3.** carbon and hydrogen **4.** carbon, hydrogen, oxygen and nitrogen
5. tin, cobalt, copper, etc.) **6.** amino acids **7.** covalent **8.** decomposition/degradation **9.** H+
10. buffers **11.** uracil **12.** slightly basic/alkaline **13.** ATP **14.** ribose **15.** phosphate
16. monosaccharide; glucose, galactose etc. **17.** DNA **18.** cation;anion **19.** triglycerides
20. saturated **21.** glycogen **22.** deoxyribose **23.** lipids **24.** 1-2 **25.** anabolism

Lost Sheep

1. carbon (not inorganic term) **2.** steroid (does not apply to protein category) **3.** OH- (only basic term) **4.** water (does not apply to lipid) **5.** chlorine (does not lose an electron in reaction) **6.** catabolism (not an anabolic term) **7.** methane (not inorganic) **8.** positive charge (only one not referring to electron loss and resulting positive ion) **9.** amino acid (does not refer to carbohydrate) **10.** phosphate (not a nitrogenous base) **11.** photon (not a subatomic particle) **12.** phosphate (not necessarily found in organic molecules) **13.** acidic (not referring to base) **14.** hydrogen (not a trace element in body) **15.** ions (does not refer to electron sharing) **16.** O_2 (not a high energy molecule) **17.** ribose (not found in DNA) **18.** thymine (not found in RNA) **19.** carbon (not a compound) **20.** alanine (not a carbohydrate)

Matching

SET 1 **1.** c **2.** e **3.** a **4.** b **5.** d SET 2 **1.** b **2.** a **3.** b **4.** b **5.** a
SET 3 **1.** c **2.** b **3.** a **4.** c **5.** c

Double Crosses

1. across:ionic; down:guanine **2.** across:aerobic; down:a. acidic **b.** ionizes **c.** carbon **3. a.** nitrogen **b.** oxygen **c.** hydrogen **d.** carbon

Sleuthing

1. more acidic **2.** H+

Word Scrambles

1. a. glucose **b.** purines **c.** water **d.** thymine **e.** phosphorus **f.** DNA TOTAL: sugar-phosphate
2. a. sodiums **b.** glycogen **c.** nitrogen **d.** carbon TOTAL:organic

Addagrams

1. a. ions **b.** acid **3.** neon **4.** cram TOTAL:monosaccharide **2. a.** atom **b.** basic **c.** salt TOTAL:catabolism

Crossword

Chapter 3

Multiple Choice

1. a **2.** b **3.** b **4.** c **5.** d **6.** d **7.** d **8.** a_**9.** b **10.** a **11.** b **12.** b **13.** c **14.** c **15.**a **16.** b **17.** d **18.** d **19.** b **20.** d **21.** d **22.** b **23.** c **24.** b **25.** a

True/False

1. F - r-RNA **2.** T **3.** T **4.** F - all **5.** T **6.** T **7.** F - Golgi complex **8.** F - mitochondrion **9.** T **10.** T **11.** F - anaphase **12.** F - lysosome **13.** T **14.** F - transcription **15.** F - mitosis **16.** T **17.** T **18.** F - integral **19.** F - internal **20.** T **21.** F - integral **22.** F - the lipid portion **23.** T **24.** F - remains stable **25.** T

Completion

1. transcription **2.** translation **3.** t-RNA **4.** interphase **5.** meiosis **6.** centriole **7.** somatic **8.** cytokinesis **9.** three **10.** DNA **11.** ribosome **12.** endoplasmic reticulum **13.** smooth ER, rough ER **14.** cytoskeleton **15.** flagella, cilia **16.** mitochondrion **17.** digestive enzymes **18.** Golgi complex **19.** nucleus **20.** plasma membrane **21.** filtration **22.** ribosomes **23.** lysosomes **24.** passive **25.** ATP

Lost Sheep

1. interphase (not a stage of mitosis) **2.** active transport (only one not passive) **3.** lysis (only term not referring to a synthesis) **4.** digestion (only term not dealing with catabolism) **5.** centrosome (not dealing with degradation) **6.** hemolysis (not having to do with hypertonicity) **7.** swell (does not occur in hypertonic solution) **8.** highly salty (only one not associated with hypotonicity) **9.** against gradient (not dealing with a passive process **10.** micrometer (not a term for cytoskeleton structure) **11.** flagella (not an inclusion) **12.** cytoplasm (not dealing with membrane) **13.** intracellular (not outside of cell) **14.** passive (not associated with active process) **15.** lipid (not a protein structure)

16. digestion (does not refer to cell division) **17.** chromomere (does not refer to genetic material) **18.** lysis (does not have to do with cell division) **19.** nucleus (does not have to do with division of cytoplasm) **20.** no contact inhibition (is not characteristic of a cancer)

Matching

SET 1 **1.** c **2.** e **3.** f **4.** d **5.** g **6.** a **7.** b **8.** e **9.** c **10.** d SET 2 **1.** c **2.** c **3.** c **4.** b **5.** b **6.** a **7.** d
SET 3 - **1.** b **2.** a **3.** c **4.** d **5.** d **6.** e **7.** c **8.** e

Double Crosses

1. across:osmosis; down:mitosis **2.** across:cilia; down:Golgi **3.** across:selective; down:reticulum **4.** across:chromatin; down:chromatid

Sleuthing

1. Both molecule C (tiniest) and molecule B (lipid soluble) **2.** Solution of 35% would crenate fastest, then 20%, then 10%; water would leave the cell and pass in to the hypertonic solutions outside by osmosis **3.** molecule C (small, with charge which would attract it to membrane); molecule B (greatest lipid solubility) **4.** molecule A (slightly larger than the average pore size and not lipid soluble)

Word Scrambles

1. **a.** somatic **b.** hemolysis TOTAL :lysosome **2.** **a.** reticulum **b.** transfer **c.** ion **d.** ATP TOTAL:transcription

Addagrams

1. **a.** lysosome **2.** metabolism **c.** pinocytosis **d.** cristae TOTAL:cytoplasmic membrane
2. **a.** ATPs **b.** hypo **c.** gas **d.** sis **e.** cyto TOTAL:phagocytosis

Crossword

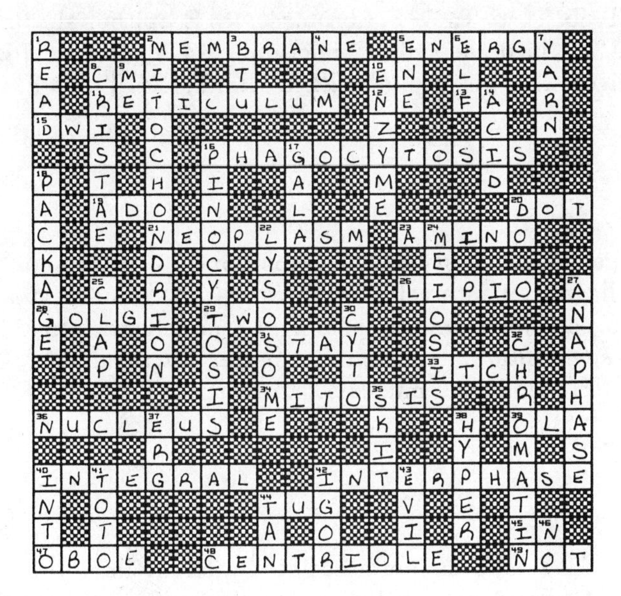

Anatomic Artwork

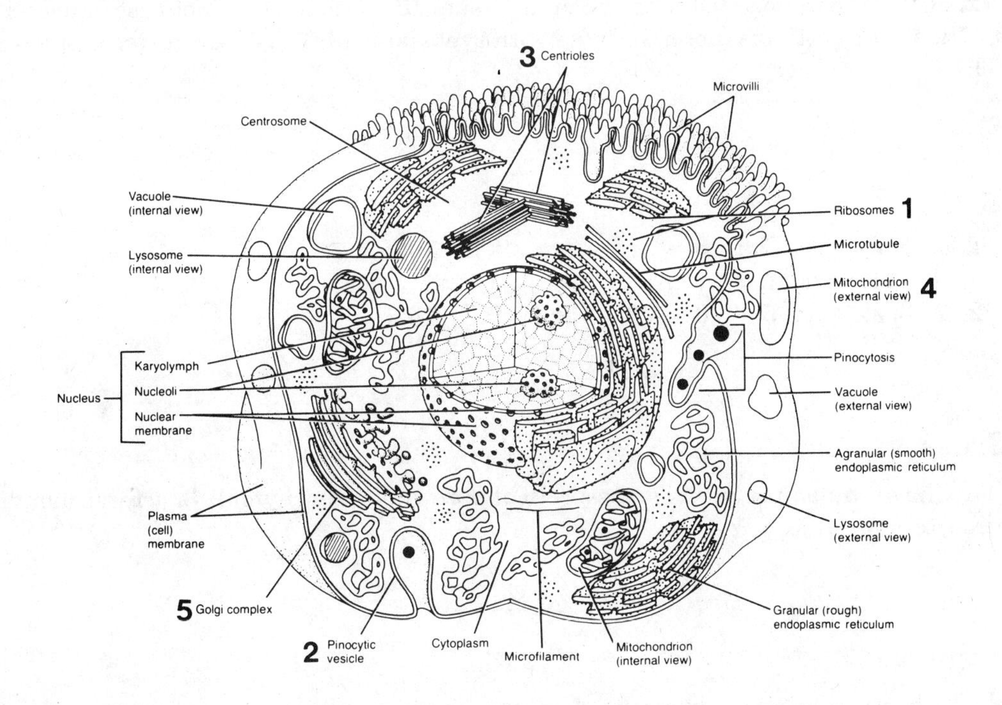

Chapter 4

Multiple Choice

1. c, **2.** b, **3.** d, **4.** b, **5.** d, **6.** d, **7.** b, **8.** c, **9.** b, **10.** b, **11.** a, **12.** d, **13.** c **14.** d, **15.** c, **16.** d, **17.** a, **18.** b, **19.** b, **20.** a, **21.** d, **22.** b, **23.** c, **24.** c, **25.** d

True/False

1. F, dense, **2.** T, **3.** T, **4.** T, **5.** T, **6.** T, **7.** F, fibroblast, **8.** F, epithelial, **9.** F, nervous, **10.** F, bone, **11.** F, plasma cells, **12.** F,'lacunae, **13.** F, endocrine, **14.** T, **15.** T, **16.** T, **17.** F, skeletal, **18.** T **19.** T, **20.** F, connective, **21.** F, epithelial, **22.** T, **23.** T, **24.** T **25.** F, synovial

Completion

1. basement membrane, **2.** goblet, **3.** stratified, **4.** transitional,_**5.** squamous, **6.** columnar, **7.** cuboidal, columnar, **8.** collagen, **9.** intercellular material, **10.** chondrocytes, osteocytes, **11.** matrix **12.** reticular, **13.** osteon/Haversian system, **14.** perichondrium **15.** connective, **16.** mesenchyme, **17.** skeletal/voluntary, **18.** cardiac muscle, **19.** dendrites, **20.** neuroglia, **21.** neurons, **22.** endocrine **23.** gland, **24.** mucus, **25.** exocrine

Lost Sheep

1. areolar (not an epithelial cell, **2.** adipose (not associated with cartilage), **3.** collagen (not associated with epithelium), **4.** transitional (not a connective tissue cell), **5.** collagen production'(not a function of epithelium), **6.** areolar (not associated with dense connective tissue), **7.** epidermis (not a connective tissue), **8.** hyaline (not part of compact bone), **9.** striated (not associated with smooth muscle),10. neuroglia (not part of neuron), **11.** Haversian canal (not found in cartilage), **12.** elastic (not associated with reticular tissue),13. mesenchyme (not associated with epithelium), **14.** elastic cartilage (not associated with synovial joints), **15.** osteocytes (not associated with mast cell)

Matching

Set 1: **1.** c, **2.** a, **3.** d, **4.** b, **5.** i, **6.** e, **7.** j, **8.** g, **9.** h, **10.** f
Set 2: **1.** b, **2.** d, **3.** e, **4.** c, **5.** e
Set 3: **1.** f, **2.** h, **3.** a, **4.** g, **5.** c, **6.** e, **7.** d, **8.** b, **9.** i, **10.** j

Double Crosses

1. across: neuroglia down: adipocyte **2.** across: reticular down: mesenchymal **3.** across: chondroitin down: chondrocyte

Word Scrambles

1. a. fibroblast, **b.** stratified, **c.** columnar, **d.** macrophage **e.** epithelium, **f.** transitional Total: basement membrane **2. a.** exocrine, **b.** cuboidal, **c.** gland, **d.** tissue, **e.** areolar Total: goblet cell

Addagrams

1. a. fiber, **b.** elastic, **c.** tendon, **d.** mucus, **e.** neuron, **f.** a Total: stratified columnar
2. a. cardiac, **b.** stratified, **c.** elastic Total: intercalated disc

Crossword

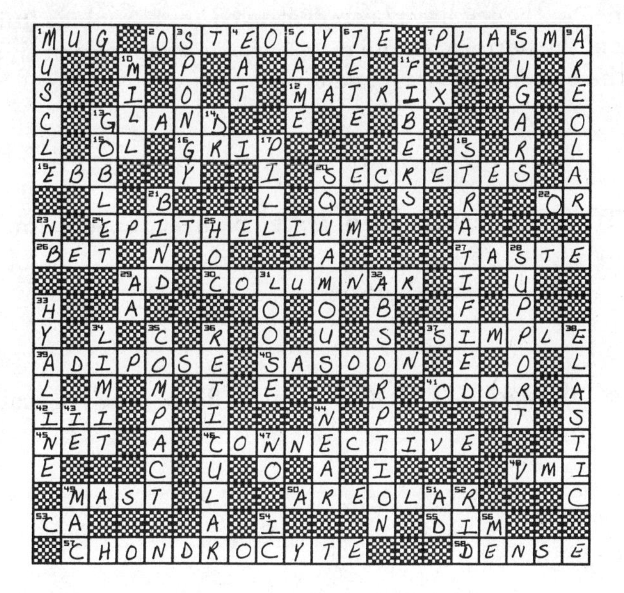

Chapter 5

Multiple Choice

1. c **2.** c **3.** c **4.** d **5.** c **6.** d **7.** b **8.** d. **9.** a **10.** a

True/False

1. F basale **2.** F melanin **3.** F basale **4.** F granulosum **5.** F inactive **6.** T **7.** F first **8.** F high **9.** T **10.** T

Completion

1. melanin **2.** dermis **3.** subcutaneous **4.** fingerprints **5.** stratum corneum **6.** melanoma **7.** oil **8.** dermis (reticular layer) **9.** germinativum (basale) **10.** papillary **11.** stratum germinativum (basale) **12.** arrector pili

Lost Sheep

1. keratin (not part of dermis) **2.** skin of forehead (does not have apocrine sweat glands) **3.** eccrine gland (not associated with hair) **4.** keratin (does not impart color to skin) **5.** artery (not part of nail) **6.** arrector pili (not associated with sweat production) **7.** corpuscle (not associated with hair) **8.** reticular (not epidermal) **9.** loose connective tissue (not associated with dermis) **10.** sweat pore (not deep in skin)

Matching

SET 1 - **1.** b **2.** e **3.** d **4.** c **5.** a SET 2 - **1.** a **2.** e **3.** b **4.** d **5.** c

Sleuthing

a. 18% **b.** the nerve endings in the leg have been destroyed **c.** to replace fluids and electrolytes lost from destroyed capillaries of the dermis **d.** the most serious complication that might be anticipated is infection due to loss of the body's first line of defense

Word Scrambles

1. a. exocrine **b.** mucus TOTAL:corneum **2. a.** trace **b.** layer **c.** subcutaneous **d.** keratin TOTAL:reticular

Addagrams

1. a. root **b.** papilla **c.** ice **d.** rear TOTAL:arrector pili **2. a.** tans **b.** scab **c.** uuu **d.** neon TOTAL:subcutaneous

Crossword

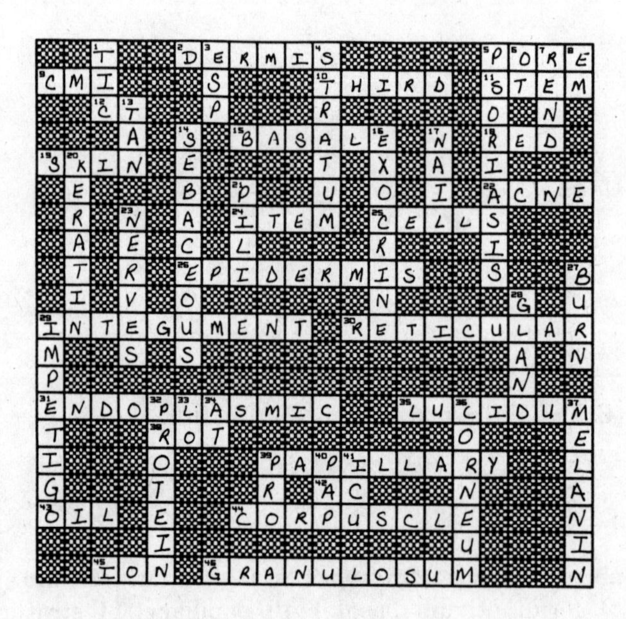

Anatomic Artwork

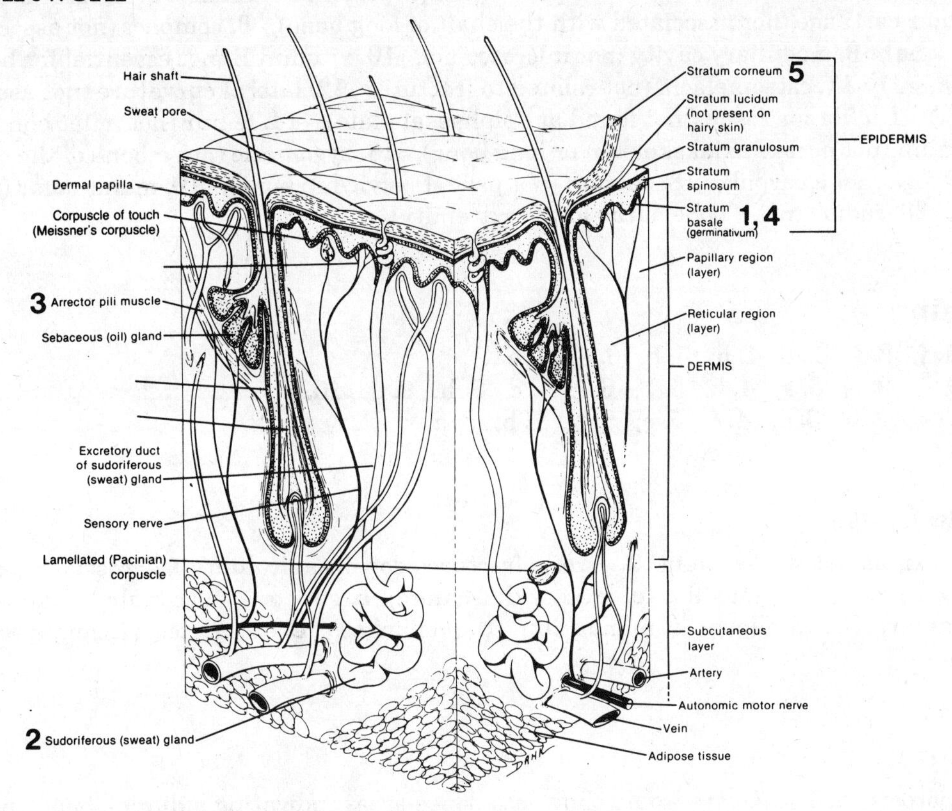

Chapter 6

Multiple Choice

1. c, **2.** b, **3.** d, **4.** c, **5.** a, **6.** c, **7.** a, **8.** b, **9.** d, **10.** d, **11.** d, **12.** c, **13.** c **14.** d, **15.** a, **16.** c, **17.** b, **18.** b, **19.** b, **20.** d, **21.** c, **22.** c, **23.** c, **24.** b, **25.** b

True/False

1. T, **2.** T, **3.** F, intramembranous, **4.** F, deposition, **5.** F, yellow, **6.** F, red, **7.** F, are not, **8.** T, **9.** F, vitamin D, **10.** T, **11.** T, **12.** F, irregular, **13.** T, **14.** T, **15.** F, trabeculae, **16.** T, **17.** T, **18.** T, **19.** **20.** F, pelvic, **21.** F, foramen, **22.** T, **23.** F, scoliosis, **24.** F, axis **25.** F, different

Completion

1. hemopoiesis, **2.** intramembranous, **3.** adipose, **4.** low, **5.** canaliculi, **6.** decreases, opposite, **7.** periosteum, **8.** piezoelectric, **9.** epiphyseal plate, **10.** osteocytes, **11.** yellow, **12.** osteomalacia, **13.** spina bifida, **14**, 65:35 (2:1), **15.** endochondral, **16.** callus, **17.** appendicular **18.** calcium salts, **19.** blood clotting, muscle contraction, nerve impulse transmission, **20.** osteoclasts, **21.** hyaline, **22.** fontanels **23.** intracartilagenous/endochondral, **24.** appendicular, **25.** hyoid

Lost Sheep

1. calcium salts (not organic), **2.** calcitonin (lowers blood calcium) **3.** collagen (does not regulate bone growth/development), **4.** adipocyte(not associated with hemopoiesis), **5.** subluxation (not a type of fracture), **6.** increased estrogen production (deposits calcium into bone from body fluids), **7.** articular cartilage (not associated with the shaft of long bone), **8.** compact (not associated with spongy bone), **9.** medullary cavity (not microscopic), **10.** vitamin B (not essential for bone development), **11.** osteomalacia (not related to fractures **12.** lateral curvature (not associated with kyphosis), **13.** fibroustissue (not found at epiphyseal plate), **14.** femur (not a flat bone) **15.** sternum (not part of axial skeleton or long bone), **16.** zygomatic (not a bone of the cranium), **17.** phalange (not a carpal), **18.** sacrum (not part of pelvic/hip bone), **19.** mandibular (not a cranial suture), **20.** radius (not a bone of the lower extremity)

Matching

Set 1: **1.** j, **2.** f, **3.** a, **4.** h, **5.** b, **6.** d, **7.** e
Set 2: **1.** i, **2.** a, **3.** a, **4.** k, **5.** b, **6.** j, **7.** c, **8.** h, **9.** d, **10.** g
Set 3: **1.** c, **2.** b, **3.** d, **4.** c, **5.** e, **6.** b, **7.** b, **8.** a

Double Crosses

1. across: xiphoid down: ischium **2.** across: fractures down: capitulum **3.** across: hemopoiesis down: **a.** humerus, **b.** maxilla, **c.** process **d.** ischium, **e.** support, **f.** scapula **4.** across: **a.** trochanter, **b.** cancellous, **c.** manubrium **d.** phalanges, **e.** osteocytes, **f.** epiphyseal down: periosteum

Sleuthing

a. osteoporosis **b.** The X-rays would show less dense areas indicating a diminished bone mass. **c.**Dietary mineral supplements to improve calcium levels; gonadal hormones to stimulate osteoblastic activity thereby decreasing the incidence of fractures; and physical therapy and exercise to strengthen the bones and muscles.

Word Scrambles

1. **a.** occipital, **b.** sphenoid, **c.** vertebrae, **d.** mandible,_**e.** endochondral, **f.** ossification, **g.** calcium Total: osteomalacia **2.** **a.** mandible, **b.** pectoral, **c.** phalanges, **d.** navicular, Total: pelvic girdle

Addagrams

1. **a.** crest, **b.** atlas, **c.** osseo, **d.** crista galli Total: costal cartilage **2.** **a.** olecranon, **b.** ilium, **c.** foramen, **f.** femur, **e.** radial TOTAL: middle cuneiform

Crossword

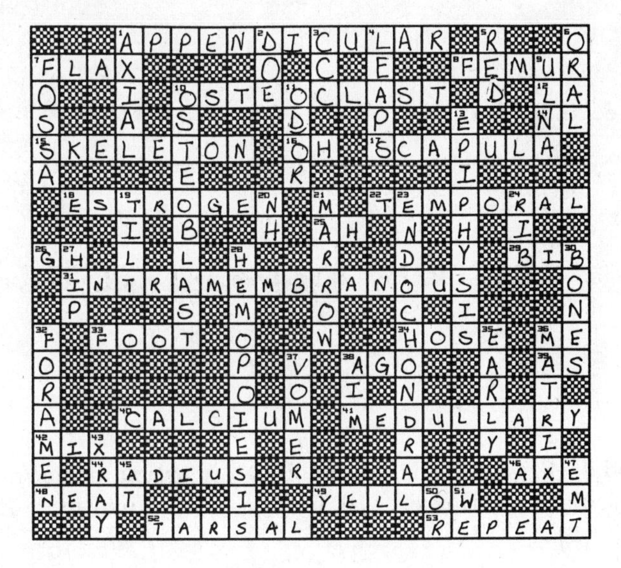

Anatomic Artwork

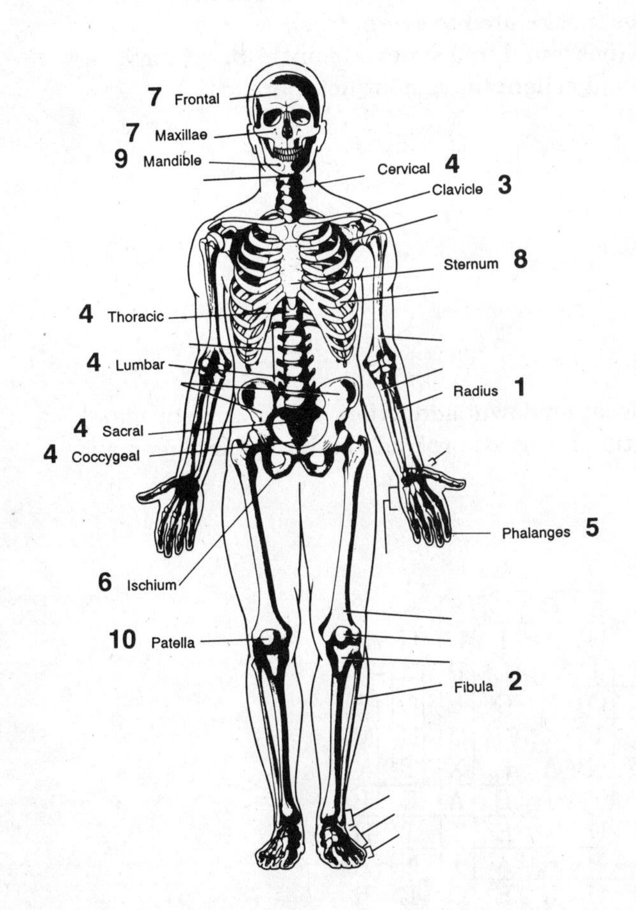

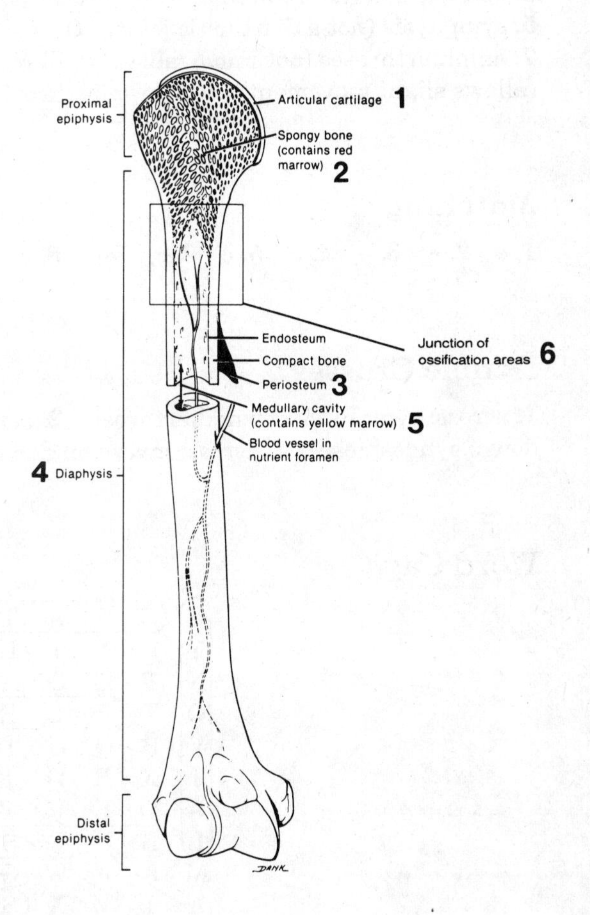

Chapter 7

Multiple Choice

1. d, **2.** b, **3.** c, **4.** b, **5.** d, **6.** d, **7.** d, **8.** c, **9.** a, **10.** b, **11.** b, **12.** c

True/False

1. T, **2.** F, hyaline cartilage, **3.** F, diarthritic, **4.** F, structural, **5.** T, **6.** F, rotation, **7.** T, **8.** F, foot, **9.** T, **10.** T

Completion

1. cartilagenous, **2.** synovial, **3.** gomphosis, **4.** mobility, **5.** articular disc/menisci, **6.** rotation, **7.** bursa, **8.** hinge, **9.** articulation/arthrosis, **10.** freely, **11.** subluxation, **12.** extension

Lost Sheep

1. bursitis (not associated with arthritis), **2.** Lyme (not related to dislocation of joint), **3.** subluxation (not related to gouty arthritis) **4.** supination (not a special movement of the foot), **5.** symphysis (not a diarthritic joint), **6.** synchondrosis (not related to symphysis), **7.** amphiarthroses (not immovable), **8.** fibrocartilage (not found in a synovial joint), **9.** symphysis (allows slight movement) **10.** amphiarthrosis (allows only slight movement/not diarthritic)

Matching

1. c, **2.** a, **3.** a, **4.** a, **5.** c, **6.** c, **7.** b, **8.** c, **9.** b, **10.** c

Double Crosses

1. across: symphysis down: diarthrosis **2.** across: dislocation down: adduction **3.** across: ellipsoidal down: syndesmoses **4.** across: inversion down: arthritis **5.** across: protraction down: cartilagenous

Word Cage

```
S C A R T I L A G E S K T
Y E N I L A Y H A I U H R
N R U O A J M J S L B G C
O T F L E X I O N O L O E
V B Q U T E R I K N U U M
I X B D I H M N A E X T Y
A O E P T J A T T B A L L
L B U R S A E D I E T I H
M A A W O N S P R A I N A
A I M X S U T U R E O L R
D R S I S O M S E D N Y S
S Y N C H O N D R O S I S
```

Sleuthing

a. Osteoarthritis usually occurs later in life after age fifty while rheumatoid arthritis generally develops between 30 - 40 years of age, but can occur in childhood. Rheumatoid arthritis is a progressive disease that can involve any joint causing severe swelling and deformity. Osteoarthritis by comparison is generally non-progressive and is found chiefly in weight bearing joints and severe swelling is not usually a symptom. **b.** The major purpose of physical therapy is to maintain, as much as possible, a useful range of motion for the joints involved and to maintain muscle tone. **c.** Gouty arthritis is due to the inability to metabolize uric acid properly.

Word Scrambles

1. a. pivot, **b.** arthritis, **c.** symphysis, **d.** rheumatism Total: amphiarthroses **2. a.** movable, **b.** menisci, **c.** synarthroses Total: synovial membrane

Crossword

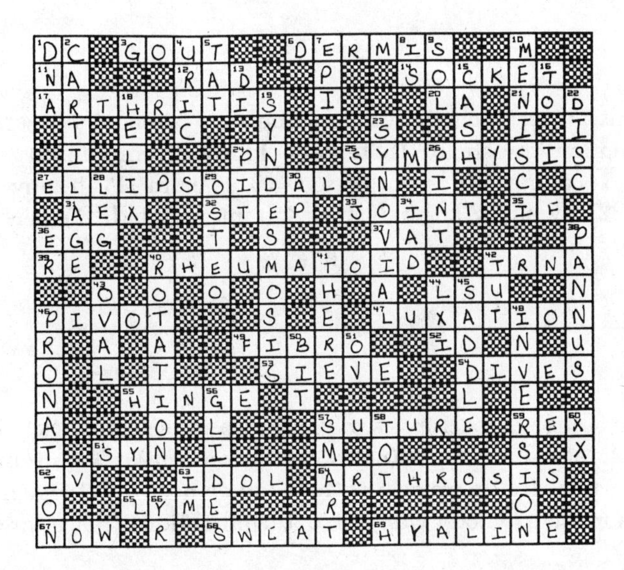

Anatomic Artwork

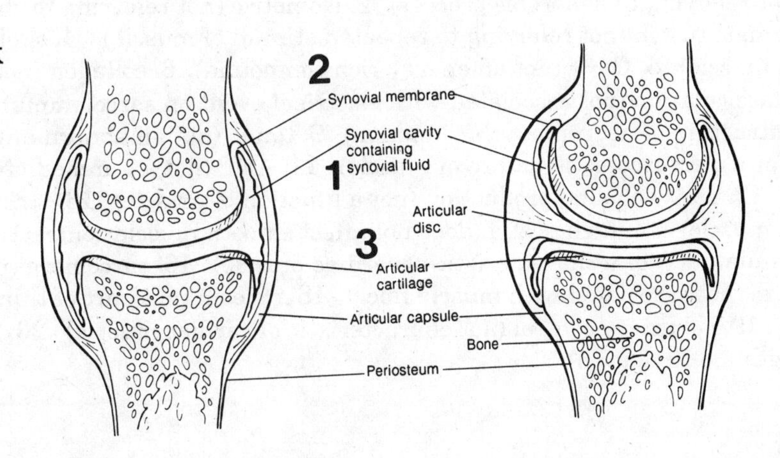

4. Synovial Joint
5. Movements: abduction; adduction; flexion; extension; circumduction; rotation
6. Shoulder, hip, elbow

Chapter 8

Multiple Choice

1. d **2.** c **3.** b **4.** c **5.** d **6.** d **7.** d **8.** b **9.** c **10.** c **11.** d **12.** c **13.** a **14.** b **15.** c **16.** d **17.** a **18.** c **19.** d **20.** c **21.** c **22.** b **23.** d **24.** c **25.** b

True/False

1. F-threshold **2.** T **3.** F-irregular pattern of myofilaments **4.** T **5.** T **6.** can **7.** F-hypertrophy **8.** F-isometric **9.** F-anaerobic **10.** T **11.** F-endomysium **12.** T **13.** F-the same strength **14.** F-myosin overlapping actin **15.** depends on nervous stimulation for the initiation of contraction **16.** T **17.** T **18.** T **19.** T **20.** F-lactic acid **21.** T **22.** T **23.** T **24.** T **25.** T

Completion

1. epimysium **b.** motor unit **3.** acetyl choline **4.** antagonists **5.** smooth/visceral **6.** myoglobin **7.** fatigue **8.** skeletal **9.** liminal/threshold **10.** tendon **11.** multiunit, unitary/visceral **12.** absolute refractory **13.** sarcomere **14.** dilate **15.** latent period **16.** calcium **17.** tetany **18.** liver **19.** calcium **20.** longer **21.** isometric **22.** myosin **23.** thin (actin) **24.** intercalated **25.** perimysium

Lost Sheep

SET 1

1. pectoralis-(not a muscle of leg) **2.** pectoralis major (not a muscle of back) biceps femoris (not a muscle of ventral thigh) **4.** latissimus dorsi (not a muscle on arm) **5.** gluteus maximus (not a muscle of abdomen) **6.** serratus(not a muscle of the head or neck) **7.** trapezius (not a muscle of face) **8.** diaphragm (not a muscle of body wall) **9.** external oblique (not a muscle of eye) **10.** bursa (not a muscle attachment)

SET 2

1. oxygen (not referring to anaerobic process) **2.** isometric (not referring to change in length of muscle) **3.** single twitch (not referring to repeated stimuli of muscle) **4.** skeletal (not associated with smooth muscle) **5.** Ca++ (not an energy-rich compound) **6.** collagen (not a muscle protein) **7.** fasciculation (not associated with electrical event on sarcolemma) **8.** epimysium (not a chemical or structure of neuromuscular junction) **9.** tonus (not an abnormality) **10.** sarcoplasmic reticulum (not a macroscopic structure of muscle) **11.** ADP (not an immediate source of energy for contraction) **12.** acetyl choline production (not a function of muscle) **13.** striations (not associated with smooth muscle) **14.** high acetyl (does not affect smooth muscle contractility) **15.** dependency on nerve stimulation (not a characteristic of cardiac muscle) **16.** sarcomere (not macroscopic) **17.** epimysium (not associated with muscle fiber) **18.** myosin (not part of thin myofilament) **19.** voluntary control (not characteristic of cardiac muscle) **20.** ACh esterase (not associated with motor neuron)

Matching

Set 1 **1.** a **2.** c **3.** b **4.** c **5.** a **6.** b **7.** b **8.** c Set 2 **1.** b **2.** g **3.** e **4.** a **5.** f **6.** c **7.** d
Set 3 **1.** g **2.** j **3.** f **4.** h **5.** i **6.** d **7.** b **8.** c **9.** a **10.** e

Double Crosses

1. across:supinator down:tensor **2.** across:intercostal down:brachii **3.** across:deltoid down:frontalis **4.** across:sartorius down:psoas **5.** across:rectus abdominus down:semimembranosus

Sleuthing

a. Carbohydrate serves as an energy source for the muscles. When it is broken down energy is produced. **b.** He relies more and more on anaerobic production of ATP. **c.** Lactic acid is accumulating and energy stores are diminishing. **d.** The muscle cells are increasing in size, not number. **e.** Both treppe and warmth may cause more calcium to be available for repeated contractions.

Word Scrambles

1. a. gluteus **2.** extension **c.** rhomboideus **d.** dystrophy TOTAL:oxygen debt **2. 2. a.** muscle **b.** antagonist **c.** refractory insertion TOTAL:synergist **3. a.** spindle **b.** epimysium **c.** myoglobin **d.** cholinesterase **e.** hypertrophy TOTAL: hamstrings

Addagrams

1. a. sarcomere **b.** dilated **c.** isotonic TOTAL:sternocleidomastoid **2. a.** actin **b.** rotation **c.** tone **d.** Ca **e.** cat TOTAL:tetanic contraction **3. a.** pronate **b.** deltoid **c.** alone TOTAL:latent period **4. a.** troponin **b.** ATPase **c.** ache **d.** heat TOTAL:cratine phosphate

Crossword

Anatomic Artwork

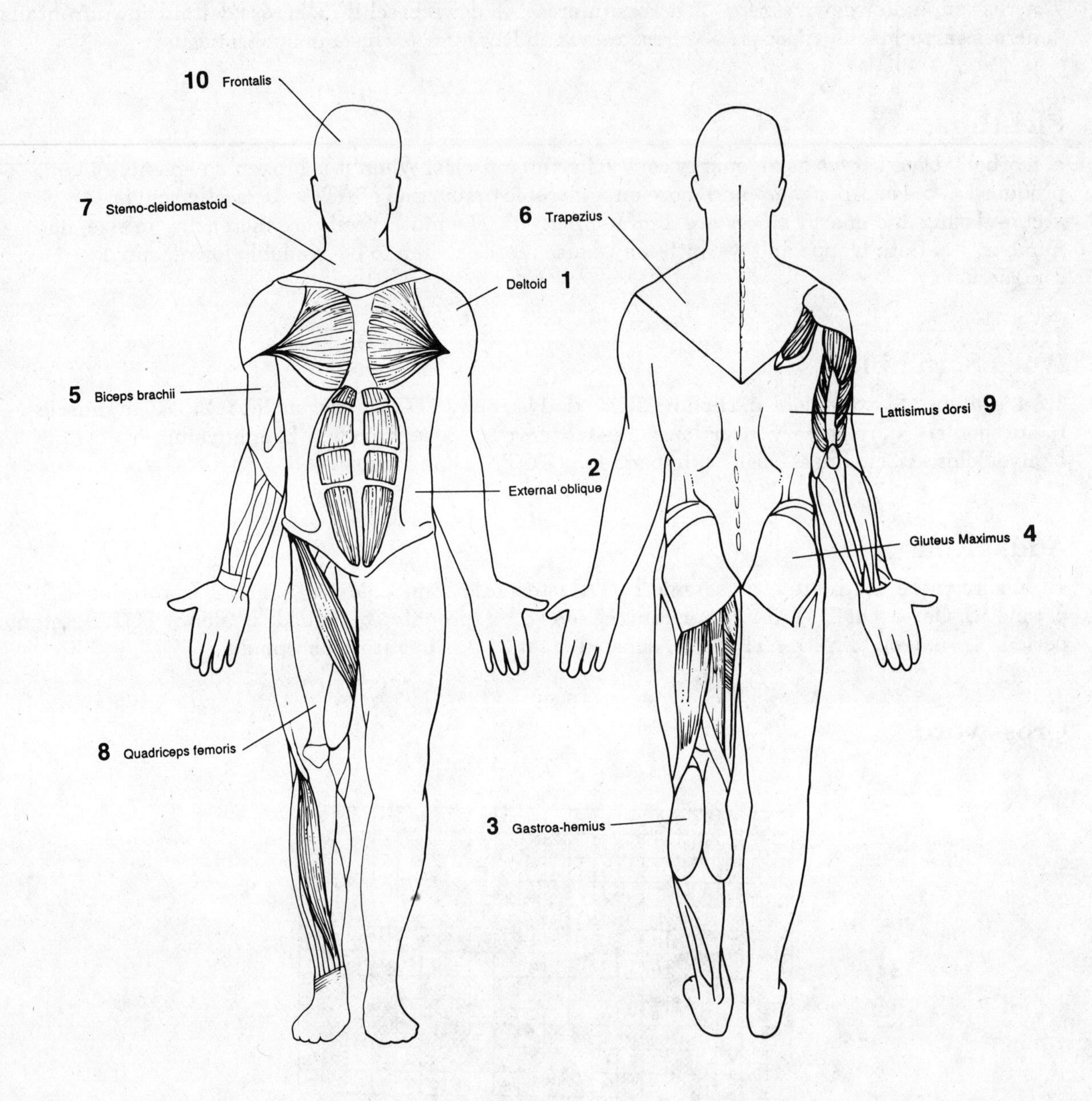

Chapter 9

Multiple Choice

1. b, **2.** a, **3.** c, **4.** c, **5.** d, **6.** d, **7.** b, **8.** a, **9.** a, **10.** b, **11.** d,**12.** c

True/False

1. T, **2.** F efferent, **3.** T, **4.** F less readily, **5.** F Central Nervous System, **6.** T, **7.** F collaterals, **8.** T, **9.** F not respond, **10.** T, **11.** T, **12.** T

Completion

1. neuroglia, **2.** Nissl Complex, **3.** motor, **4.** neurotransmitter, **5.** Nodes of Ranvier, **6.** structural,functional, **7.** gated-ion channels, **8.** polarized, **9.** repolarization, **10.** synapse, **11.** graded potentials **12.** hyperpolarized

Lost Sheep

1. myelin (not associated with C fibers), **2.** dendrite (not related to axon), **3.** Schwann (not in CNS), **4.** depolarization (not associated with inhibition of impulse transmission), **5.** Node of Ranvier (not found internally), **6.** energy expenditure elevated (not true of saltatory conduction), **7.** dendrite (not involved in transmission of impulse to postsynaptic neuron), **8.** hyperpolarized (not associated with Na increase in postsynaptic neuron), **9.** microglia (does not produce myelin sheath), **10.** efferent (not related to the sensory path)

Matching

Set 1: **1.** c, **2.** a, **3.** b, **4.** f, **5.** e, **6.** d
Set 2: **1.** a, **2.** c, **3.** b, **4.** e, **5.** d, **6.** e

Double Crosses

1. across:autonomic down:potential **2.** across:threshold down:potassium **3.** across:summation down:saltatory **4.** across:synapse down:somatic **5.** across:a. dendrite **b.** regenerate **c.** afferent **d.** ependyma **e.** facilitate f. transmit down: efferent **6.** across:refractory down: **a.** Ranvier **b.** peripheral **c.** astrocytes d. sensory

Word Cage

A D F S U B T H R E S H O L D
L T Y U E M K N B D T B K E M
O L I G O D E N D R O C Y T E
R T A S F G L P W A C T J S F
L A C T R O L O A Z F R E Y F
A X O N D S A L T A T O R Y E
R B N H I P U A D E N D N O R
E X I D E B A R B A T M O T E
H E D M U L T I P O L A R H N
P O L A R D I Z F G N L U M T
I N E U R O G E B R A I E Y I
R O L I G O D D E N D R N Z E
E F G H N I T R S Y N A P S E
P N O I T A T I L I C A F A C

Word Scramble

1. a. depolarization, **b.** potential, **c.** excitability, **d.** threshold, e. neurolemmocytes TOTAL: postsynaptic neuron **2. a.** refractory, **b.** central, **c.** potassium, **d.** dendrite, **e.** microglia TOTAL: saltatory conduction

Addagrams

1. **a.** neurolemma, **b.** dendrite, **c.** sensory TOTAL: myelinated neurons **2.** **a.** one, **b.** action, **c.** glial, **d.** end **e.** sheath TOTAL: gated ion channels

Crossword

Anatomic Artwork

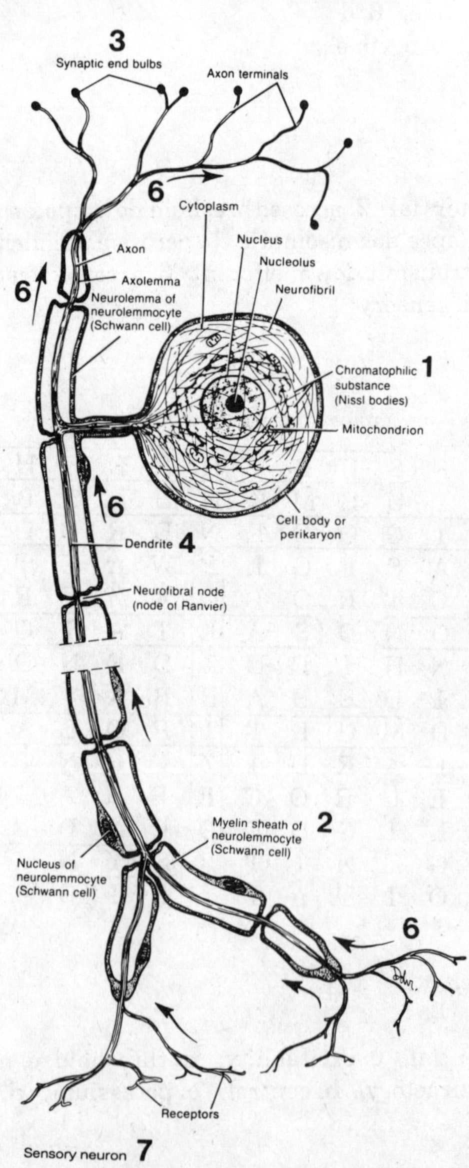

Chapter 10

Multiple Choice

1. c, **2.** a, **3.** d, **4.** c, **5.** d, **6.** b, **7.** b, **8.** a, **9.** d, **10.** b, **11.** b **12.** b, **13.** c, **14.** b, **15.** c, **16.** d, **17.** c, **18.** b, **19.** c, **20.** a, **21.** c **22.** a, **23.** b, **24.** c, **25.** b

True/False

1. F meninges, **2.** F third, **3.** F somatic, **4.** F ganglia, **5.** F medulla, **6.** T, **7.** T, **8.** T, **9.** T, **10.** F subarachnoid, **11.** F tract, **12.** T, **13.** F upper extremity and neck, **14.** T, **15.** F choroid plexus, **16.** T, **17.** F midbrain, **18.** F thalamus, **19.** F hypothalamus, **20.** T, **21.** T, **22.** T **23.** T, **24.** F sensory, **25.** T

Completion

1. meninges, 2, 31;12, **3.** hypothalamus, **4.** Willis, **5.** subarachnoid space, **6.** medulla, **7.** cerebral peduncles, **8.** REM, **9.** gyri **10.** commissural fibers, **11.** cerebellum, **12.** neurotransmitters **13.** mixed, **14.** shingles, **15.** Parkinson's disease, **16.** communicans **17.** plexi, **18.** four, **19.** choroid plexi, **20.** cerebral aqueduct/Sylvius **21.** medulla, pons, midbrain, **22.** Reticular Activating System **23.** first/olfactory, **24.** vagus, **25.** epilepsy

Lost Sheep

1. phrenic (not a cranial nerve), **2.** skeletal muscle coordination (not a function of hypothalamus) **3.** corpus callosum (not a spinal cord tract), **4.** hypothalamus (not a site where CSF flows), **5.** epidural space (not a place where CSF circulates), **6.** tract (not gray matter) **7.** postganglionic neuron (not associated with the spinal cord), **8.** ventral root (not associated with afferent path), **9.** rami (not associated with plexi), **10.** fourth ventricle (not associated with diencephalon), **11.** thin basement membrane (not part of blood-brain- barrier), **12.** regular respiration (not associated with REM sleep) **13.** right cerebral hemisphere (not generally associated with skills listed), **14.** cerebral peduncles (not associated with basal ganglia) **15.** frontal lobe (not associated with somesthetic senses)

Matching

Set 1: **1.** b, **2.** c, **3.** e, **4.** d, **5.** a, **6.** g, **7.** c, **8.** h, **9.** i, **10.** f
Set 2: **1.** c, **2.** i, **3.** h, **4.** b, **5.** f, **6.** g, **7.** j, **8.** e, **9.** a, **10.** d

Double Crosses

1. across: arachnoid, down: trochlear, **2.** across: nerve, down: gyrus **3.** across: motor, down: optic

Sleuthing

a. tremor, impaired motor performance and poor posture **b.** reduction in dopamine levels **c.** basal ganglia, substantia nigra **d.** Since basal ganglia are involved in the regulation of conscious contraction of skeletal muscle, increasing dopamine levels allows for greater control. The fat soluble L-dopa can not cross the blood-brain-barrier, be converted to dopamine and, thus, relieve the symptoms.

Word Scrambles

1. a. Willis, **b.** pia mater, **c.** reflex, **d.** hypothalamus, **e.** auditory TOTAL: extrapyramidal pathway **2. a.** cerebrum, **b.** putamen, **c.** olfactory, **d.** consciousness, **e.** TIA TOTAL: limbic system

Addagrams

1. a. peduncle, **b.** fasciculi, **c.** brain, **d.** corpus TOTAL: cerebrospinal fluid **2.** a, meninges, **b.** CVA, **c.** sulcus, **d.** tract, **e.** Reye's, **f.** sciatica, g. TIA TOTAL: reticular activating system

Crossword

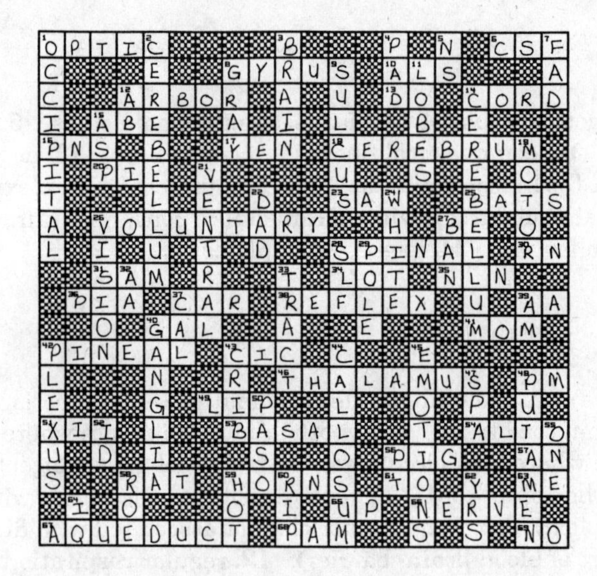

Anatomic Artwork

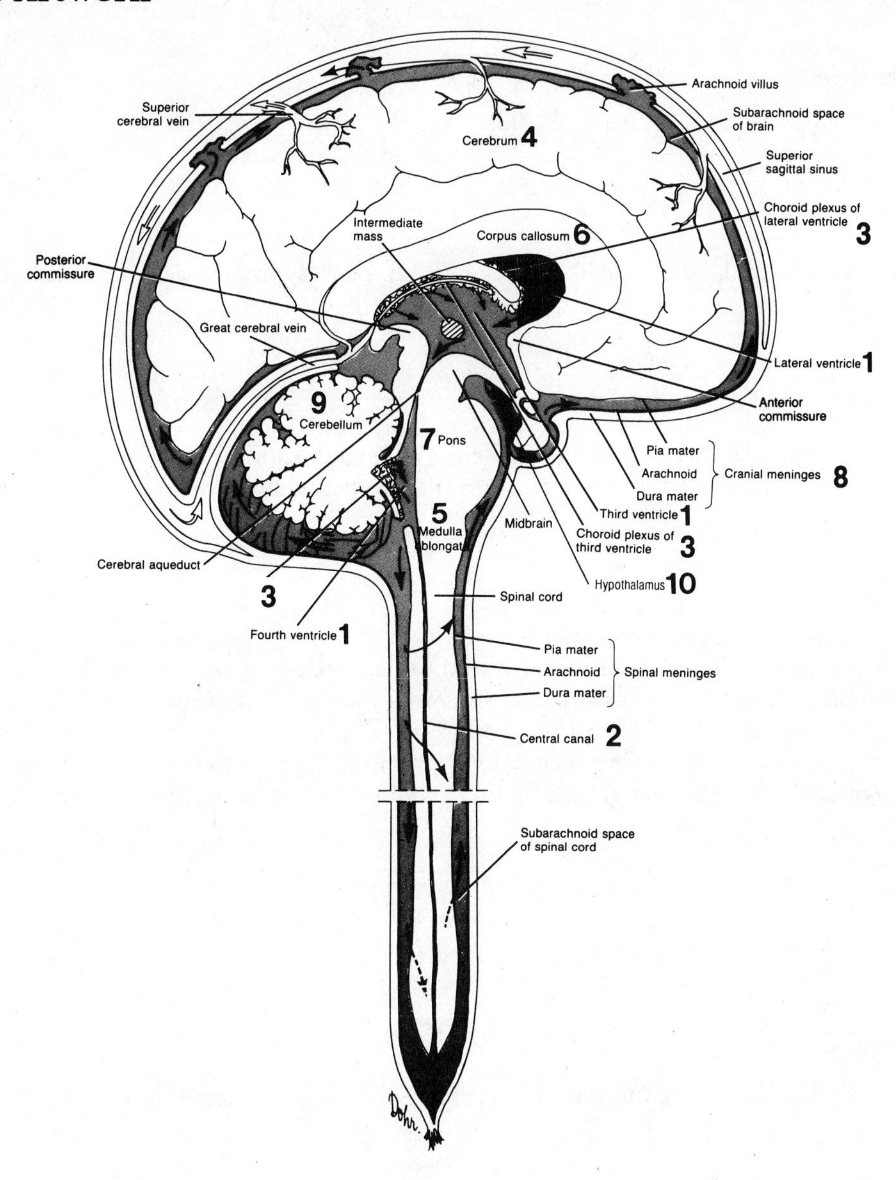

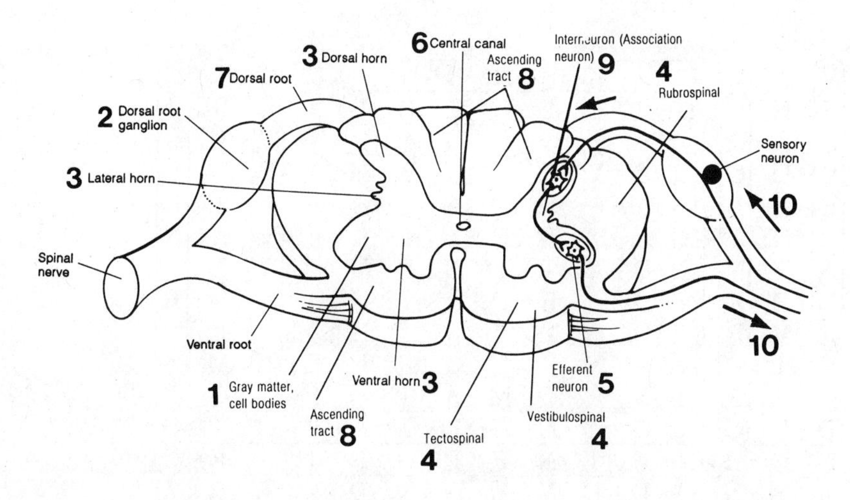

Chapter 11

Multiple Choice

1. d, **2.** c, **3.** c, **4.** c, **5.** b, **6.** c, **7.** b, **8.** d, **9.** a, **10.** c

True/False

1. T, **2.** F both, **3.** F efferent, **4.** T, **5.** T, **6.** T, **7.** T, **8.** para-sympathetic, **9.** F both stimulates and inhibits, **10.** T

Completion

1. terminal, **2.** parasympathetic, **3.** nor-epinephrine, **4.** sympathetic, **5.** increase, **6.** ganglia, **7.** sympathetic, **8.** parasympathetic, **9.** effectors, **10.** antagonistic/opposite

Lost Sheep

1. kidney (not associated with parasympathetic activity), **2.** parasympathetic (not a sympathetic function) **3.** somatic (not a visceral, efferent system), **4.** somatic (not related to visceral sensations listed), **5.** dopamine (not an ANS transmitter), **6.** dorsal root ganglion (not an ANS ganglion), **7.** terminal ganglion (not associated with the sympathetic division), **8.** parasympathetic (is not associated with the terms listed), **9.** increased digestive mobility (not part of the fight-or-flight response), **10.** sympathetic (does not generally have cholinergic postganglionic fibers)

Matching

1. d, **2.** d, **3.** b, **4.** c, **5.** b, **6.** a, **7.** b, **8.** a, **9.** b, **10.** a, **11.** b **12.** a, **13.** d, **14.** d, **15.** c

Double Crosses

across: **a.** synapse, **b.** three, **c.** ganglion, **d.** sympathetic, **e.** cholinergic down: adrenergic

Word Cage

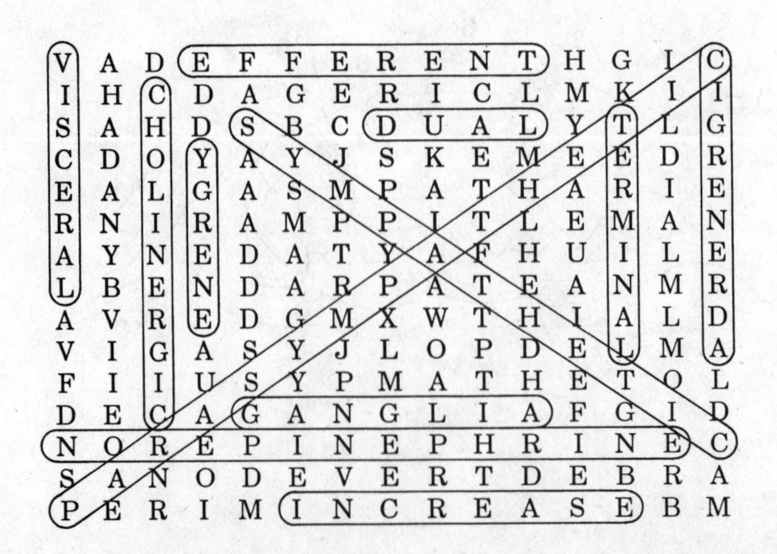

Sleuthing

a. sympathetic **b.** During stress, the sympathetic nervous system dominates and sets into motion the fight-or-flight response which causes some of the symptoms Laura was feeling. **c.** Autonomic centers in the cerebral cortex are linked to non- autonomic areas of the brain such as the thalamus and hypothalamus. This connection is the reason that anxiety, a subconscious activity can stimulate an autonomic response.

Word Scrambles

1. a. prevertebral, **b.** effectors, **c.** subconscious, **d.** parasympathetic, e. outflow TOTAL: visceral efferent fibers **2. a.** ganglion, **b.** neuroeffectors, **c.** sympathetic, **d.** norepinephrine, e. sacral TOTAL: fight-or-flight response

Addagram

a. gluconeogenesis, **b.** pancreas, **c.** constriction TOTAL: postganglionic neurons

Crossword

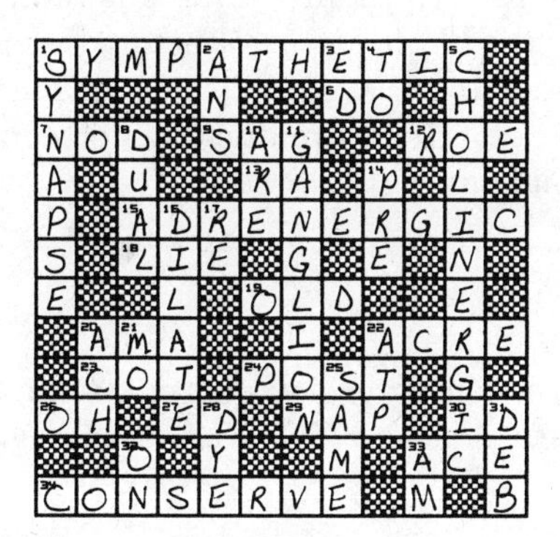

Chapter 12

Multiple Choice

1. d, **2.** b, **3.** b, **4.** a, **5.** c, **6.** d, **7.** d, **8.** c, **9.** a, **10.** d, **11.** d **12.** b, **13.** d, **14.** d, **15.** b, **16.** d, **17.** d, **18.** d, **19.** c, **20.** a, **21.** d **22.** a, **23.** b, **24.** b, **25.** b

True/False

1. T, **2.** T, **3.** F dynamic, **4.** F middle, **5.** F stapes, **6.** T, **7.** T, **8.** T, **9.** F four, **10.** F rods, **11.** F aqueous, **12.** T, **13.** T, **14.** T, **15.** T **16.** T, **17.** F dendrites, **18.** T, **19.** F unevenly, **20.** T

Completion

1. fovea centralis, **2.** A, rhodopsin, **3.** vitreous humor, **4.** pupil, **5.** optic disk/blind spot, **6.** ciliary, **7.** endolymph, **8.** basilar, **9.** outer, middle, **10.** referred pain, **11.** modality, **12.** surface **13.** pressure, **14.** optic ciasma, **15.** utricle, saccule, and semicircular canals

Lost Sheep

1. Organ of Corti (not associated with tactile sense/located in skin), **2.** 8 C (not a temperature at which thermoreceptors respond), **3.** cranial nerve II (not associated with the sense of smell) **4.** dim light (not associated with cones), **5.** ciliary body (not a refractive media of the eye), **6.** cones (not associated with night blindness), **7.** inner ear (not associated with the middle ear structures listed), **8.** crista (not associated with static equilibrium), **9.** round window (not associated with the stapes), **10.** proprioception (not a "special sense"), **11.** quick adaptation (not associated with pain or its reception), **12.** photoreceptors (do not have hair cells), **13.** trachoma (not associated with an ear infection), **14.** mechanoreceptor (not a modality or sense), **15.** cornea (not part of the vascular tunic)

Matching

Set 1: **1.** d, **2.** a, **3.** e, **4.** c, **5.** b
Set 2: **1.** c, **2.** a, **3.** b, **4.** e, **5.** d
Set 3: **1.** e, **2.** c, **3.** d, **4.** a, **5.** b

Sleuthing

1. a. The Eustachian tube leads from the throat to the middle ear and is lined with the same mucus secreting tissue as the throat. This tube was blocked and so it was difficult for her to equalize the pressure on the ear drum. **b.** A large part of the sense of taste is really related to smell, The lack of taste was due to the fact that the cold was blocking the olfactory receptors. **2. a.** The lacrimal apparatus which produces tears responds to para- sympathetic stimulation during an emotional experience. Excess tears are produced. **b.** There are ducts which convey tears away and excess production causes spillage into the nasal cavity.

Word Scrambles

1. a. glaucoma, **b.** rhodopsin, **c.** vitreous humor, **d.** choroid, e. chiasma TOTAL: accommodation
2. a. modality, **b.** fovea, **c.** retina, **g.** generator TOTAL: afterimage

Addagrams

1. a. retina, **b.** Corti, **c.** lacrimal, **d.** bilateral, **e.** mean TOTAL: tectorial membrane **2. a.** sclera, **b.** malleus, **c.** saccule, **d.** incus, **e.** iris TOTAL: semicircular canals **3. a.** basilar, **b.** craniosacral, **c.** oval, **d.** sinus TOTAL: binocular vision

Crossword

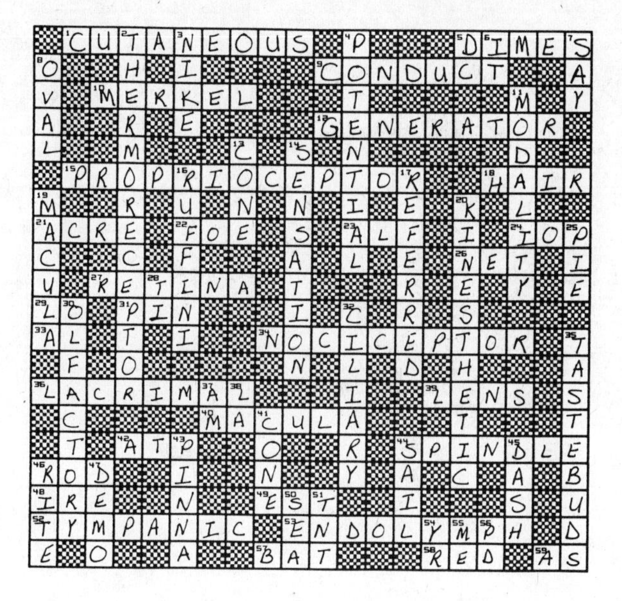

Anatomic Artwork

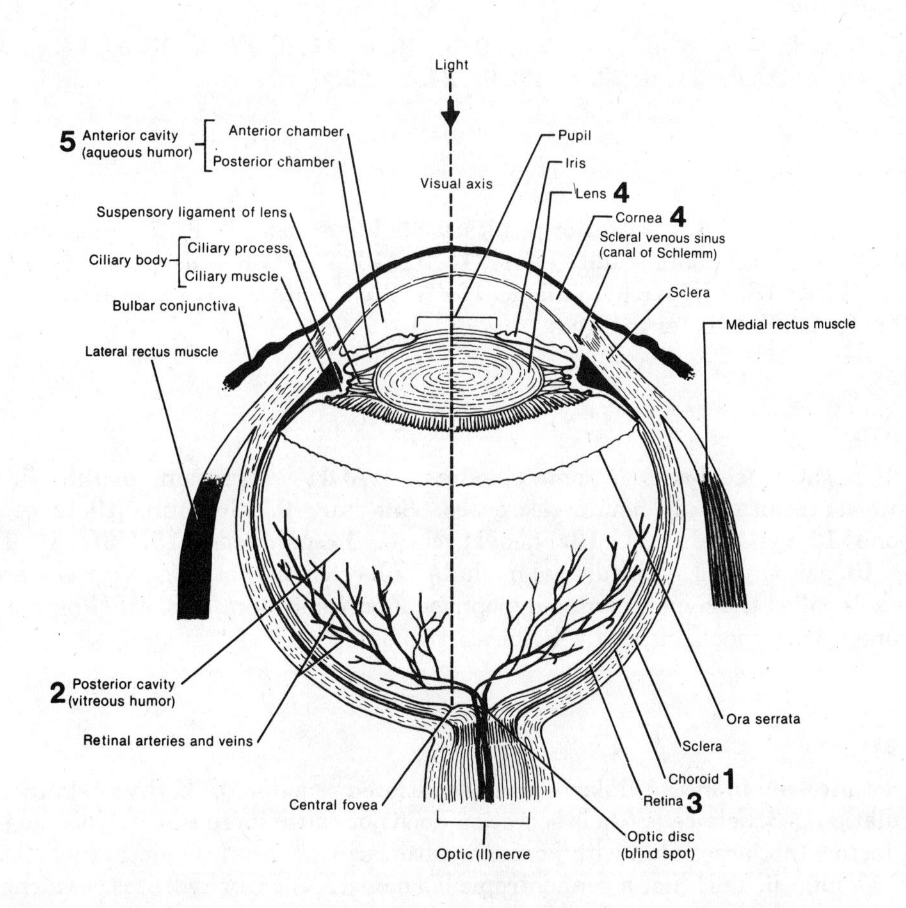

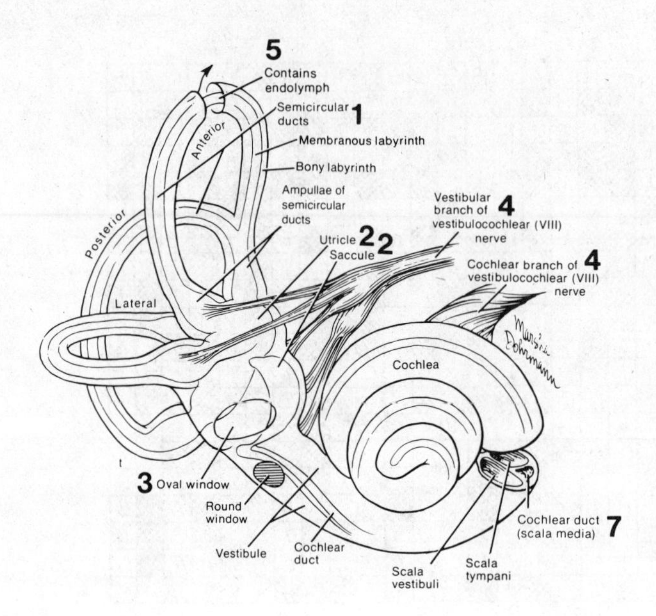

Chapter 13

Multiple Choice

1. c, **2.** b, **3.** d, **4.** d, **5.** d, **6.** c, **7.** a, **8.** d, **9.** b, **10.** c, **11.** d **12.** d, **13.** c, **14.** c, **15.** c, **16.** c, **17.** d, **18.** d, **19.** b, **20.** d, **21.** b **22.** a, **23.** b, **24.** d, **25.** d

True/False

1. T, **2.** T, **3.** F glucagon, **4.** F posterior pituitary, **5.** F steroids, **6.** T, **7.** F negative, **8.** T, **9.** F oxytocin, **10.** T, **11.** F adrenal medull **12.** T, **13.** T, **14.** T, **15.** F glucocorticoids only, **16.** F parathormon **17.** T, **18.** F hyperthyroidism, **19.** T, **20.** F water, **21.** F increase **22.** F liver, **23.** F pituitary, **24.** T, **25.** F diabetes mellitus

Completion

1. calcium, **2.** negative feedback, **3.** somatomedins, **4.** ADH/vasopressin, insulin, **5.** adrenal, **6.** ACTH, anterior pituitary, **7.** insulin, decreases, **8.** water, **9.** calcitonin, **10.** target, **11.** aldosterone **12.** hypothalamus, **13.** adenyl cyclase, **14.** pancreas, **15.** TSH **16.** TSH, **17.** calcium, **18.** parathyroid, **19.** adrenal medulla **20.** releasing factors, **21.** resistance reaction, **22.** blood vessels called the hypothalamic-hypophyseal portal system, **23.** MSH/pineal gland, **24.** aldosterone, **25.** epinephrine/cortisol/glucocorticoids.

Lost Sheep

1. AChase (not involved in intracellular hormone-initiated reactions), **2.** thymosin (not involved in calcium regulation), **3.** beta cells (their secretion does not cause increases in blood sugar levels), **4.** releasing factors (not associated with posterior pituitary), **5.** erythropoietin (not associated with the pituitary gland), **6.** TSH (not a gonadotropic hormone), **7.** carbohydrate (not a chemical class of hormones), **8.** epinephrine (not produced by the adrenal cortex), **9.** pituitary (not associated with pineal gland), **10.** Addison's disease (not a disorder associated with growth hormone), **11.** thyrotropin (not released by the posterior pituitary), **12.** goiter (not associated with adrenal cortex dysfunction), **13.** hypothyroidism (not associated with growth hormone), **14.** tyrosine (does not contain iodide), **15.** reabsorption of sodium (not a function associated with glucocorticoids), **16.** exocrine (not associated with endocrine function of pancreas), **17.** lipogenesis (not associated with glucagon function), **18.** vitamin C (not controlled by parathormone), **19.** testosterone (not a protein hormone), **20.** PTH (not associated with prostaglandins)

Matching

Set 1: **1.** d, **2.** e, **3.** b, **4.** c, **5.** a
Set 2: **1.** b, **2.** c, **3.** d, **4.** e, **5.** a
Set 3: **1.** e, **2.** g, **3.** a, **4.** k, **5.** h, **6.** m, **7.** c, **8.** j, **9.** f, **10.** b, **11.** i, **12.** d, **13.** l, **14.** o, **15.** n

Double Crosses

1. across: insulin down: medulla **2.** across: thyroid down: steroid **3.** across: endocrine down: principal **4.** across: androgens down: thyroxine **5.** across: prostaglandins down: **a.** protein, **b.** thyroid, **c.** neurons d. steroid

Sleuthing

1. a. Cortisol is an anti-inflammatory agent which inhibits fibroblast production, stabilizes lysosomal membranes and increases the body's sensitivity to vasoconstrictors. As an anti-inflammatory agent, it helps combat the inflammation of joints seen in arthritis. **b.** Some undesirable effects could include: weight gain, osteoporosis and development of diabetes, for example. **c.** Normally, decreased glucocortocoid level stimulates production of ACTH which in turn causes the release of glucocorticoids. As glucocorticoid level rises, they inhibit ACTH release by negative feedback mechanism. **2. a.** neurohypophysis **b.** ADH **c.** diabetes insipidus **d.** increased urine production and severe thirst

Word Scrambles

1. a. melatonin, **b.** oxytocin, **c.** thyroid, **d.** thymosin, **e.** goiter **f.** androgen, **g.** medulla TOTAL: myxedema **2. a.** neurohumors, **b.** negative feedback, **c.** prostaglandins, **d.** steroids TOTAL: target tissue

Addagrams

1. a. hypophysis, **b.** steroid, **c.** more TOTAL: hyperthyroidism **2. a.** cortisol, **b.** calcitonin, **c.** adrenalin, **d.** melatonin TOTAL: mineralocorticoids

Crossword

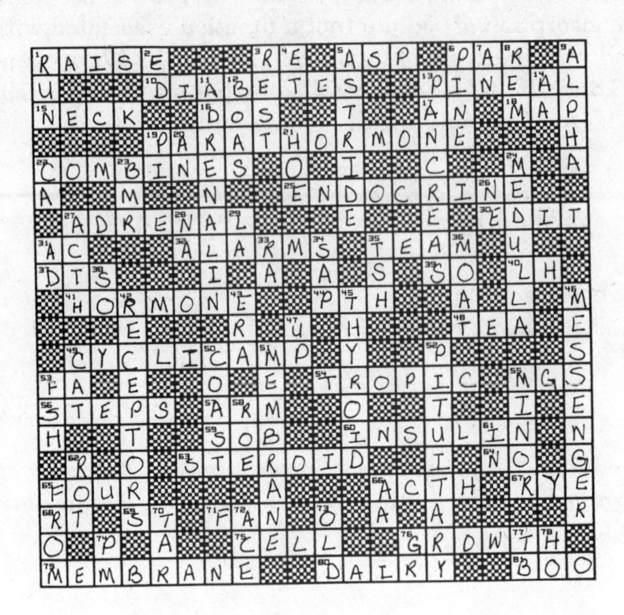

Anatomic Artwork

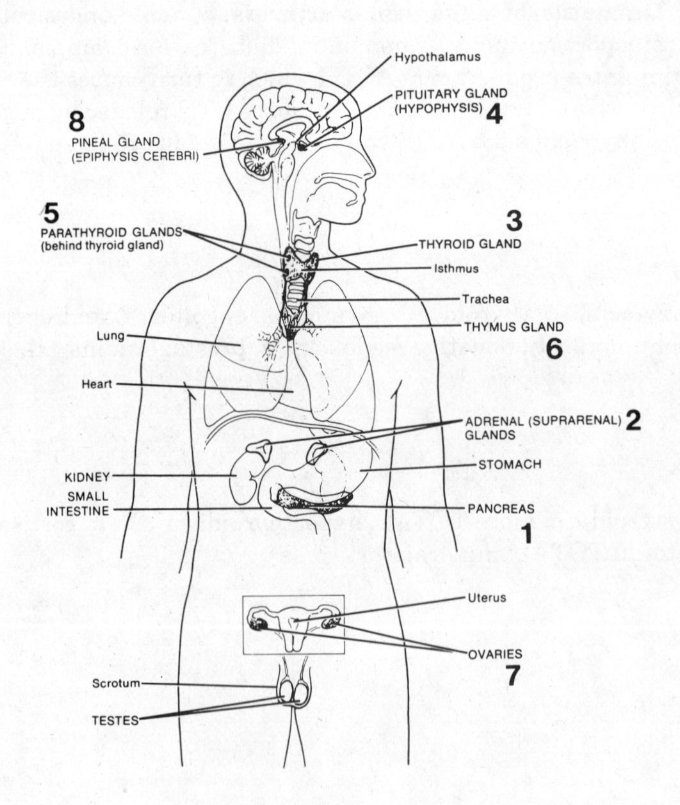

Chapter 14

Multiple Choice

1. d **2.** d **3.** a **4.** d **5.** c **6.** d **7.** d **8.** c **9.** b **10.** d **11.** d **12.** c **13.** a **14.** d **15.** b **16.** c **17.** c **18.** d **19.** c **20.** c **21.** d **22.** d **23.** d **24.** b **25.** d

True/False

1. T **2.** T **3.** T **4.** F - larger **5.** T **6.** T **7.** T **8.** T **9.** T **10.** F - diapedesis **11.** T **12.** F - whole blood sample **13.** T **14.** F - 5 million/mm3 **15.** F - T lymphocyte **16.** F - normal **17.** **18.** T **19.** F - more **20.** F - agglutinogens **21.** T **22.** F - mother Rh-, baby Rh+ **23.** T **24.** T **25.** T

Completion

1. heme **2.** fibrinolysin/plasmin **3.** clot formation **4.** basophil **5.** carbinohemoglobin **6.** **7.** 35-7. 45 **7.** megakaryocyte **8.** O **9.** agglutination, clumping, hemolysis **10.** 14-16. 5g/ml;12-15g/ml **11.** serum **12.** hemostasis **13.** liver **14.** create osmotic pressure;act as antibodies **15.** clot retraction **16.** no **17.** transport, temperature regulation, protection **18.** red cells, platelets **19.** neutrophil **20.** leukopenia **21.** polycythemia **22.** hemopoiesis **23.** calcium **24.** hemoglobin **25.** phagocytize

Lost Sheep

1. lymphocyte (not granular) **2.** 15 m(not characteristic of RBC) **3.** uremia (not a condition involving blood cells) **4.** shrink (not part of platelet response in plug formation) **5.** hemolysis (not part of WBC response) **6.** biconcave (not associated with WBC) **7.** parasitic infection (not associated with basophil) **8.** clotting (not associated with typing) **9.** seconds (not intrinsic time scale) **10.** fibrinolysin (not associated with clot formation) **11.** nitrogen (not carried by hemoglobin) **12.** biliverdin (not associated with RBC/hemoglobin formation) **13.** leukemia (not associated with red cells) **14.** type O (does not have ABO agglutinogens) **15.** hemophilia (not associated with red cells) **16.** Rh factor (not associated with clot) **17.** small (size) (not associated with proteins) **18.** fibrinogen (not normally carried by blood) **19.** hemoglobin (not associated with platelets) **20.** oxygen (not protein)

Matching

SET 1
1. **1.** a **2.** a **3.** c **4.** e **5.** d
SET 2
1. c **2.** a **3.** c **4.** a **5.** d **6.** f **7.** e **8.** b **9.** c **10.** e
SET 3
1. d **2.** c **3.** e **4.** b **5.** c **6.** d **7.** a **8.** a **9.** d **10.** c

Double Crosses

1. across:erythrocyte' down:bilirubin **2.** across:leucocyte down:hemopoiesis **3.** across a:platelets b:allergy down:reticulocytes **4.** a:albumins b:aplastic c:altitude d:allergic e:antibody f:antigens g:activate h:alkaline

Sleuthing

1. no problem - no mixing of blood till end **2.** yes - the blood she was exposed to in the first pregnancy is foreign to her **3.** she will make agglutinins against the Rh factor **4.** the agglutinins will cross the placenta into baby's blood; baby's cells will hemolyze, baby will become anemic, with large liver and spleen **5.** mother could be prevented from mounting an agglutinin response by being Rhogam (Rh agglutinins) **6.** no problem, as Rh looks familiar, and baby's blood doesn't present anything foreign

Word Scrambles

1. a. antigen **b.** hemoglobin **c.** thromboplastin **d.** albumin TOTAL:agglutination **2. a.** platelet **b.** hematocrit **c.** neutrophil **d.** antibody TOTAL:erythropoietin

Addagrams

1. a. serum **b.** hemo **c.** ag **d.** trash **e.** Ca++ **f.** poiesis TOTAL:tissue macrophages
2. a. anemia **b.** plasmin **c.** platelet **d.** pro TOTAL:plasma protein

Crossword

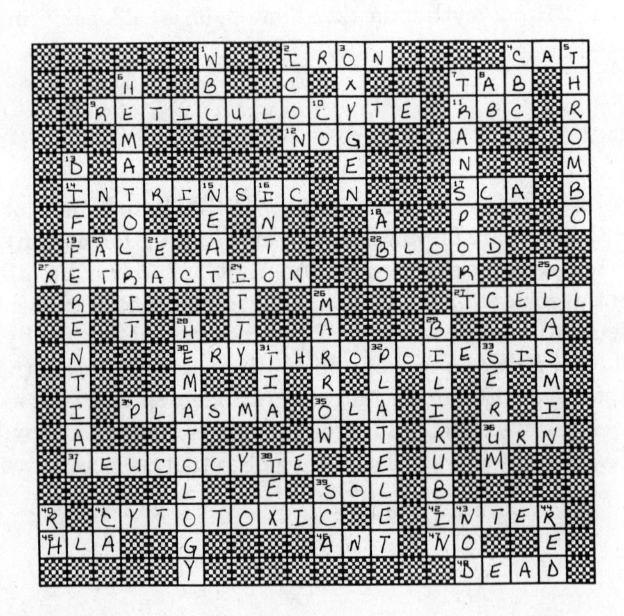

Chapter 15

Multiple Choice

1. c **2.** d **3.** a **4.** c **5.** d **6.** a **7.** d **8.** b **9.** d **10.** c **11.** d **12.** d **13.** b **14.** a **15.**b **16.** b **17.** c **18.** b **19.** d **20.** d **21.** d **22.** d **23.** b **24.** c **25.** a

True/False

1. F-70 ml **2.** T **3.** T **4.** T **5.** T **6.** F-blood pressure **7.** T **8.** T **9.** F-closed **10.** T **11.** F-aneurysm **12.** F-aorta to pulmonary artery **13.** T **14.** T **15.** F-systemic **16.** F-left **17.** F-is self-excitable **18.** F-slows **19.** F-and also **20.** T **21.** T **22.** T **23.** F-medulla **24.** T **25.** F-repolarizing

Completion

1. 70ml **2.** atherosclerosis **3.** cardiac tamponade **4.** calcium **5.** aortic semi-lunar **6.** coronary sinus **7.** left to right **8.** coronary **9.** 120 **10.** pulmonary artery **11.** cardioinhibitory center **12.** intercalated discs/gap junctions **13.** 5 **14.** relaxation **15.** cyanosis **16.** baroreceptors **17.** congestive **18.** cor pulmonale **19.** 2 **20.** chordae tendineae **21.** A-V node **22.** P wave **23.** semi-lunar **24.** right atrium **25.** striated,involuntary,self-excitable

Lost Sheep

1. left atrium (not associated with right heart) **2.** mitral valve -(not part of conductive system) **3.** vena cava (not solely in fetus) **4.** 120 mm Hg (not associated with right heart) **5.** decreased venous return (not associated with stretch and greater pumping) **6.** high BP (not a characteristic of shock) **7.** epinephrine (not connected with parasympathetic nerves) **8.** blood enters atrium (does not normally occur during ventricular systole) **9.** atrial repolarization (not associated with P wave) **10.** vagal control (not part of sympathetic control) **11.** interventricular septum (not an external structure on heart) **12.** fibrous pericardial sac (external to heart, not part of muscle of ventricle) **13.** aorta (does not carry venous blood or is it associated with right heart) **14.** cardiac skeleton (not part of muscle of heart) **15.** carotid artery (is not associated with heart) **16.** non-striated (not a characteristic of cardiac muscle tissue **17.** edema (not associated with venous return) **18.** well trained athlete (does not have a high heart rate) **19.** decreased blood pressure (not associated with high blood pressure) **20.** mitral (not associated with right heart)

Matching

SET 1 **1.** b **2.** a **3.** b **4.** c **5.** b **6.** a **7.** b **8.** b **9.** b **10.** c
SET 2 **1.** c **2.** b **3.** a **4.** b **5.** d
SET 3 **1.** d **2.** a,c **3.** d **4.** a,b **5.** a,c
SET 4 **1.** e **2.** b **3.** b **4.** e **5.** e **6.** e **7.** d **8.** c **9.** a **10.** e

Double Crosses

1. across: **a.** fibrillates **b.** chordae down:pulmonary **2.** across: **a.** tetralogy **b.** ventricular down:mediastinum **3.** across: **a.** cardiac output **b.** papillary **c.** sphygmo down:capillary **4.** **a.** angina **b.** aortic **c.** atrium **d.** artery

Sleuthing

1. death of the myocardium served by the incompetent vessel **2.** bypassing the incompetent vessel with a graft of an open vessel **3.** atherosclerosis

Word Scramble

1a. aorta **b.** reflex **c.** trunk TOTAL:flutter **2.** **a.** thymus **b.** cardiac **c.** murmur TOTAL:ductus

Addagrams

1. rate **b.** ten **c.** cardio **d.** death **e.** NE TOTAL:chordae tendinae **2. a.** arterial **b.** stenosis **c.** ovale **d.** clot **e.** rate **f.** volume **g.** ESV TOTAL:atrioventricular valves

Crossword

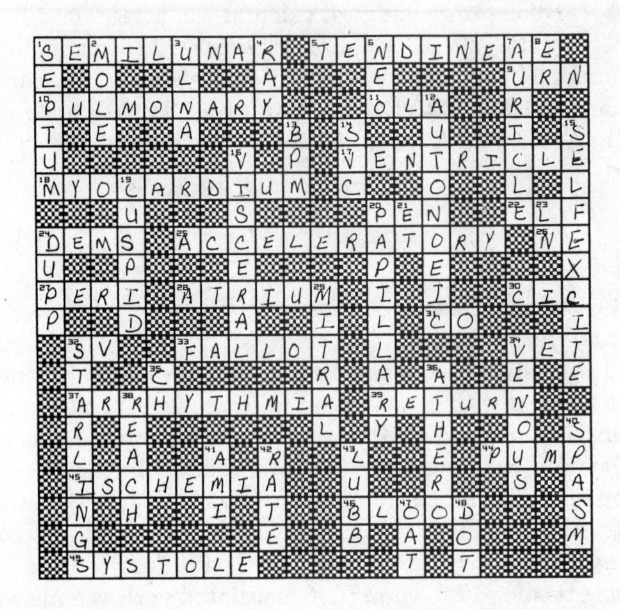

Anatomic Artwork

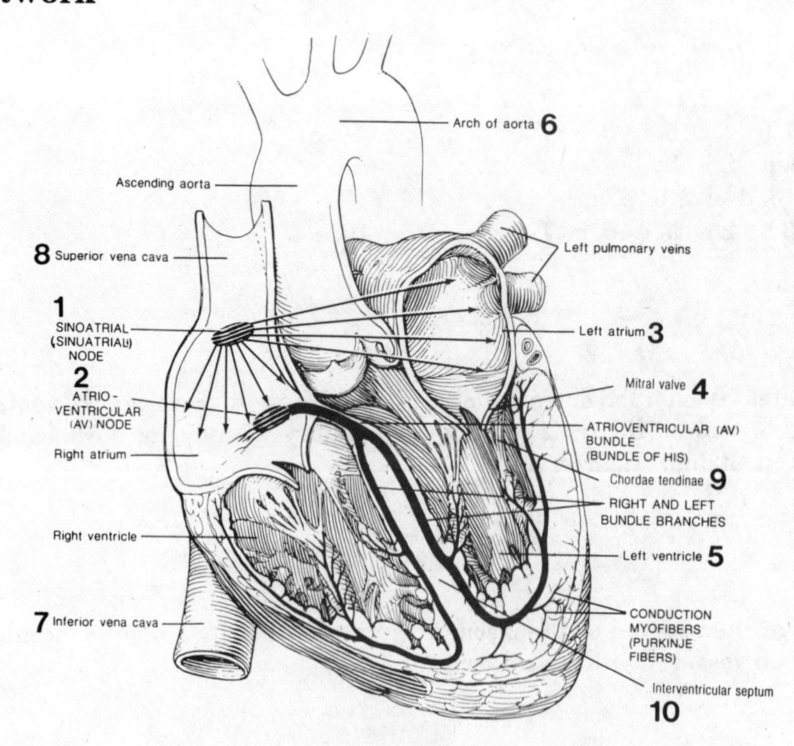

Chapter 16

Multiple Choice

1. b **2.** d **3.** a **4.** c **5.** b **6.** a **7.** d **8.** c **9.** a **10.** d

True/False

1. F - medulla **2.** F - sympathetic only **3.** T **4.** T **5.** T **6.** F - cool **7.** F - aneurysm **8.** F - least **9.** T **10.** F - cardiac output

Completion

1. dilated **2.** hepatic portal vein **3.** hypertension **4.** syncope **5.** vasomotion **6.** peripheral resistance **7.** vasomotor center **8.** baroreceptors **9.** sphygmomanometer **10.** carotid body,aortic body

Matching

SET 1 **1.** b **2.** a **3.** c,d **4.** a **5.** d SET 2 **1.** **b.** **2.** c **3.** a:b:c **4.** a **5.** b

Double Crosses

1. across:elastic down:vessels **2.** across:medulla down:solutes **3.** across:viscosity down **a.** vena cava **b.** systolic **c.** oxygen **d.** ischemia **e.** sphygmo

Word Scrambles

1. **a.** bradycardia **b.** reservoir **c.** caps **d.** carotid TOTAL:baroreceptors **2.** **a.** pulmonary **b.** vessel **c.** pale **d.** wrist **e.** bulge TOTAL:pulse pressure

Crossword

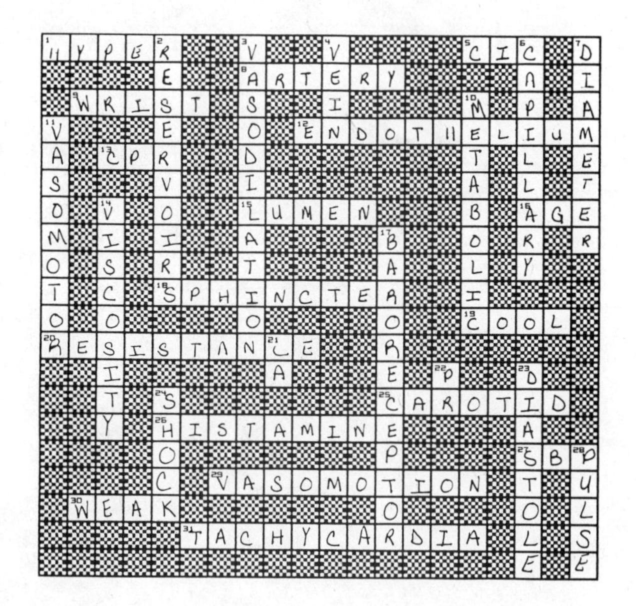

Anatomic Artwork

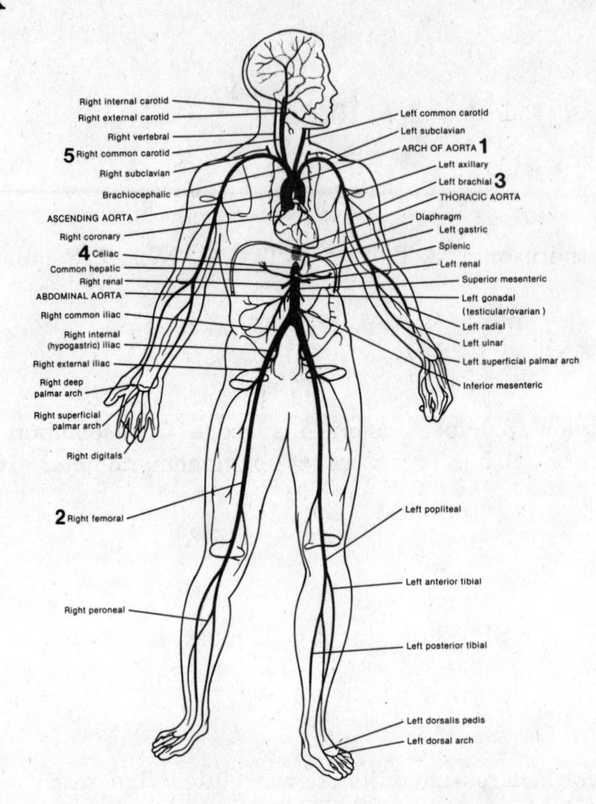

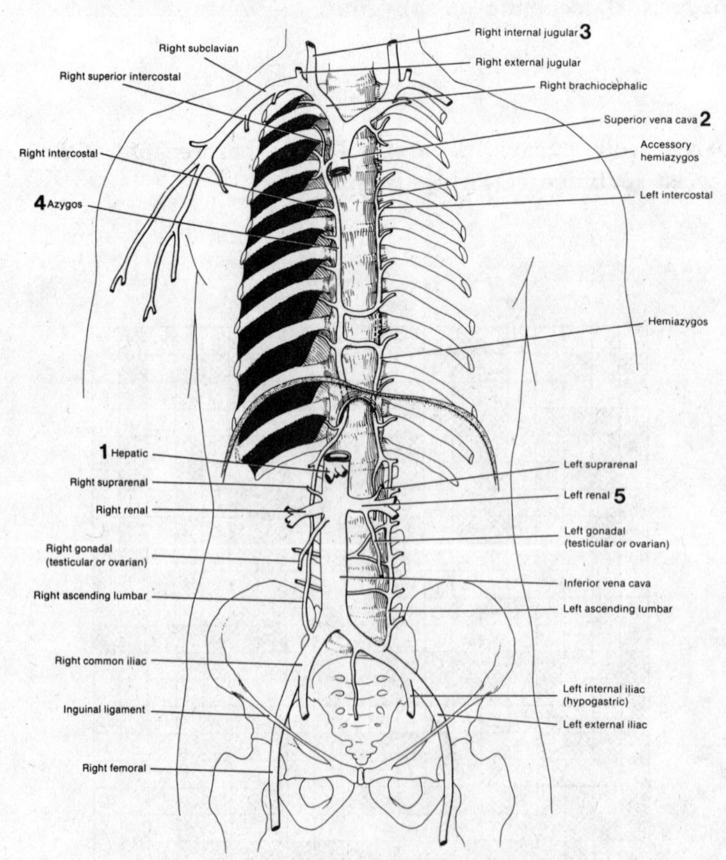

Chapter 17

Multiple Choice

1. b **2.** d **3.** d **4.** d **5.** a **6.** d **7.** d **8.** c **9.** c **10.** c **11.** b **12.** d **13.** d **14.** c **15.**c **16.** a **17.** b **18.** a **19.** a **20.** a

True/False

1. T **2.** F-lymphatic capillaries **3.** F-diapedesis **4.** F-lymph **5.** T **6.** T **7.** T **8.** F-blood only **9.** F-decreases **10.** F-the thymus **11.** T **12.** T **13.** T **14.** F-memory cell **15.** T **16.** F-has **17.** F-macrophage **18.** T **19.** F-nonspecific **20.** T

Completion

1. They are the same. **2.** protein **3.** tissue spaces **4.** leaks from capillaries **5.** reticular tissue **6.** phagocytized **7.** tonsils **8.** spleen **9.** thoracic duct **10.** edema **11.** phagocytosis **12.** immunity **13.** antigen **14.** self **15.** an antibody **16.** immunoglobulin **17.** T cells **18.** B cells **19.** general circulation **20.** macrophages

Lost Sheep

1. coolness (not characteristic of inflammation) **2.** aorta (not part of lymph system.) **3.** liver (not a lymph organ) **4.** immunity (not part of non-specific response) **5.** red cell (not part of white cell response in inflammation) **6.** neutrophil (not a lymphocyte) **7.** T cell (not involved in B cell descriptions) **8.** antibody (not a non-specific protection) **9.** histamine (not an antimicrobial chemical) **10.** diapedesis (not part of phagocytosis)

Matching

SET 1 **1.** b **2.** a **3.** c **4.** d **5.** b **6.** c **7.** b **8.** d **9.** a **10.** b
SET 2 **1.** d,f **2.** e **3.** a **4.** c **5.** b **6.** d,f **7.** d,f **8.** d,f **9.** f **10.** f

Double Crosses

1. across: macrophage **a.** microbe **b.** capsule **c.** outward **d.** antigen **e.** efferent
2. across:humoral down:compromised

Sleuthing

1. The lymph in the axillary region were removed and so there is an interruption in the draining of normal tissue fluid from her arm. **2.** The lymph nodes were removed so that the cancer would not have an easy avenue of spread in her body. Nodes can trap and subsequently release cancer cells, thus causing metastasis.

Word Scrambles

1. a. spaces **b.** ionizes **c.** body TOTAL:opsonize **2. a.** cytology **b.** macrophage **c.** sinus TOTAL:phagocytosis

Addagrams

1. **a.** stores **b.** escape **c.** cap **d.** none **e.** artificial **f.** cancer TOTAL:non-specific resistance

Crossword

Chapter 18

Multiple Choice

1. c. **2.** d **3.** b **4.** d **5.** d **6.** b **7.** b **8.** b **9.** c **10.** c **11.** c **12.** d **13.** c **14.** d **15.** d **16.** d **17.** b **18.** c **19.** d **20.** c **21.** c **22.** b **23.** c **24.** d **25.** d

True/False

1. T **2.** F - loosely **3.** T **4.** T **5.** F - 97% **6.** T **7.** T **8.** F - 200X **9.** T **10.** T **11.** T **12.** F - above the superior nasal concha **13.** T **14.** T **15.** T **16.** F - C-shaped **17.** T **18.** F - deoxygenated blood **19.** F - higher **20.** T **21.** T **22.** T **23.** T **24.** T **25.** F - speeds

Lost Sheep

1. bronchioles (not associated with alveolus) **2.** hyoid (not associated with larynx) **3.** alveolar duct (not associated with respiratory membrane) **4.** vital capacity (not a single volume of breathing) **5.** carbonic acid (not exhaled at lung) **6.** carboxyhemoglobin (not a way in which carbon dioxide travels in blood) **7.** bicarbonate (not an acid) **8.** baroreceptor (not a control center) **9.** conchae (not part of lung) **10.** cortical impulses (not part of involuntary control over respiration) **11.** pharynx (not part of nose) **12.** parathyroid (not part of larynx) **13.** diaphragm (not in mediastinum) **14.** trachea (not part of lung anatomy) **15.** expiration:duration 1 second (not a true statement related to normal breathing) **16.** pleurisy (does not deal with alveoli) **17.** higher pressure in air than in lungs (not characteristic of expiration) **18.** formation of carboxyhemoglobin (not a combination of CO_2 and hemoglobin) **19.** cerebellum (not involved in control of respiration) **20.** volume increase (not characteristic of inhalation)

Matching

SET 1

1. b **2.** c **3.** a **4.** a **5.** d **6.** d **7.** d **8.** c **9.** b **10.** a

SET 2

1. f **2.** e **3.** a **4.** d **5.** g

SET 3

1. e **2.** d **3.** b **4.** e **5.** c

Double Crosses

1. across:apneustic down:pleural **2.** across:cricoid down:olfactory **3.** across:surfactant **a.** sneeze **b.** receptor **c.** alveolus **d.** tracheal **e.** pneumonia

Sleuthing

a. hyaline membrane disease **b.** deficiency in surfactant production **c.** alveolar septal cells **d.** function is to decrease the surface tension in the alveoli so than can inflate more easily **e.** the baby's lung are too immature to produce surfactant at 23 weeks, so as the weeks ensue the, and the baby is supported by a respirator, the baby's lungs will gain the ability to produce surfactant

Word Scrambles

1. a. inhale **b.** volume **c.** dyspnea **d.** right **e.** tonsil TOTAL:hyperventilation **2. a.** capillary **b.** SIDS **c.** pure **d.** exert TOTAL:partial pressure

Addagrams

1. a. RDS **b.** dual **c.** oral **d.** more **e.** alive **f.** U **g.** U TOTAL:residual volume **2. a.** H,O **b.** air **c.** intubation **d.** itis TOTAL: Hb Saturation

Crossword

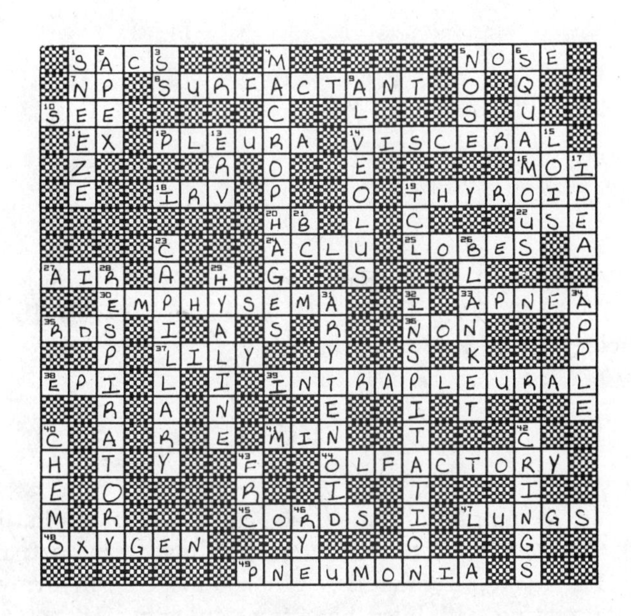

Anatomic Artwork

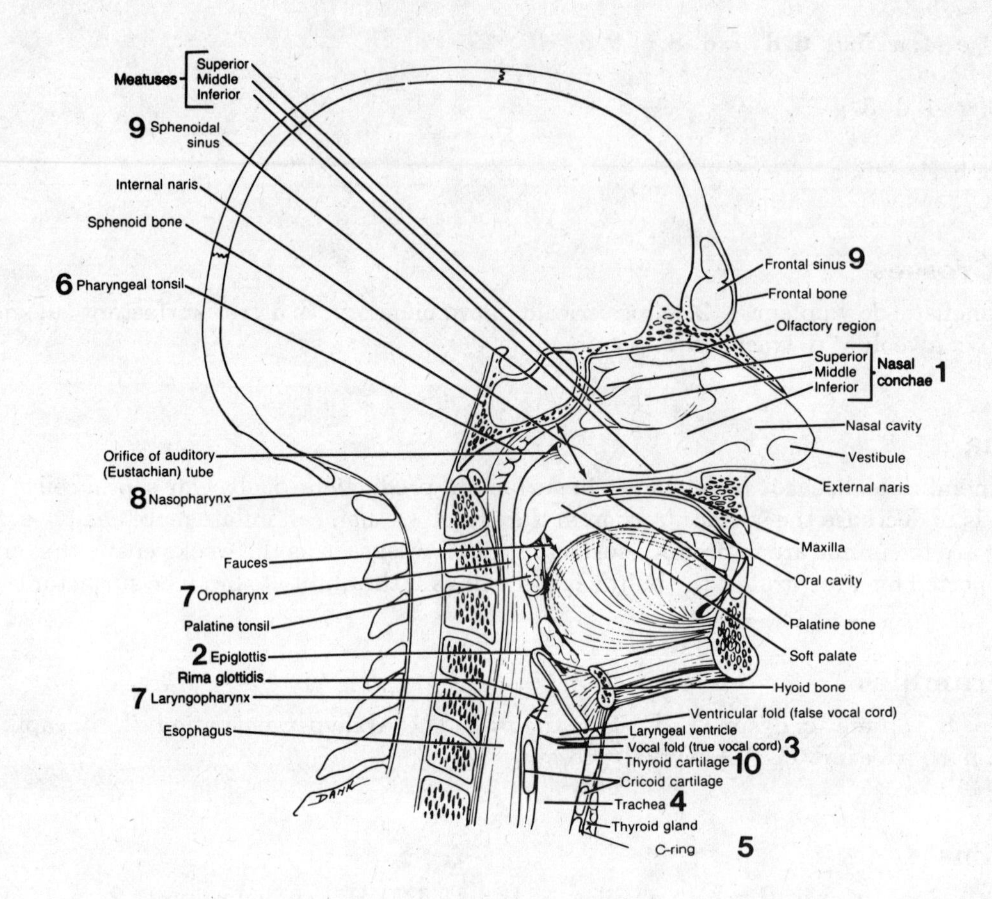

Chapter 19

Multiple Choice

1. c, **2.** d, **3.** c, **4.** d, **5.** c, **6.** b, **7.** c, **8.** c, **9.** c, **10.** c, **11.** c **12.** d, **13.** c, **14.** b, **15.** a, **16.** b, **17.** b, **18.** b, **19.** d, **20.** b, **21.** b **22.** d, **23.** a, **24.** c, **25.** c

True/False

1. T, **2.** T, **3.** T, **4.** T, **5.** F parietal, **6.** T, **7.** F microvilli, **8.** T, **9.** F emulsify, **10.** F large intestine/colon, **11.** T, **12.** T, **13.** T, **14.** **15.**NT, **16.** F filiform, **17.** T, **18.** T, **19.** F stimulates, **20.** F liver **21.** T, **22.** F cholecystokinin, **23.** T, **24.** T, **25.** T

Completion

1. cecum, **2.** cardiac sphincter, **3.** simple columnar, **4.** acini, **5.** duodenum, **6.** parotid, submandibular, sublingual, **7.** diverticula **8.** rennin, gastric lipase, **9.** gastrin, **10.** segmentation, **11.** bacterial fermentation, **12.** chyme, **13.** mouth, **14.** proteins, carbohydrates, lipids, **15.** chylomicrons, **16.** small intestine, **17.** liver, hepatic portal vein, **18.** common bile duct, **19.** cholesterol, **20.** gastric mucosa, **21.** peptic, **22.** rennin, **23.** vagus, **24**, A, **25.** peristalsis

Lost Sheep

1. glucose (not a disaccharide), **2.** amylase (not a nutrient), **3.** appendix (does not pertain to the liver), **4.** dipeptide (does not refer to a lipid), **5.** rugae (not a division of the small intestine), **6.** gastrin (not an enzyme), **7.** parietal (not a salivary gland), **8.** haustra (not associated with the stomach), **9.** amylase (not related to the stomach), **10.** mesenteric (not a sphincter), **11.** proteins (not a term referring to carbohydrate), **12.** cecum (not an accessory gland of the digestive tract), **13.** antrum (not part of the small intestine) **14.** pepsin (not a constituent of saliva), **15.** secretin (does not stimulate gastric secretions), **16.** acid secretion (does not occur in the mouth), **17.** cystic duct (not associated with the pancreas) **18.** rugae (not a part of the liver), **19.** churning (not a liver function), **20.** circular folds (not associated with the large intestine

Matching

Set 1: **1.** c, **2.** e, **3.** d, **4.** d, **5.** a
Set 2: **1.** e, **2.** c, d, **3.** c, **4.** b, d, **5.** a
Set 3: **1.** h, **2.** c, **3.** d, **4.** h, **5.** b, **6.** a, **7.** d, **8.** e, **9.** a, **10.** h, **11.** h, **12.** f, **13.** c, **14.** d, **15.** h

Double Crosses

1. a. peristalsis, **b.** proteolytic, **c.** polypeptide **2. a.** masticate, **b.** fungiform, **c.** vermiform, **d.** falciform **3. a.** sphincters, **b.** saccharide, **c.** sublingual **4.** across: **a.** bolus, **b.** sphincter, **c.** jejunum, **d.** valve, **e.** bilirubin, **f.** voluntary, **g.** colon, **h.** chylomicron down: lingual frenulum

Sleuthing

a. In this case intolerance refers to the inability of the individual to digest lactose due to the absence of the enzyme lactase. **b.** Lactose is a disaccharide found in milk and if there were no intolerance the secretion of the small intestines containing lactase would break the disaccharide into two monosaccharide units, glucose and galactose. **c.** The end products would be absorbed into the blood stream and sent directly to the liver via the hepatic portal vein. **d.** Undigested carbohydrate is being digested in the large intestine by bacteria and this causes the production of gas.

Word Scrambles

1. a. esophagus, **b.** pharynx, **c.** sigmoid, **d.** portal TOTAL: peristalsis **2. a.** bolus, **b.** villi, **c.** sodium bicarbonate TOTAL: bilirubin **3. a.** colon, **b.** gall bladder, **c.** enzyme, **d.** mucus, **e.** fats, **f.** acini TOTAL: common bile duct

Addagrams

1. a. cephalic, **b.** pyloric region, **c.** carbohydrates, **d.** hard, **e.** gland **f.** dentin TOTAL: hydrochloric acid and pepsinogen **2. a.** bulimia, **b.** left, **c.** saliva, **d.** intestines, **e.** amino TOTAL: fat soluble vitamins 3. **a.** Meissner's, **b.** peritoneal, **c.** gastrin, **d.** salts, **e.** digestion TOTAL: segmentation and peristalsis

Crossword

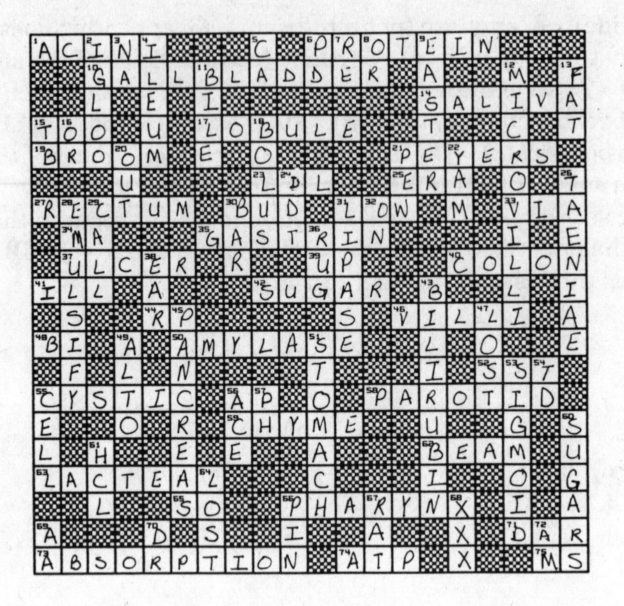

Anatomic Artwork

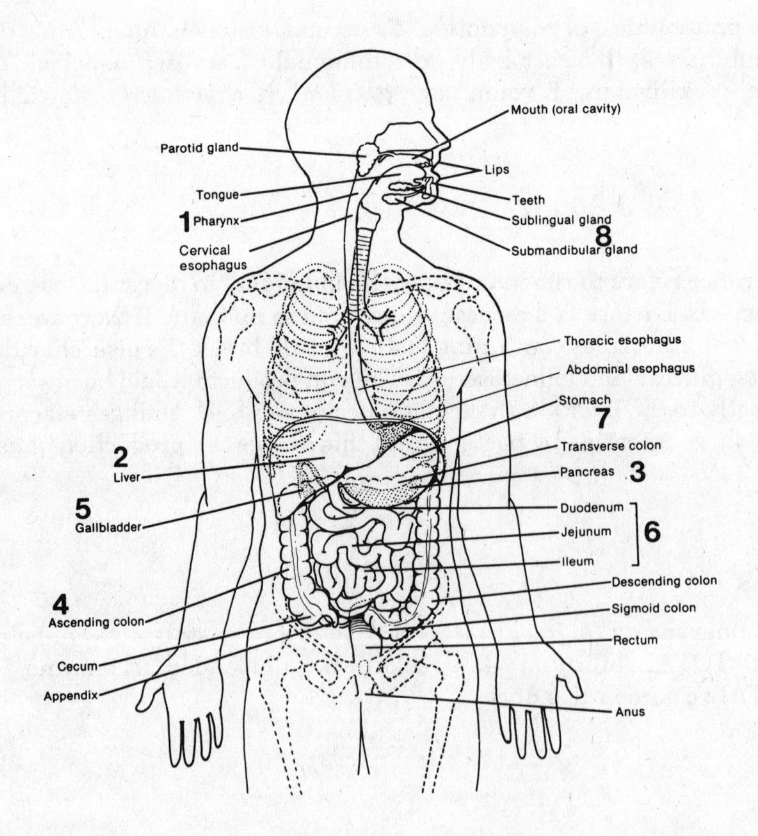

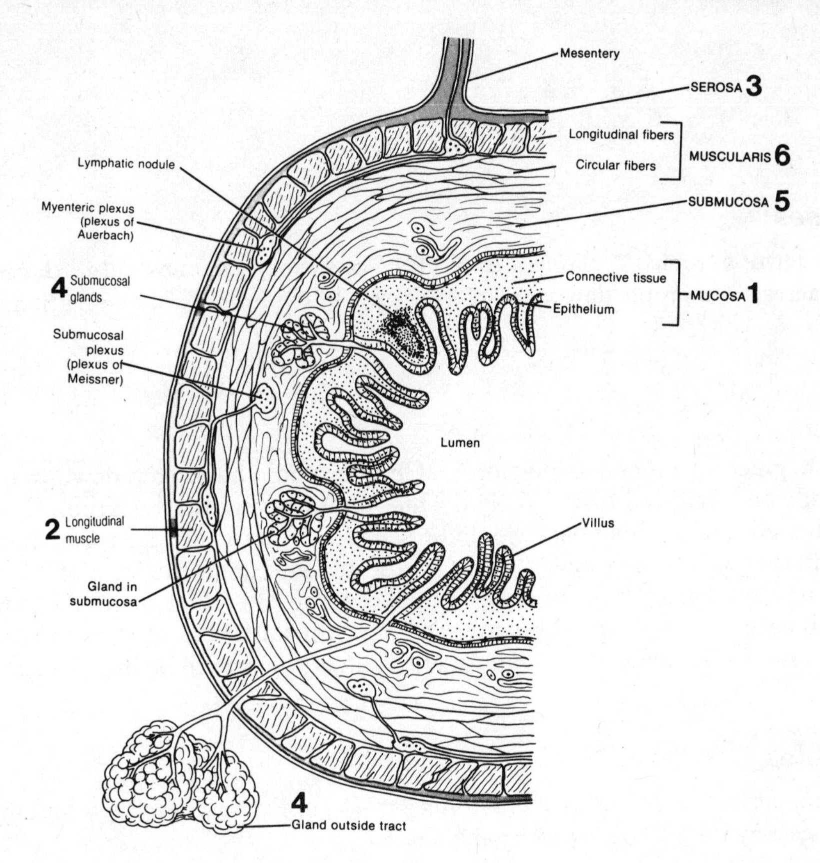

Chapter 20

Multiple Choice

1. c, **2.** d, **3.** d, **4.** d, **5.** a, **6.** d, **7.** c, **8.** a, **9.** d, **10.** c, **11.** a, **12.** d

True/False

1. T, **2.** T, **3.** T, **4.** F aerobic, **5.** F anabolic, **6.** T, **7.** T, **8.** F glycolysis, **9.** F conduction, **10.** T, **11.** T, **12.** T

Completion

1. water, **2.** hypothalamus, decreased/halted, **3.** metabolism. **4.** enzyme, **5.** respiration, **6.** gluconeogenesis, **7.** kilocalories, **8.** acetyl CoA, **9.** deamination, **10.** 60 - 75 , **11.** thyroxine, **12.** radiation

Lost Sheep

1. vitamin B (not a fat soluble vitamin), **2.** lipogenesis (not acatabolic reaction involving the fats), **3.** aerobic (not associated with the anaerobic items listed), **4.** protein synthesis (not associated with the breakdown of amino acids), **5.** beta oxidation (not an anabolic reaction), **6.** secretes bicarbonate (not a liver function), **7.** vitamin E (not associated with phenylketonuria), **8.** magnesium (not a micromineral), **9.** carbohydrate (not associated with deamination) **10.** hunger center inhibition (does not result in satiety center activity)

Matching

Set 1: **1.** b, **2.** a, **3.** b, **4.** a, **5.** b, **6.** a, **7.** a, **8.** b
Set 2: **1.** d, **2.** f, **3.** c, **4.** b, **5.** a, **6.** e, **7.** g

Double Crosses

1. across: ketosis down: aerobic **2.** across: **a.** Kreb's, **b.** satiety, **c.** substrate, **d.** essential down: deamination **3.** across: anaerobic down: **a.** anabolism, **b.** oxidation, **c.** respiration, **d.** catabolic, **e.** coenzymes

Sleuthing

a. carbohydrate **b.** glucose, a monosaccharide **c.** Glucose would be broken down to pyruvic acid which in turn would be converted to acetyl CoA. This would travel into the Kreb's Cycle and the hydrogen ions released during this phase would be sent through the electron transport chain. The products formed in this would be carbon dioxide, water and energy in the form of ATP. **d.** Insulin would be called upon to facilitate the transport of glucose into the cells for metabolism. **e.** Jim would be better off eating a snack that had glucose for a "quick lift" but also contained starch which would release glucose more slowly over time. An example of this would be bread with jelly.

Word Scrambles

1. a. niacin, **b.** tomatoes, **c.** D TOTAL: deamination **2. a.** thyroid, **b.** hypothalamus, **c.** convection, **d.** coenzyme TOTAL: cytochrome chain

Addagrams

1. a. niacin, **b.** anemia, **c.** diets, **d.** acidosis, **e.** osteomalacia TOTAL: essential amino acids **2. a.** catabolism, **b.** ascorbic, **c.** steroid, **d.** aerobic, **e.** lose TOTAL: basal metabolic rate

Crossword

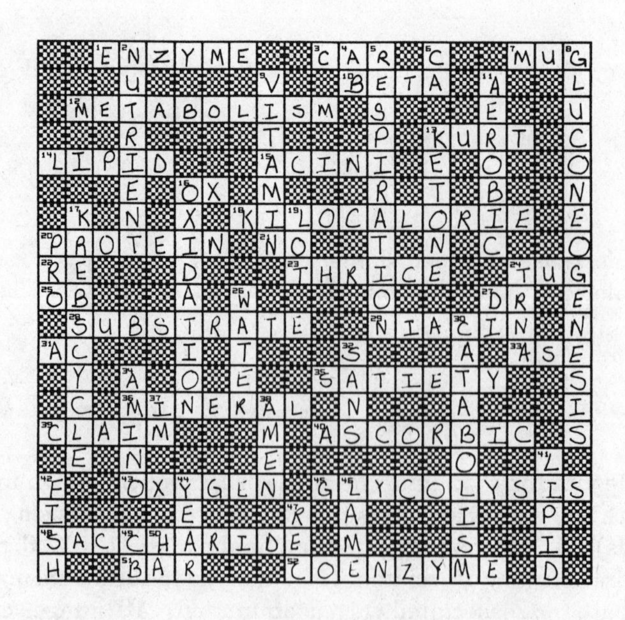

Chapter 21

Multiple Choice

1. c **2.** c **3.** a **4.** b **5.** c **6.** c **7.** d **8.** d **9.** c **10.** b **11.** d **12.** a **13.** d **14.** d **15.**d **16.** b **17.** d **18.** b **19.** b **20.** a **21.** a **22.** d **23.** b **24.** b **25.** d

True/False

1. T **2.** F-excess **3.** T **4.** F-99 **5.** T **6.** T **7.** T **8.** T **9.** F-efferent **10.** F-10 **11.** T **12.** T **13.** T **14.** F-mellitus **15.** T **16.** F-ureter **17.** T **18.** F-part of **19.** T **20.** T **21.** T **22.** T **23.** T **24.** T **25.** T

Completion

1. parasympathetic **2.** glucose/ketones **3.** hemodialysis **4.** water, chloride, sodium **5.** secreted **6.** 700-800 **7.** endothelial-capillary **8.** osmosis **9.** obligatory **10.** facultative **11.** 200-400 **12.** urethra **13.** 25% **14.** active transport **15.** secreted **16.** endothelial-capillary membrane **17.** greater **18.** nephron **19.** uremia **20.** 4. 6-8. 0 **21.** H+ **22.** 95% **23.** aldosterone,ADH **24.** afferent arterioles **25.** transitional epithelium

Lost Sheep

1. calyx (not part of nephron) **2.** PCT (not part of blood supply of nephron) **3.** glucose (not a normal constituent of urine) **4.** capillaries (not a macroscopic structure of kidney) **5.** blood colloidal pressure (not a pressure which pushes outward) **6.** sebaceous glands (not an auxiliary means of fluid loss) **7.** vasodilation (not associated with vasoconstriction) **8.** secretion (not associated with filtration process) **9.** pelvis (not a microscopic structure) **10.** major calyx (not associated with nephron) **11.** renal artery (not a microscopic blood vessel) **12.** sodium (not associated with glucose) **13.** ADH (not associated with JGA) **14.** constriction of afferent arteriole (not associated with increase in GFR) **15.** red blood cells (not a normal constituent of urine) **16.** glucose (does not normally appear in urine) **17.** capsule (not part of microscopic structure of nephron) **18.** calyx (not associated with bladder) **19.** FSH (hormone is not associated with normal water control) **20.** thyroid (gland not associated with water balance)

Matching

SET 1
1. b,c **2.** b,c **3.** a,c **4.** a,c **5.** a
SET 2
1. a **2.** e **3.** b **4.** a **5.** d **6.** c **7.** b **8.** a **9.** e **10.** e
SET 3
1. b **2.** c **3.** c **4.** e **5.** f **6.** d **7.** a **8.** a **9.** d **10.** b

Double Crosses

1. across:secretion down:suppression **2.** across:a. medullary **b.** micturition **c.** angiotensin down:filtration **3.** across:hypotonic down:hematuria **4.** across:a. ammonia **b.** retroperitoneal down:juxtaglomerular

Sleuthing

a. It has exceeded the Tm. **b.** Ketone bodies from the breakdown of fats. **c.** There is an osmotic of water into the kidney tubules because of the high concentration of glucose.

Word Scrambles

1. a. blood **b.** water **c.** membrane **d.** glomerulus TOTAL:Bowman's **2. a.** penis **b.** vein **c.** salts TOTAL:pelvis **3. a.** urethra **b.** cortex **c.** renal TOTAL:ureter **4. a.** coxae **b.** glucose **c.** medulla **d.** jaundice **e.** arteriole TOTAL:juxtaglomerular cells

Addagrams

1. a. secretion **b.** tubule **c.** lobar **d.** medulla **e.** Henle **f.** macula **7.** plasma TOTAL:endothelial-capsular membrane

Crossword

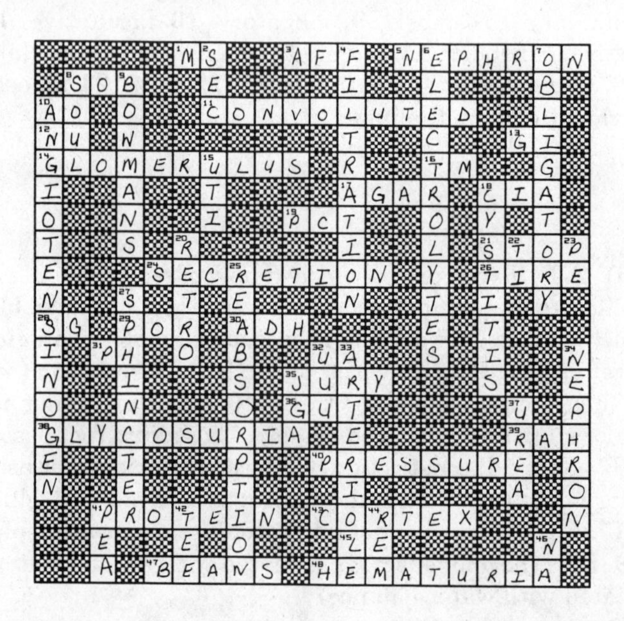

Anatomic Artwork

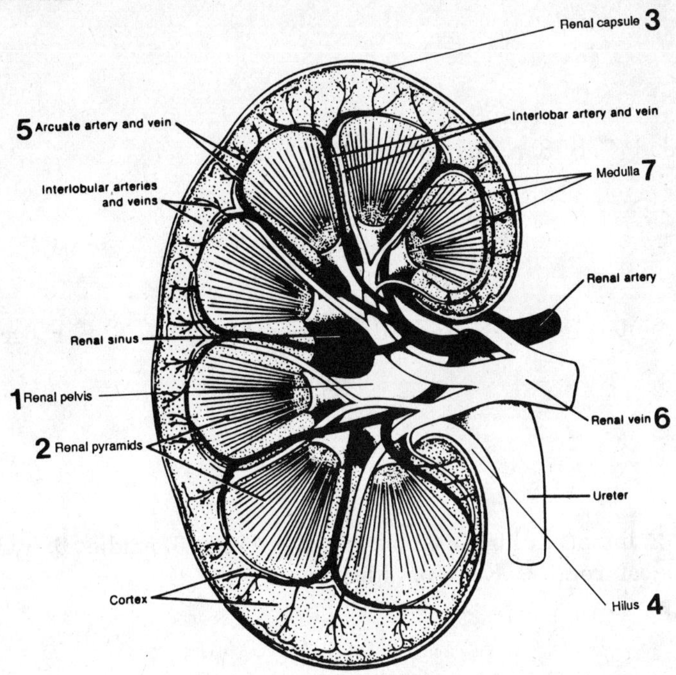

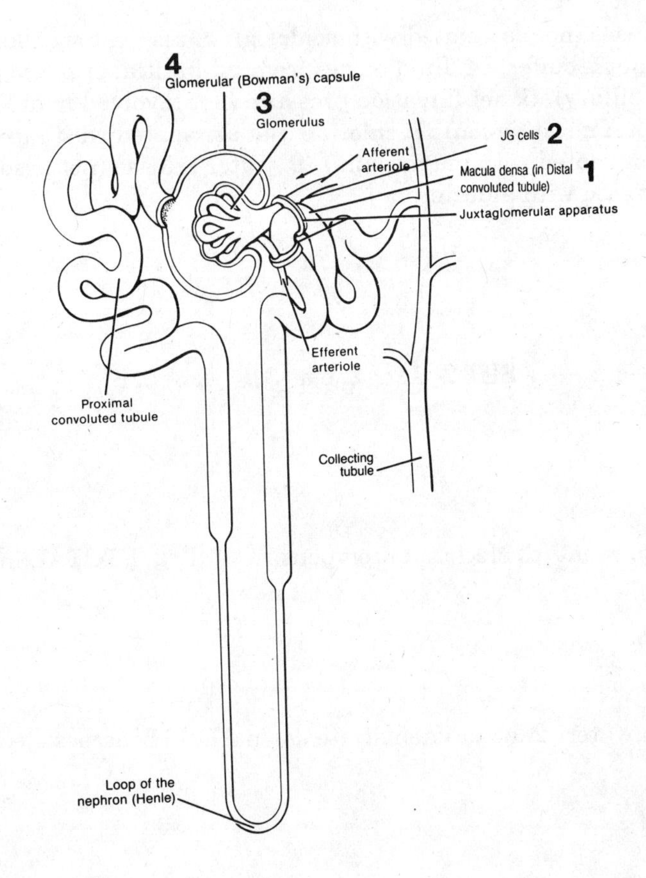

Chapter 22

Multiple Choice

1. d **2.** d **3.** b **4.** a **5.** d **6.** d **7.** b **8.** a **9.** b **10.** c

True/False

1. T **2.** T **3.** F - distal convoluted tubule **4.** T **5.** T **6.** F - PTH, CT **7.** T **8.** F - greater with NaCl **9.** T **10.** F

Completion

1. ingested food and drink **2.** intracellular **3.** osmosis **4.** buffers **5.** acidic **6.** hydrostatic **7.** cation, anion **8.** ADH, aldosterone **9.** K+ **10.** aldosterone

Lost Sheep

1. hyperventilation causes (not associated with acidosis) **2.** respiratory alkalosis (not associated with acidosis) **3.** ketones (not a buffer) **4.** fluid moves from interstitial space to plasma (not the direction of movement out of capillary) **5.** net filtration pressure (not involved in aldosterone regulation) **6.** anion (calcium is not a negative ion) **7.** chloride (not associated with intracellular) **8.** extracellular (not associated with magnesium) **9.** water release (not associated with ADH) **10.** marrow (not associated with aldosterone)

Matching

SET 1 **a.** d **2.** e **3.** c **4.** a **5.** a SET 2 **1.** c **2.** b,a **3.** b **4.** c **5.** c

Addagram

1a. ATP **b.** osmotic **c.** renal **d.** electro **e.** compensate **f.** I **g.** I TOTAL:interstitial compartment

Double Crosses

1. across:protein down:water **2.** across:dehydrate down:ADH **3.** across:electrolyte down:chlorides

Sleuthing

1-3 her belly is swollen because she is not taking a sufficient amount of protein in her diet; she does not have the "building blocks" to make her own protein, and therefore her blood osmotic pressure is low; fluid flows out to the tissue spaces in her abdominal cavity

Crossword

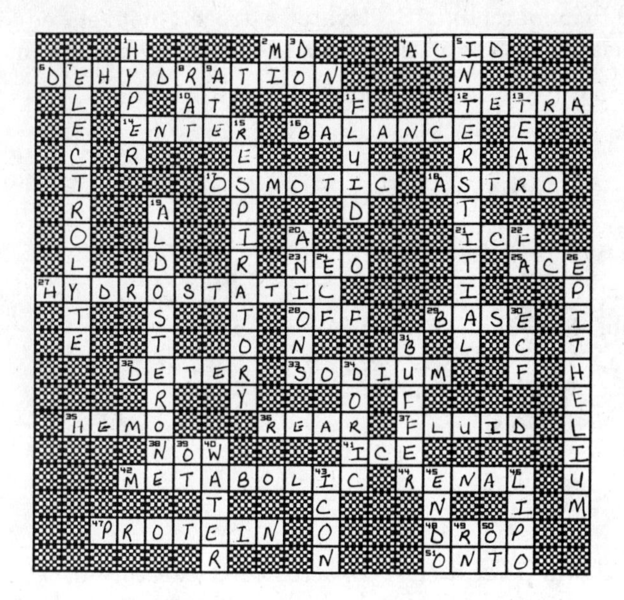

Chapter 23

Multiple Choice

1. d, **2.** c, **3.** b, **4.** d, **5.** a, **6.** c, **7.** c, **8.** a, **9.** c, **10.** d, **11.** a **12.** c, **13.** d, **14.** b, **15.** d, **16.** a, **17.** b, **18.** b, **19.** a, **20.** c, **21.** c **22.** d, **23.** d, **24.** b, **25.** c

True/False

1. F Staphylococcus aureus, **2.** T, **3.** T, **4.** F Fallopian tube, **5.** F‘endometrium, **6.** F tunica albuginea, **7.** T, **8.** T, **9.** T, **10.** T, **11.** F inguinal canal, **12.** F urethra, **13.** T, **14.** T, **15.** F homologous, **16.** T **17.** T, **18.** F cryptorchidism, **19.** F Meiosis I, **20.** F Sertoli, **21.** T **22.** T, **23.** T, **24.** F seminal vesicle, **25.** T

Completion

1. spermiogenesis, **2.** midpiece, **3.** FSH, **4.** puberty, **5.** ejaculatory duct, **6.** 48, **7.** fructose, seminal vesicle, **8.** glans penis, **9.** 7. 20-7. 6 **10.** oogenesis, **11.** secondary oocyte, polar body, **12.** fimbriae **13.** cervix, **14.** endometrium, **15.** rugae, **16.** areola, **17.** climateric, estrogen, **18.** corpus albicans, **19.** 14, **20.** FSH, **21.** venereal **22.** syphilis, **23.** amenorrhea, **24.** impotence, **25.** pelvic inflammatory disease

Lost Sheep

1. prostate (not associated with the testes), **2.** Graafian cells (not associated with the testes), **3.** spermatids (not associated with Meiosis I), **4.** age 8 (not associated with puberty), **5.** rete testes (not part of the spermatic cord), **6.** 30% volume (not associated with the seminal vesicles), **7.** Cowper's glands (not part of the penis), **8.** bulbourethral (not associated with the penis), **9.** epididymis (not associated with production of seminal fluid), **10.** Leydig cell (not associated with inhibin), **11.** Fallopian tube (not a part of the ovary), **12.** haploid (does not have 46 chromosomes), **13.** fundus (not part of the Fallopian tube), **14.** fornix (not associated with the uterus), **15.** Cowper's glands (not a female structure), **16.** perineum (not associated with mammary glands), **17.** menopause (not related to menarche), **18.** elevated estrogen (not related to menstrual phase) **19.** menstruation (not associated with ovulation), **20.** preovulatory phase (not associated with post ovulatory phase)

Matching

Set 1: **1.** d, **2.** g, **3.** c, **4.** a, **5.** f, **6.** e, **7.** b, **8.** c
Set 2: **1.** e, **2.** a, **3.** c, **4.** b, **5.** g, **6.** d, **7.** f
Set 3: **1.** f, **2.** g, **3.** d, **4.** a, **5.** b, **6.** c, **7.** i, **8.** j, **9.** h, **10.** e

Double Crosses

1. across: spermatogenesis down: **a.** testicles, **b.** acrosome, **c.** scrotum, **d.** Fallopian, **e.** relaxin, **f.** deferens, **g.** follicles, **h.** puberty, **i.** radiata, **j.** anterior, **k.** elevated, **l.** ectopic, **m.** inhibin
2. across: endometrium down: **a.** oogenesis, **b.** estrogens, **c.** gametes, **d.** cavernosa, **e.** pellucida
3. across: ovulation down: **a.** chromosomes, **b.** lobules, **c.** spermatozoa, **d.** seminal, **e.** examination

Sleuthing

1. a. The cyst is composed of tissue that is identical to the lining of the uterus. **b.** Each month the tissue proliferates and bleeds in response to estrogens and progesterone, like the endometrium of the uterus. **c.** (1) menstrual phase - the stratum functionalis is shed with a discharge of blood, tissue fluid and epithelial cells. (2) preovulatory phase - endometrial repair occurs. (3) ovulation - a uterine phase, endometrial growth continues (4) postovulatory phase - the endometrium is undergoing an increase in size, vascularity and glandular secretion due to presence of progesterone which is secreted by the corpus luteum which was formed in the ruptured follicle. **2. a.** vas deferens **b.** Yes, the process of erection depends on neural and vascular mechanisms. **c.** Yes, seminal fluid production by accessory glands is not affected by a vasectomy. Sperm form a minimal percentage of semen. **d.** No, the union of the two systems, excretory and reproductive, is t the urethra

Word Scrambles

1. a. cervix, **b.** foreskin, **c.** spermatocyte, **d.** gonad, **e.** urethra TOTAL: Toxic Shock Syndrome
2. a. acrosome, **b.** Cowper's, **c.** epididymis, **d.** functionalis, TOTAL: spermatic cord

Addagrams

1. a. sustentacular, **b.** cyst, **c.** luteum, TOTAL: menstrual cycle **2. a.** infertility, **b.** follicles, **c.** glans, **d.** tetrad, TOTAL: interstitial cells of Leydig

Crossword

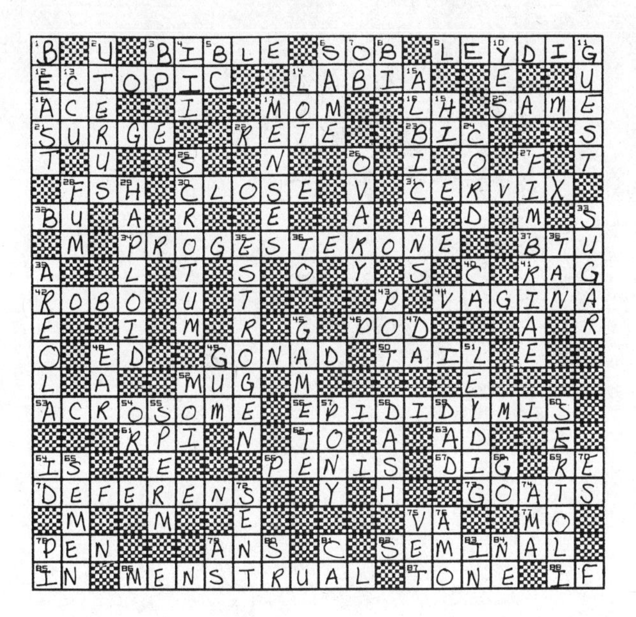

Anatomic Artwork

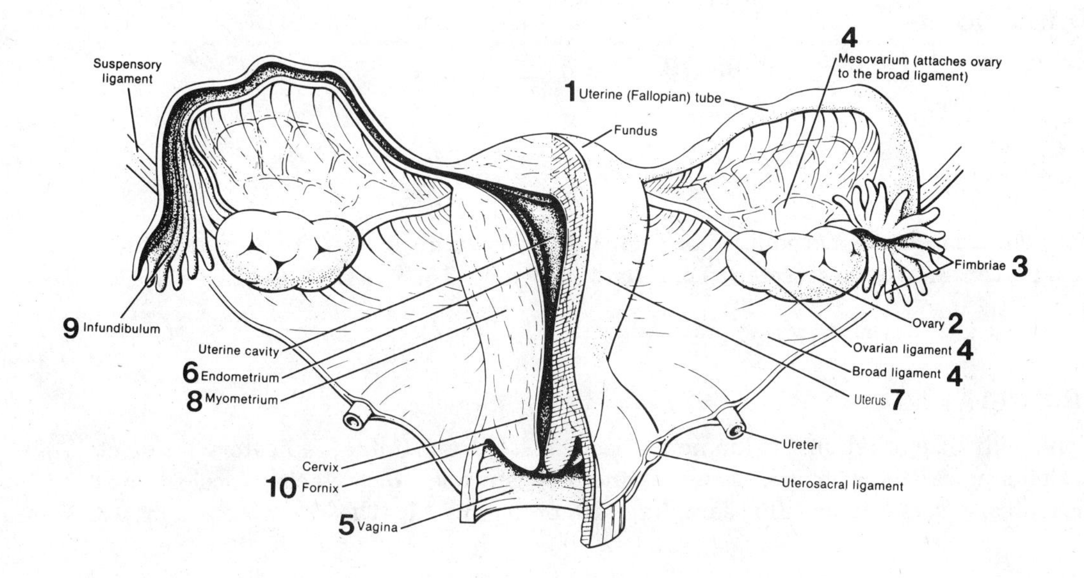

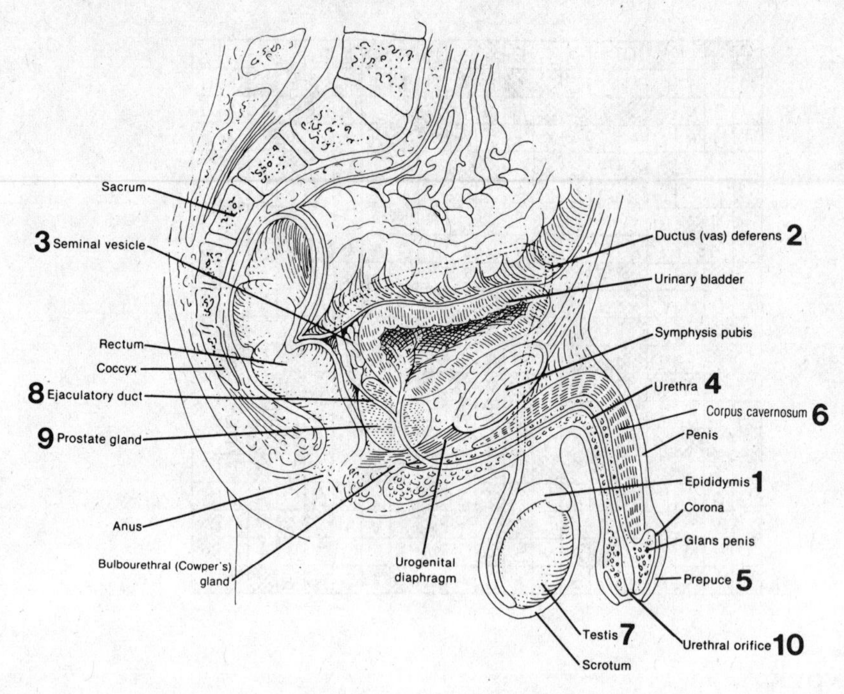

Chapter 24

Multiple Choice

1. c **2.** d **3.** c **4.** c **5.** d **6.** c **7.** d **8.** b **9.** c **10.** d **11.** b **12.** c

True/False

1. F-zona pellucida **2.** F-endothelium **3.** T **4.** T **5.** F-morula **6.** T **7.** T **8.** F-do not mix **9.** T **10.** T **11.** F - continue to be secreted **12.** F-umbilical vein **13.** T **14.** T **15.** T

Completion

1. chorionic villi **2.** epithelium of skin,nerve tissue etc. **3.** gut lining, repiratory lining etc. **4.** most muscle, skeleton, cartilage etc. **5.** estrogens and progesterone **6.** oxytocin **7.** colostrum **8.** amniocentesis **9.** chorionic villus sampling **10.** prolactin **11.** IUD **12.** diaphragm **13.** corona radiata

Lost Sheep

1. morula (not hollow) **2.** muscle (does not come from ectoderm) **3.** epithelium of skin (does not develop from endoderm) **4.** mesoderm (is not part of placenta) **5.** umbilical cord (is not one of extraembryonic membranes) **6.** glucagon (is not a hormone of pregnancy or labor) **7.** FSH (is not a hormone of pregnancy) **8.** estrogens (are not involved with labor) **9.** estrogens (is not involved in milk production or letdown) **10.** fossa ovale (is not an embryonic structure) **11.** rhythm (not an artificial means of birth control) **12.** ovulation (not associated with pregnancy)

Matching

SET **1.** **1.** c **2.** a **3.** a **4.** b **5.** b
SET **2.** **1.** c **2.** b **3.** d **4.** a **5.** e

Double Crosses

1. across:a. teratogen **b.** umbilical **c.** lactation down:fertilization **2.** across:**a.** primary **b.** meiosis **c.** pregnancy **d.** colostrum **e.** down:amniocentesis **3.** **a.** endoderm **b.** ectoderm **c.** embryonic **d.** estrogens

Sleuthing

1. usually 14-16 weeks after conception **2.** a puncture of the uterus and the removal via a hypodermic needle of 10-20 ml of fluid which contains sloughed off fetal cells;chromosomal disorders and biochemical defects are then examined **3.** Down's syndrome, hemophilia, Tay-Sachs disease, sickle cell anemia **4.** Chorionic villus sampling, done at 8-10 weeks of gestation; done through the vagina

Word Scrambles

1. **a.** vagina **b.** alleles **c.** ligation **d.** rhythm **e.** ejaculation **f.** condom TOTAL:chorionic villi
2. **a.** placenta **b.** sponge **c.** rhythm TOTAL:genotype

Addagrams

1. **a.** labor **b.** PRL **c.** oral **d.** births **e.** T TOTAL:trophoblast **2.** **a.** morula **b.** IUD **c.** blood **d.** color **e.** crib TOTAL:umbilical cord

Crossword

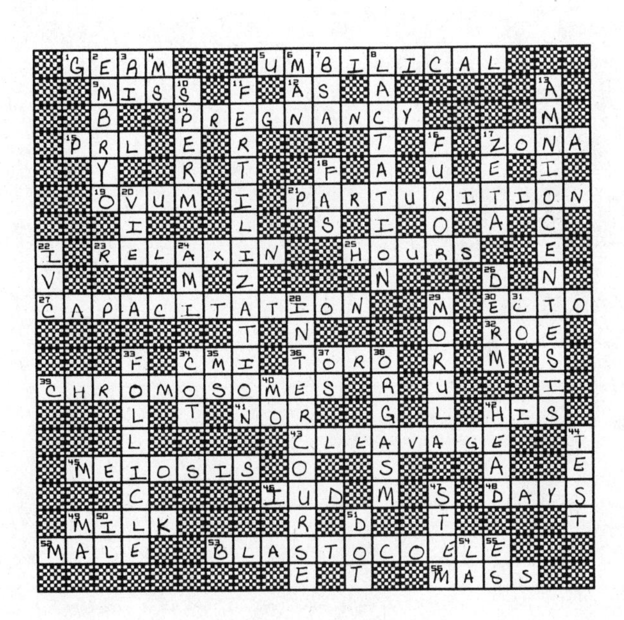